Return to an Order of the Honourable the House of Commons date

Report, evidence and supporting papers of the Inquiry into identification of Bovine Spongiform Encephalopathy (BSE) and v Disease (vCJD) and the action taken in response to it up t

The BSE Inquiry

Volume 1

Findings and Conclusions

Lord Phillips of Worth Matravers
Mrs June Bridgeman CB
Professor Malcolm Ferguson-Smith FRS

Ordered by The House of Commons *to be printed October 2000*

887–I   LONDON: THE STATIONERY OFFICE   £29.50 (inc VAT)

© Crown copyright 2000
First published 2000

Volume 1
Findings and Conclusions

Terms of Reference, Committee Members and Report Volumes — xiii
A note on the footnotes — xiv

Executive Summary of the Report of the Inquiry — xvii

Introduction — xvii
1. Key conclusions — xvii
2. The identification of the emergence of BSE — xviii
3. The cause of BSE — xix
4. Assessment of risk posed by BSE to humans — xx
5. Communication of the risk posed by BSE to humans — xxi
6. Measures to eradicate the disease in cattle — xxi
7. Measures to address the risks posed by BSE to humans — xxii
    Slaughter and compensation — xxii
    Food risks — xxiii
8. Medicines — xxiv
9. Cosmetics — xxvi
10. Occupational risk — xxvii
11. Other pathways of infection — xxviii
12. Pollution and waste control — xxviii
13. The identification of vCJD — xxviii
14. Victims and their families — xxix
15. Research — xxx
16. Some general lessons — xxxi

1 Introduction — 1

Our task — 3
The structure of the Report — 7
Transmissible Spongiform Encephalopathies — 10
Transmission to humans — 11
The story in a nutshell — 13
    What happened? — 13
    Why did it happen? — 20

2 Setting the context — 23

The cattle industry — 23
Slaughterhouses — 24
Renderers — 25
The animal feed industry — 27

The meat industry	27
The pharmaceutical industry	28
Other uses of bovine products	28
Government and BSE	29
Handling risk	31
Risk evaluation	31
Risk management	31
BSE and risk	32

3 The early years, 1986–88   33

Identification of a new disease in cattle	33
Restraints on information	34
What was the cause of BSE?	36
The scrapie theory	37
The ruminant feed ban	38
Exports	41
Human health implications	42
Mr MacGregor's reaction	45
Sir Donald Acheson's advice	46

4 The Southwood Working Party and other scientific advisory committees   48

The Southwood Working Party	48
Epidemiology	49
Risk to humans	50
Other scientific advisory committees	55
The Consultative Committee on Research into SEs (The Tyrrell Committee)	55
The Spongiform Encephalopathy Advisory Committee (SEAC)	57

5 The animal health story   58

Ruminant feed ban	59
The first BAB	61
UKASTA's information about breaches of the ban	62
Cross-contamination in feedmills	63
What went wrong?	66
Introduction of the animal SBO ban	68
The voluntary animal SBO ban	69
The cat	70
The pig	71
The statutory animal SBO ban	72

The operation of the statutory animal SBO ban	73
Before the ban	74
The human SBO Regulations	75
Enforcement	75
The voluntary animal SBO ban	77
The statutory animal SBO ban	77
Reliance on the voluntary animal SBO ban	79
Reliance on the human SBO ban	79
Knacker's yards and hunt kennels	80
SBO in transit	81
Responsibility	81
Monitoring	82
Renderers	84
Slaughterhouses	85
Knacker's yards and hunt kennels	86
'Cradle to grave' reviews	86
The truth emerges	86
The penny drops	87
The Meat Hygiene Service takes over and a new SBO stain is introduced	88
More shortcomings revealed	88
The new Order	90
Did the provisions of the animal SBO ban matter?	92
Why did it take so long?	93
Two fundamental issues	94
Conclusions	95
Cattle-tracking	96
Breeding	96

6 Protecting human health9   98

Introduction	98
CJD surveillance	99
Surveillance recommended by the Southwood Working Party and the Tyrrell Committee	99
The CJD Surveillance Unit established	99
How the surveillance system worked	99
PHLS excluded from CJD surveillance	100
Slaughter and compensation	101
Was compensation too low?	104
Ante-mortem inspection	105
Compensation changed again	105
Unanticipated burdens	105
Introduction of the ban on Specified Bovine Offal (SBO) in human food	106
Government response to the *Southwood Report*	106

The decision to introduce the human SBO ban	110
Preparation of the Regulations	113
Brain, spinal cord, thymus, spleen and tonsils	115
Tripe and rennet	115
Mesenteric fat	115
Casings	115
Calves under 6 months of age	116
Mechanically recovered meat (MRM)	117
BSE and human health in 1990	121
Implementation, enforcement and monitoring of the human SBO ban	121
Bovine brains	122
Slaughterhouse practices and mechanically recovered meat	123
Europe and lymphoid tissue	126
Alarms and reassurances	127
The cat	128
The Agriculture Committee	130
SEAC considers the safety of beef	131
A look ahead	132
The false peace – 1 January 1991 to 31 March 1995	133
Slaughterhouse standards	134
History of the setting up of the Meat Hygiene Service	135
Monitoring compliance with the SBO Regulations	136
MRM on the agenda again	137
The distal ileum of calves	137
Advances in knowledge of BSE	139
Knowledge about dose	141
Two dairy farmers die from CJD	141
Vicky Rimmer	143
Chinks in the armour – April–December 1995	143
Action at last on MRM	146
Cause for concern	147
Public debate	148
A campaign of reassurance	150
The final months	151
Mr Hogg's questions	152
SEAC's meetings on 5 January and 1 February 1996	154
The storm clouds gather	156
Rumbles of thunder	156
The storm breaks	157
Postscript	160
Contingency planning	161
What would contingency planning have achieved?	164

7 Medicines and cosmetics — 166

Medicines — 166
The medicines licensing system — 167
Medical devices — 168
Phase 1: the initial response on veterinary medicines — 169
Phase 1: the initial response on human medicines — 170
    The period up to March 1988 — 170
    March–December 1988 — 171
        Initial action by the CMO and MD — 171
        The NIBSC discussion — 171
        Galvanising MD — 172
        The paper for the BSC — 172
        Sir Richard Southwood's concerns about biologicals — 173
Phase 2: preparing joint guidelines, January–March 1989 — 174
    The final draft of the *Southwood Report* — 174
    The continuing concern on vaccines — 176
    CSM and VPC approval and the issue of the guidelines — 176
    Was the action taken adequate? — 177
        The Southwood message and how it was interpreted — 178
        Were non-binding guidelines appropriate? — 178
        Was the scope of the guidelines adequate? — 178
        Were existing stocks of injected products treated appropriately? — 179
Phase 3: implementing the guidelines after March 1989 — 181
    The context for handling matters — 181
    Collecting and analysing the information — 181
    The SBO ban and pharmaceuticals — 182
    How the BSEWG operated — 183
    First meeting of the BSEWG on 6 September 1989 — 183
        The follow-up to the first meeting — 184
    Second meeting of the BSEWG on 10 January 1990 — 185
        The follow-up to the second meeting — 186
    Third meeting of the BSEWG on 4 July 1990 — 186
    Fourth meeting of the BSEWG on 31 October 1990 — 187
        Veterinary products — 187
    Final meeting of the BSEWG in July 1992 — 187
Overview of the way the guidelines were implemented — 188
    Veterinary medicines — 188
    Human medicines — 189
Research into pharmaceuticals — 190
Cosmetics and toiletries — 192
    The main products — 192
    Regulation — 192

The Tyrrell recommendation on cosmetics	193
Was the initial action adequate?	193
Was DTI action adequate?	194
Action taken thereafter	194
The adequacy of the response	196

8 Occupational risk — 198

Those at risk	198
Chronology of occupational safety advice	198
ACDP advice to laboratories, medical workers and undertakers	200
Chronology of drafting of ACDPWG advice	201
The issue of guidance to schools about dissecting bovine eyeballs	203
Chronology of guidance on bovine eyeball dissection	203
Overview of occupational health	205

9 Potential pathways of infection — 207

Consideration of an audit of the uses of cattle tissues	207
The Tyrrell recommendation	207
Reasons for this outcome	210
Where responsibility lay	210

10 Pollution and waste control — 212

11 Wales, Scotland and Northern Ireland — 215

Wales	216
Scotland	216
Northern Ireland	217
Collective government and working relationships	218

12 Science and research — 219

Scientific conclusions about BSE	219
Alternative theories	222
The organophosphate theory	222
The autoimmune theory	222
Research	222

13 What went right and what went wrong? — 226

A recipe for disaster	226
The identification of the disease and its cause	227
The Government's response	228
Eradication of BSE	228
Possible transmissibility to other animals	229

Possible transmissibility to humans	229
Shortcomings and possible reasons for them	231
Was there a conflict of interest in MAFF?	231
Other conflicts of interest	232
Perception of risk	232
Ignorance and failures of communication	235
Ignorance of views as to the minimum infective dose for cattle	236
Ignorance of views as to the minimum infective dose for humans	237
Ignorance of pathways of infection	237
Failures of communication	238
Between the Southwood Working Party, the Government and the public	238
Between SEAC, the Government and the public	238
Lack of rigorous consideration when giving effect to policy	239
The best being the enemy of the good	239
Inappropriate use of advisory committees	240
Administrative structures	242
Interdepartmental structures	242
DH role	243
Structure within MAFF	243
Chief Medical Officers and Chief Veterinary Officers	244
Central and local government	244
Central government and the Territorial Departments	246
Individual criticisms: redressing the balance	246

14 Lessons to be learned — 249

Episodes in the BSE story	249
Lessons from the fact that BSE emerged	249
Commentary	249
Lessons	250
Lessons from the transmissions of BSE	250
Commentary	250
Lessons	250
Lessons from the spread of the BSE epidemic	250
Commentary	250
Lessons	251
Lessons from the identification of BSE	251
Commentary	251
Lessons	251
Lessons from the consideration of the nature and implications of BSE	252
Commentary	252
Lessons	252
Lessons from the investigation of the cause of BSE	252

Commentary	252
Lessons	253
Lessons from the introduction of the ruminant feed ban	253
Commentary	253
Lessons	253
Lessons from the introduction of slaughter with compensation	254
Commentary	254
Lessons	254
Lessons from the *Southwood Report*	254
Commentary	254
Lessons	254
Lessons from the introduction of the animal SBO ban	255
Commentary	255
Lessons	255
Lessons from the implementation and enforcement of the animal SBO ban	255
Commentary	255
Lessons	255
Lessons from the introduction of the human SBO ban	256
Commentary	256
Lessons	256
Lessons from the final months	256
Commentary	256
Lessons	257
Lessons in respect of Wales, Scotland and Northern Ireland	257
Commentary	257
Lessons	257
Lessons from the emergence of vCJD	258
Commentary	258
Lessons	258
Lessons from the handling of non-food routes of transmission to humans	258
Commentary	258
Lessons	258
Lessons from the approach to BSE and medicines	259
Commentary	259
Lessons	259
Lesson from the approach to BSE and cosmetics	260
Commentary	260
Lesson	260
Lesson from the approach to BSE and occupational risk	260
Commentary	260
Lesson	260
Lesson in relation to pollution and waste control	260

Commentary	260
Lesson	260
Lessons in relation to research	261
Commentary	261
Lessons	261
The use of scientific advisory committees	261
Commentary	261
Lessons	262
Dealing with uncertainty and the communication of risk	264
Commentary	264
Lessons	266
The legislative framework	266
Commentary	266
The problem	267
Power to order the slaughter of animals	267
Power to order the destruction of parts of an animal	268
Power to ban the use of material for specified purposes	269
Legislative constraints in relation to medicines	270
Legislative constraints in relation to cosmetics	271
General constraints of European law	271
Lessons	272
The experience of vCJD victims and their families	272
Commentary	272
Lessons	273

Annex 1: Procedures adopted by the BSE Inquiry — 275

Thoroughness and openness	275
Fairness	278

Annex 2: Individual criticisms — 280

The early years	280
The Southwood Working Party	281
Protection of animal health, 1989–96	281
Protection of human health, 1989–96	281
Medicines and cosmetics	283
Potential pathways of infection	284

Glossary — 285

Who's who — 289

Index — 297

The BSE Inquiry Terms of Reference

To establish and review the history of the emergence and identification of BSE and new variant CJD in the United Kingdom, and of the action taken in response to it up to 20 March 1996; to reach conclusions on the adequacy of that response, taking into account the state of knowledge at the time; and to report on these matters to the Minister of Agriculture, Fisheries and Food, the Secretary of State for Health and the Secretaries of State for Scotland, Wales and Northern Ireland.

The Members of the Committee

Lord Phillips of Worth Matravers, Master of the Rolls
Mrs June Bridgeman CB
Professor Malcolm Ferguson-Smith MBChB, FRCPath, FRCP(Glasg.), FMedSci, FRSE, FRS

The Volumes of the Report

1 Findings and Conclusions

2 Science

3 The Early Years, 1986–88
4 The Southwood Working Party, 1988–89
5 Animal Health, 1989–96
6 Human Health, 1989–96
7 Medicines and Cosmetics
8 Variant CJD
9 Wales, Scotland and Northern Ireland
10 Economic Impact and International Trade
11 Scientists after Southwood

12 Livestock Farming
13 Industry Processes and Controls
14 Responsibilities for Human and Animal Health
15 Government and Public Administration

16 Reference Material

A note on the footnotes

During the course of its deliberations, the BSE Inquiry published many thousands of documents, along with transcripts of its oral hearings. These formed the evidence on which the Inquiry based its chronological accounts, discussions and conclusions. When footnotes refer to these sources, they are coded according to the Inquiry's filing system, which can be consulted by the public in two ways:

- either through the Public Record Office, which has a copy of all the evidence in electronic form on a series of CD-ROMs; or
- on the BSE website (www.bseinquiry.gov.uk).

YB codes: eg, YB88/12.22/4.1

YB refers to Year Books. These are documents collected in chronological order, year by year. They come from a variety of sources, but many of them are letters, memoranda and minutes of departmental meetings. For example, the one mentioned above refers to a document dated 22 December 1988 (YB88/12.22), which is the fourth document filed for that day and, specifically, its first page (4.1).

S codes (written witness statements): eg, S387 Tomlinson para. 6

A witness statement is written evidence supplied to the Inquiry. In the example above, the 'S' classifies the evidence as a witness statement, and it is number 387, paragraph 6. 'Tomlinson' shows that it was written by Sir Bernard Tomlinson. When people have sent in more than one witness statement, these statements are classified S387, S387A, etc.

T codes (transcripts of oral hearings): eg, T40 pp. 121–2

A number of witnesses gave oral evidence to the Inquiry, and the 'T' references indicate transcripts of the relevant hearings. The example above refers to day 40 of the oral hearings, pages 121–2.

IBD codes: eg, IBD1 tab 2 para. 5.3.5

These are Initial Background Documents – a selection of published material that was supplied by the Ministry of Agriculture, Fisheries and Food at the start of the Inquiry. The example refers to the first file, or 'bundle', of such background documents, and to the second document in that bundle. In this case, it is the Report of the Southwood Working Party on Bovine Spongiform Encephalopathy, paragraph 5.3.5.

M codes: eg, M29 tab 3

These are further bulky documents from a variety of sources ('M' stands for 'Materials'). They have been filed in series of bundles in the same way as the Initial Background Documents and the other series of bundles described below.

L codes: eg, L3 tab 6

These refer to legislation (ie, Statutory Instruments – Regulations, Orders, etc – and Acts), which is generally available in published form. For convenience the legislation most frequently referred to at hearings was filed in a series of L bundles.

DM codes: eg, DM01

Documents from the Ministry of Agriculture, Fisheries and Food (MAFF)

DH codes: eg, DH01

Documents from the Department of Health

DW codes: eg, DW01

Documents from the Welsh Office

DS codes: eg, DS01

Documents from the Scottish Office

DN codes: eg, DN01

Documents from Northern Ireland Departments

DO codes: eg, DO01

Documents from other Departments

SEAC codes: eg, SEAC1

Documents relating to the Spongiform Encephalopathy Advisory Committee

FEG codes: eg, FEG1

Documents relating to the Lamming Committee (the Expert Group on Animal Feedingstuffs)

Tyrrell codes: eg, Tyrrell1

Documents relating to the Consultative Committee on Research into Spongiform Encephalopathies chaired by Dr David Tyrrell

Executive Summary of the Report of the Inquiry

Introduction

By our terms of reference, we have been required:

> To establish and review the history of the emergence and identification of BSE and variant CJD in the United Kingdom, and of the action taken in response to it up to 20 March 1996; to reach conclusions on the adequacy of that response, taking into account the state of knowledge at the time; and to report on these matters to the Minister of Agriculture, Fisheries and Food, the Secretary of State for Health and the Secretaries of State for Scotland, Wales and Northern Ireland.

In this Executive Summary, we give an overview of our key findings and conclusions. We refer to things that went right as well as to some of the errors, inadequacies and shortcomings that we have identified in the response to BSE. We do not attempt here to explain or even list all of these. In particular we do not explain the criticisms of individuals that appear in our Report. These need as a matter of fairness to be read in their proper context, as we explain at paragraph 30 of this volume.

1. Key conclusions

- BSE has caused a harrowing fatal disease for humans. As we sign this Report the number of people dead and thought to be dying stands at over 80, most of them young. They and their families have suffered terribly. Families all over the UK have been left wondering whether the same fate awaits them.

- A vital industry has been dealt a body blow, inflicting misery on tens of thousands for whom livestock farming is their way of life. They have seen over 170,000 of their animals dying or having to be destroyed, and the precautionary slaughter and destruction within the United Kingdom of very many more.

- BSE developed into an epidemic as a consequence of an intensive farming practice – the recycling of animal protein in ruminant feed. This practice, unchallenged over decades, proved a recipe for disaster.

- In the years up to March 1996 most of those responsible for responding to the challenge posed by BSE emerge with credit. However, there were a number of shortcomings in the way things were done.

- At the heart of the BSE story lie questions of how to handle hazard – a known hazard to cattle and an unknown hazard to humans. The Government took measures to address both hazards. They were sensible measures, but they were not always timely nor adequately implemented and enforced.

- The rigour with which policy measures were implemented for the protection of human health was affected by the belief of many prior to early 1996 that BSE was not a potential threat to human life.

- The Government was anxious to act in the best interests of human and animal health. To this end it sought and followed the advice of independent scientific experts – sometimes when decisions could have been reached more swiftly and satisfactorily within government.

- In dealing with BSE, it was not MAFF's policy to lean in favour of the agricultural producers to the detriment of the consumer.

- At times officials showed a lack of rigour in considering how policy should be turned into practice, to the detriment of the efficacy of the measures taken.

- At times bureaucratic processes resulted in unacceptable delay in giving effect to policy.

- The Government introduced measures to guard against the risk that BSE might be a matter of life and death not merely for cattle but also for humans, but the possibility of a risk to humans was not communicated to the public or to those whose job it was to implement and enforce the precautionary measures.

- The Government did not lie to the public about BSE. It believed that the risks posed by BSE to humans were remote. The Government was preoccupied with preventing an alarmist over-reaction to BSE because it believed that the risk was remote. It is now clear that this campaign of reassurance was a mistake. When on 20 March 1996 the Government announced that BSE had probably been transmitted to humans, the public felt that they had been betrayed. Confidence in government pronouncements about risk was a further casualty of BSE.

- Cases of a new variant of CJD (vCJD) were identified by the CJD Surveillance Unit and the conclusion that they were probably linked to BSE was reached as early as was reasonably possible. The link between BSE and vCJD is now clearly established, though the manner of infection is not clear.

2. The identification of the emergence of BSE

- Individual cattle were probably first infected by BSE in the 1970s. If some lived long enough to develop signs of disease, these were not reported to or subject to investigation by the Central Veterinary Laboratory (CVL) of the State Veterinary Service (SVS).

- The Pathology Department of the CVL first investigated the death of a cow that had succumbed to BSE in September 1985, but the nature of the disease that had caused its death was masked by other factors and was not recognised at the time. This is not a matter for criticism.

- The Pathology Department considered two further cases of BSE at the end of 1986 and identified these as being likely to be a Transmissible

Spongiform Encephalopathy (TSE) in cattle. This identification was commendable.

- This part of the story demonstrates both the benefits and the limitations of the passive surveillance system operated by the SVS.

3. The cause of BSE

- Gathering of data about the extent of the spread of BSE was impeded in the first half of 1987 by an embargo within the SVS on making information about the new disease public. This should not have occurred.

- By the end of 1987 Mr John Wilesmith, the Head of the CVL Epidemiology Department, had concluded that the cause of the reported cases of BSE was the consumption of meat and bone meal (MBM), which was made from animal carcasses and incorporated in cattle feed. This conclusion was correct. It had been reached with commendable speed.

- The following provisional conclusions of Mr Wilesmith, which were generally accepted at the time as a basis for action, were reasonable but fallacious:

    – the cases identified between 1986 and 1988 were index (ie, first generation) cases of BSE;

    – the source of infection in the MBM was tissues derived from sheep infected with conventional scrapie;

    – the MBM had become infectious because rendering methods which had previously inactivated the conventional scrapie agent had been changed.

- The cases of BSE identified between 1986 and 1988 were not index cases, nor were they the result of the transmission of scrapie. They were the consequences of recycling of cattle infected with BSE itself. The BSE agent was spread in MBM.

- BSE probably originated from a novel source early in the 1970s, possibly a cow or other animal that developed disease as a consequence of a gene mutation. The origin of the disease will probably never be known with certainty.

- The theory that BSE resulted from changes in rendering methods has no validity. Rendering methods have never been capable of completely inactivating TSEs.

- The theory that BSE is caused by the application to cattle of organophosphorus pesticides is not viable, although there is a possibility that these can increase the susceptibility of cattle to BSE.

- The theory that BSE is caused by an autoimmune reaction is not viable.

4. Assessment of risk posed by BSE to humans

- One of the most significant features of BSE and other TSEs is the fact that they are diseases with very long incubation periods. Thus the question whether BSE was transmissible to humans was unlikely to be answered with any certainty for many years, and scientific experiments were bound to take a long time. The Government had to deal with BSE against this background of uncertainty as to the transmissibility of the disease.

- MAFF officials appreciated from the outset the possibility that BSE might have implications for human health.

- By the end of 1987 MAFF officials had become concerned as to whether it was acceptable for cattle showing signs of BSE to be slaughtered for human consumption. However, the Department of Health (DH) was not asked to collaborate with MAFF in considering the implications that BSE had for human health. It should have been.

- Only in March 1988, by which time MAFF officials had advised their Minister that animals showing signs of BSE should be destroyed and compensation paid, did MAFF advise the Chief Medical Officer (CMO) Sir Donald Acheson of the emergence of BSE and ask him for his view of the possible human health implications.

- On Sir Donald's advice, an expert working party, chaired by Sir Richard Southwood, was set up to advise on the implications of BSE. After their first meeting in June 1988, the Southwood Working Party advised that cattle showing signs of BSE should be slaughtered and destroyed. This advice was of crucial importance in safeguarding human health. The Working Party had concerns about some occupational health risks in relation to BSE and some risks posed by medicinal products. They notified the responsible authorities of these concerns. On 9 February 1989 they submitted a Report to the Government in the knowledge that it would be published. The report concluded that the risk of transmission of BSE to humans appeared remote and that 'it was most unlikely that BSE would have any implications for human health'.

- This assessment of risk was made on the following basis:

  - BSE was probably derived from scrapie and could be expected to behave like scrapie. Scrapie had not been transmitted to humans in over 200 years and so BSE was not likely to transmit either.

  - So far as occupational and medicinal risks were concerned, the authorities which had been notified about these could be relied upon to take appropriate measures to address them.

- The Report did not, as it should have done, make clear the basis for its assessment of risk. It did comment that if the assessment was incorrect the implications would be extremely serious. This warning was lost from sight. The *Southwood Report* was, in years to come, repeatedly cited as constituting a scientific appraisal that the risks posed by BSE to humans were remote and that no precautionary measures were needed other than those recommended by the Working Party.

- Precautionary measures were nonetheless put in place that went beyond those recommended by the Working Party. The wisdom of those measures was demonstrated as the years went by and facts were learned about BSE which threw doubt on the theory both that it was derived from scrapie and that it would behave like scrapie.

- In May 1990 a domestic cat was diagnosed as suffering from a 'scrapie-like' spongiform encephalopathy. This generated widespread public and media concern that BSE had been transmitted to the cat and might also be transmissible to humans. Subsequently, more domestic cats were similarly diagnosed. These events shifted the perception of some scientists of the likelihood that BSE might be transmissible to humans. By 1994 the Spongiform Encephalopathy Advisory Committee (SEAC) evaluated the risk of transmissibility to humans as remote only because precautionary measures had been put in place.

5. Communication of the risk posed by BSE to humans

- The increasing knowledge about BSE over the years, which threw doubt on the theory that it would behave like scrapie, was not concealed from the public. However, the public was not informed of any change in the perceived likelihood that BSE might be transmissible to humans.

- The public was repeatedly reassured that it was safe to eat beef. Some statements failed to explain that the views expressed were subject to proper observance of the precautionary measures which had been introduced to protect human health against the possibility that BSE might be transmissible. These statements conveyed the message not merely that beef was safe but that BSE was not transmissible.

- The impression thus given to the public that BSE was not transmissible to humans was a significant factor leading to the public feeling of betrayal when it was announced on 20 March 1996 that BSE was likely to have been transmitted to people.

6. Measures to eradicate the disease in cattle

- Once Mr Wilesmith had identified MBM as the probable vector of BSE, the Government introduced the appropriate measure to prevent further infection and to stop the spread of the BSE agent – a ban on incorporating ruminant protein in ruminant feed. This had a dramatic effect in reducing to a fraction what had been an escalating rate of infection. It did not, however, bring infection to an end.

- The manner in which the Government introduced the ruminant feed ban was influenced by misconceptions as to:

    – the scale of the infection;

    – the amount of infective material needed to transmit the disease.

- Ignorant of the fact that the rate of infection had escalated to thousands of cases a week, the Government gave the animal feed trade a 'period of grace' of some five weeks to clear existing stocks of feed before the ban took effect. Some members of the feed trade, being given an inch, felt free to take a yard and continued to clear stocks after the ban came into force. Farmers in their turn used up the stocks that they had purchased. This led to thousands of animals being infected after the ruminant feed ban came into force on 18 July 1988.

- More serious was a failure to give rigorous consideration to the amount of infective material that was proving capable of transmitting the disease. The false assumption was made that any cross-contamination of cattle feed in feedmills from pig or poultry feed containing ruminant protein would be on too small a scale to matter.

- In fact, as subsequent experiments were to demonstrate, a cow can become infected with BSE as a result of eating an amount of infectious tissue as small as a peppercorn. Cross-contamination in feedmills resulted in the continued infection of thousands of cattle. Because it takes, on average, five years after initial infection for the clinical signs of BSE to become apparent, this was not appreciated until 1994.

- From September 1990 contamination of cattle feed with pig and poultry feed should not have resulted in infection. This was because, following the experimental transmission of BSE to a pig, MAFF on the advice of SEAC introduced a measure in September 1990 aimed at protecting pigs and poultry from BSE. This was a ban on the inclusion in pig and poultry feed of MBM derived from the parts of the cow that might be expected to carry high infectivity if an animal were incubating or suffering from the disease – 'Specified Bovine Offal' or SBO.

- However, there was a failure to give proper thought to the terms of this measure when it was introduced. The animal SBO ban was unenforceable and widely disregarded. Infectious bovine offal continued to find its way into pig and poultry feed and then, by cross-contamination, into cattle feed.

- Only in 1994 did the fact of the continuing infection and the reasons for it become appreciated. Regulations were revised and a rigorous enforcement campaign launched to coincide with the takeover in 1995 by a new national Meat Hygiene Service (MHS) of the enforcement duties in slaughterhouses, previously carried out by local authorities. The success of these measures is now becoming apparent. They were replaced after 20 March 1996 by the radical step of banning the incorporation of all animal protein in animal feed.

7. Measures to address the risks posed by BSE to humans

Slaughter and compensation

- Compulsory slaughter and destruction of all animals showing signs of BSE was a crucial measure to protect human health and, incidentally, animal health. It prevented the use, for any purposes, of sick animals, which could otherwise have been sent to the slaughterhouse for human consumption.

- A compulsory slaughter and compensation scheme was introduced in August 1988, following the commendable interim advice of the Southwood Working Party. Had there been prompt and adequate collaboration between MAFF and DH, this measure could and should have been introduced months earlier.

- Levels of compensation to farmers were adjusted on two occasions, but at no time did they lead to any significant failure to comply with the duty to notify the SVS of animals showing signs of BSE.

Food risks

- The Southwood Working Party considered that all reasonably practicable precautions should be taken to reduce the risks that would exist should BSE prove to be transmissible to humans. However, they did not make this plain in their Report and did not recommend that the possible risks from eating animals incubating BSE but not yet showing signs of the disease ('subclinical cases') called for any precautions, other than a recommendation that manufacturers should not include ruminant offal and thymus in baby food. This was a shortcoming in their Report.

- Because of a failure to subject the *Southwood Report* to an adequate review, MAFF and DH failed to identify this shortcoming. Concern about the food risks posed by subclinical cases was, however, expressed by some scientists, by the media and by the public. With the agreement of DH, MAFF reacted by announcing in June 1989 that those categories of offal of cattle most likely to be infectious (SBO) were to be banned from use in human food. The introduction of this vital precautionary measure was commendable. However, this ban was presented to the public in terms that underplayed its importance as a public health measure.

- Careful consideration was given by MAFF and DH in 1989 to the terms of the human SBO ban, with one important exception. During the consultation process, concerns were raised about the practicality of ensuring the removal of all of the spinal cord during abattoir processes, and about the practice of mechanical recovery of scraps left attached to the vertebral column for use in human food ('mechanically recovered meat' or MRM). However, MAFF officials discounted these concerns without subjecting them to rigorous consideration – in particular no advice was sought as to the minimum quantity of spinal cord that might transmit the disease in food.

- MAFF gave detailed consideration to spinal cord and MRM in 1990. A lengthy paper was submitted to SEAC, the Government's new expert advisory committee on TSEs. Unhappily, as a result of a breakdown of communications, MAFF officials understood that the members of SEAC were not concerned about the inclusion in human food of an occasional scrap of spinal cord, so that no action was called for. In fact the advice of some, at least, of the members of SEAC was premised on the false assumption that spinal cord could readily be removed from the carcass in its entirety, and would be so removed.

- This was one of a number of occasions that has given rise to lessons for the future about the proper use of expert committees by the Government.

- Not until 1995 was action taken in relation to MRM. Following the takeover by the Meat Hygiene Service of the enforcement of Regulations in slaughterhouses, occasional instances were discovered of failure to remove all spinal cord from the carcass. Strenuous and successful steps were taken to improve standards of compliance with the Regulations in slaughterhouses. Eventually, in December 1995, on SEAC's advice the extraction of MRM from the spinal column of cattle was banned.

- Up to 1995, MRM was a potential pathway to the infection of humans with BSE, not merely because of the risk of inclusion of the occasional portion of spinal cord, but because the material recovered by the MRM process included dorsal root ganglia. These were peripheral nervous tissues which were not thought to be infectious at the time, but which have since been demonstrated to be infectious in the late stages of incubation.

8. Medicines

- Despite the highly regulated licensing regime for medicines, systematic records of the action taken in response to BSE in respect of individual medical products are lacking.

- Past experience of the transmission of animal disease through vaccines, and of transmission of CJD through medication and through the contamination of surgical instruments, showed that minute particles of infected tissue from an apparently healthy donor could transmit a TSE.

- MAFF officials recognised in 1987 that there was a risk that BSE might be transmitted through veterinary products and began to take steps to address this risk which were commendable. They failed, however, to share their concerns with those in DH who were responsible for handling human medicinal products. This was inadequate interdepartmental liaison.

- On learning of BSE in March 1988 the CMO, Sir Donald Acheson, sought to ensure that the potential risks that the disease posed in relation to human medicinal products were addressed. However, Medicines Division (MD) did not bring the matter before their advisory committees until November 1988. Of this period, two months' delay was attributable to a failure to accord the matter appropriate priority.

- MD did not appreciate the extent of the concern felt by the Southwood Working Party about medicines administered by injection and about the existing stocks of these. This was compounded by the wording of the *Southwood Report*, which described the risk posed by medicines as remote without making it plain that this risk assessment was predicated on the assumption that remedial measures were being taken to address the risk.

- Having regard to the legislative constraints, it was reasonable to issue guidelines in relation to both human and veterinary medicinal products rather than resort to direct regulatory action.

- Production of the relevant human and veterinary medicines involved similar raw materials and processes. The approach in respect of each needed to be consistent. Yet DH and MAFF did not discuss joint guidelines until January 1989. Once again this reflected inadequate interdepartmental liaison.

- The decision to continue to use existing vaccine stocks until these could be replaced was reasonable. Vaccines cannot be produced overnight. An embargo on existing stocks would have led to interruptions, potentially lengthy, in vaccination programmes. The overwhelming professional opinion at the time was that there was bound to be death and disablement in the event of breaks in the vaccination programmes, on a scale which far outweighed the potential risks from BSE. Some comfort can be derived from the 1993 results of tests carried out on bovine serum by the Neuropathogenesis Unit (NPU), which failed to lead to infection in mice.

- The task of identifying medicinal products to which the guidelines applied was made more difficult and protracted by:

  – the inadequate database of licensed products;

  – the need to make case-by-case enquiries in relation to thousands of products;

  – inadequate staffing;

  – unclear management responsibilities; and

  – the administrative dislocation involved in reorganisation at the time of the relevant DH and MAFF divisions as Executive Agencies.

  Staff from the two new Agencies – the Medicines Control Agency (MCA) and the Veterinary Medicines Directorate (VMD) – worked diligently to overcome these difficulties.

- The establishment of the BSE Working Group with a high-powered membership to advise all of the section 4 committees on human medicinal products thought to pose a potential risk was a sound decision.

- The small number of products that included high-risk tissues as an ingredient was identified and dealt with reasonably promptly.

- The role of the BSE Working Group, like that of the Committee on Safety of Medicines (CSM) and Veterinary Products Committee (VPC), was purely advisory. The task of identifying individual products for consideration by the Group and following up recommendations made by the Group was for officials.

- Decisions taken in relation to individual medicinal products were reasonable, but the speed with which decisions were taken and followed up suffered from lack of clear and purposeful leadership in the MCA.

- More effective handling arrangements were adopted within DH's Procurement Division (serving the National Health Service) to review medical devices.

- Existing stocks of a small number of human vaccines prepared using bovine tissues may have been used up to 1992 and of animal vaccines for even longer.

- The decision to continue using existing stocks of vaccines was not considered to be one that needed to be taken or approved by Ministers. Had it been, we consider that Ministers would have accepted the overwhelming professional advice, but would have been concerned to see that the process of phasing out these stocks was more vigorously pursued.

- Officials in the MCA and VMD do not appear to have been systematically accountable to anyone for the manner in which the phasing out exercise was handled. Nor, given the low-profile handling, was there any parliamentary or public scrutiny of their actions.

9. Cosmetics

- Cosmetics, like topically applied medicines, might be applied to the skin, eye or mucous membranes but were covered by a less stringent regulatory regime under the aegis of the Department of Trade and Industry (DTI). The category presenting the highest risk comprised 'exotica' or 'premium products', such as anti-ageing creams, which might contain lightly processed brain extracts, placental material, spleen and thymus.

- MAFF and DH failed to alert DTI to the need to consider the risk through cosmetics from BSE despite this having been identified by the *Tyrrell Report* in June 1989. This contributed to several months' delay in the start of action to secure their safety.

- Guidance was provided to the industry in February 1990 on the initiative of DTI, but was made available only to members of the cosmetics and toiletries trade association. This was the most significant single action to address the risk from cosmetics.

- Thereafter no further initiative was taken by DTI. A muddled situation developed about lead responsibility for action. Responsibility for taking action should have been clearly understood to rest with DTI with professional advice from DH.

- Following a request from SEAC in July 1991 for the cosmetics guidance to be updated, DH omitted to advise DTI about this and subsequently made its own unsuccessful approach to the trade association in April 1992 seeking detailed information. DTI was brought back into the picture only in September 1992 at a meeting between DH, MAFF and the trade association.

- The confusion about lead responsibility both between Departments and within DH continued thereafter, and responsibility for updated UK guidance was effectively left with the trade association. The topic became embroiled in protracted negotiations at European level on EU guidelines, and the trade association UK guidance did not emerge until 1994.

- The hallmarks of the handling of BSE in relation to cosmetics were lack of purposeful leadership and an absence of a sense of urgency. Manufacturers were left to use up stocks, and checks were not made to ensure they reformulated their products. This has left unanswered questions both about what material was being used, and about how long production continued and on what scale.

10. Occupational risk

- The possibility of contracting illness from contact with diseased animals or their tissues was a well-recognised occupational hazard. Workers in a wide range of occupations were potentially in contact with the tissues of BSE-infected cattle or with those of human victims. All of these occupations needed to be identified and to receive appropriate guidance about the precautions to reduce risk in respect of BSE and other TSEs.

- The delays in issuing advice to many of those concerned were unacceptable. Ultimately the main occupations at risk were identified and advice given. But a detailed chronology shows that it took over three years to complete the task of issuing simple warnings and basic advice to the most obvious high-risk trades.

- Work began in 1991 on guidance to those handling risk tissues in laboratories, hospitals and mortuaries. This took until September 1994 to be completed and issued. During that process a so-called 'fast track' professional letter took 14 months to prepare.

- In a different field, it took two-and-a-half years for advice to be issued to schools about risks from dissecting bovine eyeballs, though SEAC had asked in June 1990 for this to be done.

- The slow and erratic responses have indicated weaknesses in the standard system for handling a wide-ranging disease threat. The slow tempo of action, in part attributable to time spent on polishing and refining advice, stemmed from three factors:

  – a failure in communication: the perception that the *Southwood Report* had indicated that the risk to humans from BSE was remote even without any further action, and a belief in the Health and Safety Executive (HSE) that action was being taken simply as a response to political and media pressures;

  – the absence of a comprehensive review of pathways of transmission, which might have helped pinpoint where the issue of urgent advice could not wait;

  – the decision to use the slow-paced existing consultative and drafting arrangements. This ought not to have been at the expense of prompt and straightforward interim warnings.

- The mistakes made in handling the occupational threats from BSE and the questions raised by them need to be carefully considered by the HSE.

11. Other pathways of infection

- There was a need to establish all the pathways by which bovine products or by-products might come into contact with humans or other animals. This need was recognised by MAFF officials at an early stage and also by the Government's expert advisers on BSE. However, the exercise was never carried out prior to March 1996. As a result, no coordinated or comprehensive consideration was given to the various routes by which BSE might infect human beings or other animals.

12. Pollution and waste control

- MAFF was directly responsible for disposing of cattle carcasses from the compulsory slaughter scheme. Major problems included the large volume of carcasses and initial serious underestimation of the numbers that would arise. MAFF handled this difficult and unpopular disposal task energetically and competently.

- The disposal of SBO material was not MAFF's direct responsibility and was less straightforward to manage. Initially this material did not constitute waste as such because it was a marketable product for rendering into tallow and MBM. It did not become controlled waste, to be disposed of only at a licensed destination, until after the animal SBO ban and SEAC advice that the protein product of SBO should not be used as an agricultural fertiliser.

- Other forms of waste included effluent passing down drains to sewers and rivers. None of the usual precautions or conditions attached by water authorities to discharges would have inactivated the BSE agent.

- Blood, slaughterhouse and rendering plant waste, including that from plants that rendered SBO, and sewage sludge from works handling their effluents, might lawfully be spread as agricultural fertiliser.

- Some of the failures to identify and address these matters promptly can be attributed to the defective state of environmental regulatory action at the time, and the transitional turmoil of measures to rectify this.

- General waste disposal systems as a potential transmission pathway for BSE received scant attention from those handling BSE prior to 1996. The matter was not referred to or addressed by the Southwood Working Party, the Tyrrell Committee or SEAC. All of them advocated a systematic review of the destination of all bovine materials. Had this been carried out, it might have identified waste disposal issues.

13. The identification of vCJD

- The Southwood Working Party noted that if BSE were to be transmitted to humans it would be likely to resemble CJD and suggested that surveillance be put in place to identify atypical cases or changing patterns of the disease.

- The task of detecting any variation in the characteristics of cases of CJD which might indicate infection with BSE was entrusted to the CJD Surveillance Unit (CJDSU), a research team of dedicated medical scientists headed by Dr Robert Will, a neurologist with extensive experience of CJD.

- No role in this was given to the Public Health Laboratory Service (PHLS), an established service for the surveillance of new and existing disease, among other things.

- The decision to establish a new team specifically for this purpose was vindicated by the prompt detection of the emergence of vCJD by the CJDSU.

- The conclusion reached by SEAC on 16 March 1996 that the most likely explanation for the cases of a new variant of CJD in young people was exposure to BSE has since been compellingly supported by scientific evidence.

- It should have been apparent to both MAFF and DH by early February 1996 at the latest that there was a serious possibility that the scientists would conclude that it was likely that BSE had been transmitted to humans. The two Departments should have worked together, in consultation with SEAC, to explore the possible policy options that would be available should this occur.

- There was no interdepartmental discussion or consideration of policy options within either Department until the middle of March 1996. The views of SEAC were awaited, both as to whether the cases of vCJD were linked with BSE, and as to what action should be taken if they were. This was an inadequate response.

- Under intense pressure from the Government, on 20 March 1996 SEAC advised among other things that the appropriate course was that carcasses from cattle over 30 months old should be deboned in licensed plants supervised by the Meat Hygiene Service and the trimmings classified as SBO.

- The Government immediately announced that it was accepting this advice. In doing so it was wrong-footed, for this course proved neither practicable nor acceptable to the public. A policy of banning consumption of cattle over 30 months had to be introduced instead.

14. Victims and their families

- The unusual problems of the diagnosis, treatment and care of the early cases of vCJD meant that for some of the victims and their families the tragic horror of the disease was made the more difficult to bear by lack of the appropriate treatment, assistance and support.

- Victims of vCJD and their families have special needs which should be addressed.

15. Research

- The Southwood Working Party made wise recommendations in relation to research, not least that an expert committee be set up to advise on this.
- That committee, the Tyrrell Committee, rapidly recommended research priorities which formed the basis of much of the research that followed.
- After some initial delay, BSE research was adequately funded by the Government.
- Attempts to agree that a director, or 'supremo', should oversee and coordinate research were initiated by Sir Donald Acheson but foundered in the face of concerns on the part of the Research Councils and MAFF for their independence.
- Coordination of research effort is desirable in order to achieve:

    – identification of gaps in research;

    – determination of research priorities;

    – identification of the best sources of expert assistance;

    – a well-constructed plan for funding from the outset;

    – competition for research projects;

    – peer review of projects; and

    – efficient arrangements for provision of clinical material to researchers.

- A research supremo might have identified the following areas where research could profitably have been started earlier or pursued with more vigour:

    – experiments to transmit scrapie to cattle to test the scrapie origin assumption;

    – tests for BSE in sheep;

    – identification of the minimum infective dose which could transmit BSE orally to cattle;

    – assessment of the sensitivity of mice to BSE for use in experiments;

    – ante- and post-mortem tests for BSE;

    – a test for ruminant protein in compound feed;

    – epidemiology.

16. Some general lessons

- The lessons to be learned from the BSE story are set out in Chapter 14 of this volume.

1. Introduction

1 In December 1986 a new animal disease was discovered by the State Veterinary Service. It quickly became known as Bovine Spongiform Encephalopathy or BSE. It caused irreversible 'spongy' changes to the brains of cattle and was invariably fatal. The public called it 'mad cow disease'.

2 For ten years the Government told the people:

- there is no evidence that BSE can be transmitted to humans;
- it is most unlikely that BSE poses any risk to humans; and
- it is safe to eat beef.

3 Then, on 20 March 1996, Mr Stephen Dorrell, the Secretary of State for Health, stood up in Parliament and announced that ten young people had contracted a new variant of the harrowing, and invariably fatal, Creutzfeldt-Jakob disease – vCJD – and that it was probable that they had caught BSE. Further cases of vCJD were to follow. By September 2000 there had been over 80[1] cases and the frequency with which they were being reported seemed to be growing.

4 For nearly three years we have been examining all that is known about the history of BSE and vCJD and looking at how these diseases were handled by the Government and by others in the period between December 1986 and 20 March 1996. This Report sets out what we have found.

5 In 1986 the United Kingdom had a worldwide reputation for competence and efficiency in animal health and welfare matters, and in the handling of outbreaks of serious animal diseases. Its skilled veterinarians and scientists, with the State Veterinary Service and veterinary laboratories in the forefront, operated established processes to identify, contain and eradicate animal diseases. They worked closely with farmers, veterinarians in private practice, public health professionals and the relevant industrial sectors. They raised awareness, gave advice, and recommended statutory regulation where appropriate and compensation if need be. The process required well-established communication between advisers and practitioners, effective systems of animal surveillance and information-gathering, programmes of research, and detailed shared understanding of the links between animal and human health in all its aspects, including the food chain.

6 The UK also had highly regarded public health processes of long standing to handle outbreaks of human disease. These included surveillance, preventive action, such as immunisation and advice, and treatment. The health of the nation was at the heart of the remit of the Health Ministers and the professional responsibility of the four Chief Medical Officers, one for each part of the UK, who advised the Government.

7 What went wrong after the new fatal degenerative brain disease of cattle, BSE, emerged in 1986? Why did the announcement in 1996 that humans had probably

---

[1] Including probable cases who were still alive

been struck down by this particular brain disease find the guardians of public health and the world at large so shocked, and apparently unprepared, and leave the public so disillusioned? Our remit does not extend to the frantic diplomatic activity and other events after that date, but the consequences are still bearing heavily on the British economy and have inflicted tragedy on some families and left blighting uncertainty and fear hanging over many more.

8 The full extent and effects of the human disease will not be discernible for many years to come. Baffling questions include the unusual nature of Transmissible Spongiform Encephalopathies (TSEs), the reasons why specific people have become prey to the human version of BSE, and the extent to which others, particularly those exposed to the agent in the 1980s, may yet develop it. These difficult and still unresolved questions have hampered and bedevilled the whole course of events. What we do know is that as of September 2000, shortly before publication of this Report, over 80 victims of vCJD, most of them young, had had their lives destroyed and their families' happiness and hopes had been irreparably damaged.

9 BSE has been a peculiarly British disaster. Almost all the victims of vCJD have been in the United Kingdom. Only four other human victims of vCJD have been diagnosed elsewhere.[2] Over 170,000 cattle have been diagnosed with BSE here compared with fewer than 1,500 abroad, mostly it would appear traceable to British-sourced animals or infected feed at the beginning of the British epidemic. So far, over 4.7 million British cattle have had to be slaughtered, and their carcasses burned or buried as potentially dangerous waste.[3] A thriving high-quality cattle and meat export industry has been wiped out. The livelihood of thousands of farmers and businesses has been damaged. Even at this tail-end of the animal epidemic there were still over 2,000 cases of BSE notified in 1999 and cases continue to be reported as we write.

10 Small wonder that people want to know why it happened and whether it was handled wisely and well. In particular:

- What was the cause of BSE emerging and spreading country-wide? Was it as a result of intensive modern farming practices? Was it a result of inadequate regulation or lowered standards? Why is it so overwhelmingly the UK that has been afflicted?

- Seventy-four victims, mostly young people, have died of a new variant of CJD. Is it certain that they contracted this dreadful disease as a result of some form of connection with BSE? If so, why was it that they were struck down?

- Was the emergence of BSE and its threat to human health effectively handled by those whose responsibility it was to do so?

- Did individuals respond as they should have done, having regard to the state of knowledge at the time?

- Was the truth about the nature of BSE and the threat it posed concealed from the public? Has there been a cover-up?

- Did we make proper use of our scientists?

---

[2] This represents two confirmed and one probable case in France and one confirmed case in the Republic of Ireland.
Source: CJD Surveillance Unit, 20 September 2000
[3] Figures up to 30 June 2000. Source: MAFF

- Did our health and welfare services adequately cater for the special needs of those who contracted vCJD and their families?
- What lessons does the catastrophic course of events hold for public policy and the way we do things in the future?

11 These questions have been very much in our minds throughout this Inquiry, as we have explored exactly what happened day by day during the ten years that led up to the announcement of 20 March 1996 that BSE had probably generated a new and fatal human disease. Some questions, such as the numbers who are likely to succumb to the human disease, we are not in a position to answer. Our remit is to report to Ministers on the course of events and the adequacy of the responses to them in the light of knowledge at the time. We have sought to do so thoroughly and fairly. We have reviewed not only the years since BSE first emerged, but the events that led up to it. We have read a large number of scientific publications. We have sifted 3,000 files of documents, and have studied 1,200 statements and many contributions from the public, whom we have sought to keep fully up to date with every stage of our proceedings. We have listened to 138 days of public oral evidence from 333 witnesses.

12 A recurring theme in the BSE story – a point we look at in detail later in our Report – has been growing public suspicion and dissatisfaction that important information was not being shared and discussed openly so that people were denied proper choices in matters that deeply affected them and their families. One of our goals in settling the conduct of our Inquiry was to make our investigations as open as good practice and modern technology could ensure, with any significant material we received made freely available to all. Witnesses' statements and transcripts of our hearings have been made available free of charge to all with access to the Internet. Hundreds of fuller dossiers of assembled factual material have throughout been available in more conventional form for those who wish to inspect them at our offices. We have placed in the public domain a unique corpus of official documents, and we have sought to throw light on a range of normally internal public policy processes. Our aim has been to be as thorough, open and fair as we could possibly be. Annex 1 to this volume describes the procedures we adopted for this purpose.

13 We have welcomed the spirit of cooperation we have been shown by the previous and current administrations and many other organisations in opening their archives to us. As some of our witnesses pointed out, they too are consumers of animal products and they too have children and grandchildren whom they cherish. We have made heavy demands for information on many witnesses and the voluntary response has been remarkable.

Our task

14 Our Terms of Reference require us:

To establish and review the history of the emergence and identification of BSE and variant CJD in the United Kingdom, and of the action taken in response to it up to 20 March 1996; to reach conclusions on the adequacy of that response, taking into account the state of knowledge at the time; and to

FINDINGS AND CONCLUSIONS

> report on these matters to the Minister of Agriculture, Fisheries and Food, the Secretary of State for Health and the Secretaries of State for Scotland, Wales and Northern Ireland.

15 Establishing and reviewing the history of the emergence of BSE and vCJD requires us to consider what occurred and why. Ascertaining what occurred is not straightforward, for we believe that the initial emergence of BSE was neither recorded nor appreciated, and the aid of the epidemiologist is needed to try to reconstruct what happened. Ascertaining why BSE and vCJD occurred is even more difficult. Many scientists around the world have been conducting research which bears on these questions. We have reviewed the results of this research to see what, at the time of writing our Report, can be said with a reasonable degree of confidence about the causes of BSE and vCJD. Many questions remain unanswered, but we believe that a number of widely held beliefs can be shown to be misconceptions.

16 Next we are required to establish and review the history of the response to the emergence of BSE and vCJD up to 20 March 1996. That was the day on which the Government announced the identification of a new variant of CJD and the conclusion that the cases were probably linked to exposure to BSE.

17 Establishing the response to the emergence of vCJD involves focusing on the few months leading up to 20 March 1996, during which the emergence of the disease was identified. In contrast, considering the action taken in response to the emergence of BSE has been a massive exercise. That action spanned a period of nearly ten years, starting in December 1986, when the emergence of a new disease in cattle was first suspected. The action involved the five Government Departments to which this Report is addressed, and on occasion other Departments, the Prime Minister and Cabinet. It involved local authorities throughout the United Kingdom charged with enforcing Regulations introduced to deal with BSE. It involved many other public bodies. It involved the rendering industry, the animal feed industry, the food industry, the pharmaceutical industry, and, of course, the farming industry. It involved the media. It involved the consumer and it involved the public.

18 When we speak of the consumer, we do not refer simply to those who ate beef. Products derived from the cow enter the food chain in a variety of guises. Tallow, the fat that is extracted by the rendering process, and gelatine, derived from the skin and bones of cattle, are used in a wide variety of foodstuffs. But the public was involved not merely as consumers of food. Bovine tissues and fluids are used in, or in the production of, medicinal products swallowed, injected or inoculated. They are used in the manufacture of surgical devices. They are incorporated in cosmetics. The emergence of BSE put in question the safety of each of these products. It also raised questions about the handling of waste derived from the manufacture of these products or directly from carcasses.

19 Not only have we been required to establish the action taken in response to the emergence of BSE, we have been asked to reach conclusions on the adequacy of the response, taking into account the state of knowledge at the time.

INTRODUCTION

20 On the last day of the hearings we made the following observations about this part of our task:

> The mechanisms by which policy decisions in Government are taken are complex. The important decisions involve preparation of information and advice to submit to a Minister, preparation that often involves a number of different officials. It is easy with hindsight to assert that an assumption should not have been made, or that a decision was inadequate, misguided or dilatory, or that there was a culpable failure to take action that the situation required. Public opinion, as events unfolded and reached crisis point, has made many such value judgements. Hardly a day goes by today without BSE being referred to in the media as epitomising maladministration, usually by the use of an epithet such as 'the BSE scandal'. We believe that we have been asked to consider the adequacy of the response to BSE so that these accusations, insofar as they relate to the period with which we are concerned, can receive a fair and dispassionate consideration.

21 As we shall shortly explain, in the years with which we are concerned, most of those responsible for responding to the challenge posed by BSE emerge with credit. But we have found that a number of aspects of the response to BSE were inadequate. There are lessons to be learned from the events of those years. We stress that identifying those lessons is more important than examining whether individuals should be criticised. Nevertheless, any description of inadequacies is bound to lead people to ask whether individuals are to be criticised. We have given anxious consideration to that question.

22 A finding that an action constituted an inadequate response to BSE does not necessarily mean that those responsible for the action should be criticised. An action may not have been adequate because it did not satisfactorily deal with things that were known about a problem at the time. But it would not be right to criticise an individual unless, given the knowledge of that particular individual, he or she should have acted differently.

23 We have approached our task on the premise that it ought to be possible to identify those with responsibility for the policy decisions, the actions to implement policy and the public communications that together made up the response to BSE.

24 In practice we have found allocation of individual responsibility difficult. In part this has been due to the passage of time, which has rendered individual recollection of material facts at least unreliable and frequently non-existent. In part this has been due to the complexity of the administrative processes. The willingness of those concerned to give us unrestricted access to internal papers, and to disclose these to the public, has enabled us and the media and the public to gain an insight into those processes which we believe to be unprecedented.

25 Our Inquiry has led us to consider in depth:

- the relationship between Ministers and officials;
- the relationship between Government Departments;
- the relationship between administrators and professionals within Departments;

- the relationship between public authorities and expert advisers; and
- the relationship between central and local government.

26 These relationships formed the structure within which major and minor decisions of policy came to be taken and implemented.

27 When considering individual responsibility we have had to bear in mind this structure. We have had to bear in mind the way in which the public administrative system works. Many decisions are the product of a team effort to which individuals have made different contributions. A faulty decision may be the result of an error of judgement in assessing the available scientific and other data, or it may have resulted from an individual failure or failures in the provision of data, or the provision of expert advice in relation to it.

28 We have had to bear in mind the constraints on advisers and decision-makers: constraints of law, constraints of resources, constraints of established government policy; and constraints of the legitimate interests of the agricultural and other industries as well as those of the consumer. The background volumes of our Report (which, as we explain below, have been prepared by Inquiry staff) contain information about these constraints.

29 We describe in Annex 1 to this volume the procedures that we adopted to ensure that this Inquiry was thorough, open and fair. These included particular procedures adopted in Phase 2 of the Inquiry for those areas which we considered might give rise to criticisms of individuals. Fairness demanded that individuals be given notice of any potential criticisms. Such a course had its costs. Those notified of potential criticisms, and the lawyers advising them, naturally devoted and diverted their efforts to attempting to meet the criticisms. This tended to focus attention on the areas to which the potential criticisms related, albeit that these were not necessarily the most important areas of the Inquiry, and thus to unbalance the process.

30 In considering the adequacy of the action of individuals we have kept in the forefront of our minds the dangers of hindsight. We have had regard to all the surrounding circumstances which have often explained and excused action which at first blush seemed open to criticism. We have had well in mind that in any situation there is likely to be a range of responses from the inspired to the unimaginative, all of which fall within the compass of a reasonable response. Only where, having regard to all the relevant circumstances, we have concluded that the response of an individual fell below the standard to be expected of a person holding his or her position, have we indicated that the individual was at fault. We have done so in clear language, stating that the individual 'should' or 'should not' have acted in a particular way. Where we have not made an express criticism, none should be implied. So as to avoid any misunderstanding, a list of individual criticisms can be found in Annex 2 to this volume, with cross-references to locations in the Report where the matter is discussed.

31 Consistently with this approach, when considering the actions of Government Ministers, we have not adopted the traditional convention whereby Ministers are held accountable for the actions of those in their Department, regardless of their personal level of involvement. As with other individuals, we have only criticised a Minister where we have concluded that, in all the circumstances, his or her response fell below the standard to be expected of that Minister in the light of his or her knowledge at the time.

32 This is not to say that we have proceeded on the basis that a Minister should never be criticised for following advice from officials. The fact that a Minister has followed this course cannot preclude the conclusion that he or she should have acted differently. It is, however, an important factor when considering whether a Minister should be criticised.

33 There are some instances where we have found the response inadequate, but have not identified failings on the part of specific individuals. These are usually cases where we have felt that, having regard to the constraints on our time and resources, an attempt to identify individual responsibility could not be justified. In all such instances we would emphasise that it would be wrong and unfair to infer fault on the part of any individual.

The structure of the Report

34 Almost every aspect of the BSE story takes us into territory that may well be unfamiliar to the average reader of this Report. Anyone who wishes to follow the story fully will need to understand:

- the involvement of government in UK agriculture during and after the Second World War;
- the influence of the Common Agricultural Policy on agricultural production;
- the digestive system of the cow;
- intensive feeding methods designed to boost milk production;
- feed compounding;
- rendering;
- slaughterhouse techniques;
- the administrative structure of the Government Departments and local authorities involved;
- the powers available to government to regulate and enforce;
- the use made by government of advisory committees;
- basic human and animal biology;
- genetics; and
- current scientific knowledge in relation to the nature of Transmissible Spongiform Encephalopathies (TSEs).

35 These topics form the background to ten years of activity in response to the emergence of BSE. We must review that activity in context. A key consideration in an exercise as far-ranging and complex as this Inquiry is how best to present and make widely available the significant material and findings we have assembled.

36 We are conscious that while some will wish to follow, in detail, our examination of the BSE story, or some specific parts of it, most will not have the time or the energy for such an exercise. The majority will wish to read, in simple language, a summary account of the emergence of BSE and how it was handled, with particular reference to its implications for human health. More particularly, the majority will be looking to us to answer, as best we can, a number of questions about BSE, vCJD and the conduct of government in relation to them over the period with which we are concerned. This volume aims to meet those wishes of the majority.

37 The emergence of BSE called for responses of different kinds and in relation to different areas of activity. In this volume we propose to follow a topic-based approach. At the outset we shall explain the nature of Transmissible Spongiform Encephalopathies and examine the assumption which lies at the root of this Inquiry: that the variant of the human disease CJD is a consequence of the emergence of BSE. We conclude this chapter by setting out the BSE story in a nutshell. In the next chapter we have included sections about the industries which feature in the BSE story; how government was set up to handle an issue like BSE; and handling risk. We aim in that chapter to give much of the background that will enable the reader to follow the story in the rest of this volume.

38 Chapters 3 to 6 contain a narrative of a part of the BSE story which, for the most part, has been in the public eye:

- the emergence of BSE;
- the theories as to its cause;
- the measures taken to try to eradicate it;
- the concerns that humans might be able to catch BSE and worries about the safety of beef;
- the official reassurances about the risk to humans and the safety of beef; and
- the dreadful discovery that BSE had probably been transmitted to humans after all.

39 In Chapters 7 to 9 we turn to parts of the story of which the public was generally not aware at the time. As a result of recent media coverage, the subject matter of Chapter 7 – steps taken to address the possibility that BSE might have infected medicines, vaccines and cosmetics that used bovine products as ingredients or in the manufacturing process – has now become public. But Chapters 8, 9 and 10 deal with the less familiar topics of guidance given to occupational groups which may have been at risk from handling potentially infected tissues at work; the consideration given to tracing all the uses of bovine tissue and thus all possible pathways along which infection may have been transmitted; and the impact of BSE on pollution and waste control. In Chapter 11, we summarise our main findings about the part played by the Territorial Departments, as they then were, in Wales, Scotland and Northern Ireland.

40 In Chapter 12 we set out the conclusions we have been able to draw about the scientific response to BSE, dealing with some important questions, such as the origin of the BSE agent.

INTRODUCTION

41 We conclude with two chapters which fulfil what we believe to be the essence of our remit, that is, to understand why things happened in the way they did and to suggest how lessons may be learned from the BSE story for the benefit of those facing similarly difficult situations in future.

42 In summarising our findings and conclusions in a manner and at a length which we hope will make them accessible to all, we have had to paint with a broad brush and to leave untouched some parts of the gigantic canvas. The picture is painted in greater detail in the remaining 15 volumes, starting with Volume 2, which contains an analysis of the scientific evidence. Volumes 3 to 9 contain a detailed description and analysis of the events which are summarised in this volume.

43 Volume 10, which is a background volume, describes the impact of BSE on the economy and looks at how international trade was affected. Before BSE emerged, the majority of exports from the UK, of both live cattle and beef, went to the European Union (EU).[4] After BSE emerged, these exports were subjected to restrictions that were imposed under European law. They did, however, benefit from the protection of the Single European Market, which made it unlawful for individual members of the EU to impose more stringent requirements on UK exports. Our Terms of Reference require us to consider the response to the emergence of BSE in the UK. We have not traced the deliberations that took place in Europe – in which representations of the UK played a key role – which determined the extent of the restrictions consequent upon BSE that were placed on our trade with the EU.

44 So far as the export of live cattle was concerned, the EU response was to restrict this to cattle of a BSE-free provenance which, after 1990, were aged less than six months. So far as beef was concerned, exports were restricted to beef on the bone of a BSE-free provenance, or beef off the bone from which all obvious nervous and lymphatic tissue had been trimmed. From December 1994 there were exemptions in respect of beef from younger cattle.

45 Statistics of exports of cattle and beef during the period with which our Inquiry is concerned are set out in Chapter 5 of Volume 10. They make interesting reading. Despite the EU restrictions, our exports of live cattle to the EU climbed steadily between 1988 and 1994, dropping only slightly in 1995. Outside the EU, sales of live cattle slumped to negligible proportions after 1989. The value of exports of beef on and off the bone to the EU climbed by 1995 to well over double their value in 1987. Outside the EU, sales of beef off the bone slumped between 1986 and 1993, before recovering to close to previous levels. Sales of beef on the bone reduced to negligible proportions after 1987.

46 Volume 11 looks at the important role in the BSE story played by scientific committees and independent scientists. It forms the basis for a large number of lessons to be learned about the use of expert scientific committees which are set out in the final chapter of this volume.

47 The factual parts of these volumes have been based in large measure on 'draft factual accounts', which were collated from the evidence, were published as the

---

[4] The European Union (EU) came into existence on 1 November 1993 as a result of the Maastricht Treaty. It incorporated but did not replace the European Community. Throughout the volumes of this Report, the term EU is generally used for consistency's sake (even if sometimes chronologically incorrect), except where specific reference is made to the functions conferred by the European Community Treaty or to its legal effect

Findings and Conclusions

Inquiry progressed, and have been revised on the basis of comments received and additional evidence. To these we have added, in Volumes 2 to 9 and 11, sections of comment and discussion in which we have considered conflicts of evidence and explained the conclusions that we have drawn from the facts. Readers who want detailed explanations for the findings and conclusions set out in this volume will find them in those volumes. They will also find an abundance of references to source material, which will remain accessible to the public. In this volume we have sought to keep references to a minimum.

48 Volumes 10 and 12 to 15 contain background material which provides a detailed context in which the BSE story is set. Volume 16 contains relevant reference material. It should be noted that Volumes 10 and 12 to 16 are background volumes which have been prepared by researchers on the Inquiry team under our supervision and guidance. Conclusions of the Committee are not to be found in these volumes.

49 It has been clear that speedy access to Inquiry material through the Internet has been widely appreciated, and we have therefore cast and referenced our Report and its supporting material in a form immediately transmissible through this medium. We hope that it will thus prove another example of open practice on matters of legitimate public concern.

Transmissible Spongiform Encephalopathies

50 Our Terms of Reference speak of two diseases: BSE, a disease of cattle; and variant CJD, a human disease. These are varieties from a rare group of diseases known as Transmissible Spongiform Encephalopathies (TSEs). TSEs cause the appearance of microscopic holes in the brain, giving it a sponge-like appearance – hence the term 'spongiform'. They are invariably fatal and affect both humans and animals. In 1986 a number of TSEs had been identified both in animals – scrapie in sheep and goats, Chronic Wasting Disease (CWD) in wild deer in North America and Transmissible Mink Encephalopathy (TME); and in humans – Creutzfeldt-Jakob Disease (CJD), Gerstmann-Sträussler Syndrome (GSS), kuru and Fatal Familial Insomnia (FFI). Although a signal feature of these diseases is that they are transmissible in the manner described in paragraph 52 below, they can occur, at least in humans and probably in other species, as a result of a genetic mutation that is inherited or, in some cases, that may arise spontaneously.

51 When BSE was first identified, the nature of the infectious agents causing TSEs was a matter of controversy. It was known that the agents were extremely difficult to inactivate – they could withstand treatments commonly used to disinfect virus-contaminated materials – and that researchers had failed to detect an immune response in hosts to their presence in a variety of experiments. Although these features suggested that TSEs were not caused by conventional viruses, some believed that they must be caused by an unconventional virus. This belief was challenged by those who thought that TSEs were transmitted as a result of a reaction between proteins. This theory has now won general, though not universal, acceptance.

52 How, under this theory, does transmission of these diseases occur? Let us take BSE as an example. The building blocks of every animal, including the human animal, are proteins. These are minute particles which have different chemical compositions. BSE involves the deformation of one of these proteins (prion protein)[5] in very large numbers within the brain of the cow, until the brain develops a spongy appearance and is fatally damaged. The same deformation of this protein takes place in other specific tissues in the cow. If some of the deformed proteins of an animal suffering from BSE are introduced into the body of another animal or into a human ('the host'), they may induce similar proteins that are found in the host to deform in the same way. By a kind of chain reaction, deformation of these proteins may spread to and within the brain of the host, until finally the brain is so damaged that the host is taken ill and dies.

53 The prion protein exists in its normal form in all animals, but its chemical composition is not precisely the same in each. It can even have slight variations in animals of the same species as a result of minor variations of the prion gene. The more similar the prion protein in infected animals to that in the host animal, the easier the transmission of a TSE appears to be. Thus transmission is easiest between animals of the same species. When the animals are of different species, the 'species barrier' will sometimes prevent transmission altogether.

54 The obvious way in which deformed protein from an animal incubating a TSE may be introduced into another animal is as food. There are, however, other possibilities. For instance, medical products administered by injection are sometimes derived from animal tissues or fluids. Experiments have shown that it is very much easier to transmit a TSE to an animal by injecting infected tissue directly into the brain than by feeding it to the animal. A minute quantity will suffice for such intracerebral transmission; indeed CJD has sometimes been transmitted on surgical instruments used in neuro-surgery despite their sterilisation.

Transmission to humans

55 The two most worrying questions people ask about BSE are:

- Is it certain that the victims of the variant form of CJD have caught BSE?
- And, if so, how many victims are there likely to be?

56 We shall here summarise our conclusions about the link between BSE and vCJD, which are the subject of more detailed coverage in vol. 2: *Science* and in vol. 8: *Variant CJD*.

57 The unusual clinical features and novel pathology of the early cases of CJD in young people suggested this was a new variant of the disease. Much experimental work has been done to investigate whether there is a link between this new variant of CJD and BSE, and we believe there is now sufficient evidence to be confident that vCJD is caused by the transmission of BSE to humans. In outline, the main evidence, in addition to the temporal and geographical association of the two diseases, which leads us to reach this conclusion is as follows:

---

[5] Professor Stanley Prusiner, who coined the term 'prion protein' and who was awarded a Nobel prize for his work in this field, assisted us with a presentation of the prion theory in Phase 1 of the Inquiry

i. in strain-typing studies in both mice and primates the disease patterns (incubation period and disease pathology) of BSE, vCJD, feline spongiform encephalopathy (FSE) and TSEs of exotic ruminants were shown to be extremely similar,[6] while differing from those of scrapie and sporadic CJD;

ii. patterns known as glycosylation patterns, produced by analysing samples of brain using a technique called western blotting, are the same for BSE and vCJD. The patterns for BSE and vCJD are different from those for other TSEs such as sporadic CJD and iatrogenic CJD; and

iii. in transgenic mice in which the mouse prion gene has been replaced by the bovine prion gene, inoculation with tissue derived from BSE-infected cattle produces the same disease pattern and incubation period as inoculation with tissue derived from patients with vCJD.

58 It is not possible to say whether BSE was transmitted to humans through consumption of beef or beef products, or by some other means; nor is it possible to say when individual infection occurred. There are a number of other unanswered questions:

- *Why does vCJD affect young people?* Possible explanations meriting further investigation include: the possible disproportionate consumption by young people of beefburgers, some of which contained high-risk material; higher incidence of infections such as tonsillitis or gastroenteritis in children than adults, giving rise to transmission through broken skin or mucous membranes; infection through gum lesions associated with eruption of teeth; and transmission via childhood vaccines prepared in cultures containing bovine constituents.

- *How many more people will succumb to vCJD?* To attempt to answer this question is not required by our Terms of Reference, nor would we feel able to do so. Estimates of the possible size of a vCJD epidemic are made difficult by the many variables associated with the disease. Many important factors in determining the likelihood of BSE transmission to an individual are unknown, such as dose, route of exposure, incubation period, genetic susceptibility and scale of the species barrier between cattle and humans. Nevertheless, several groups of epidemiologists and statisticians have attempted to predict the possible number of cases. Projections have in the past ranged from small numbers to many millions and it is not possible at this stage to reach a firm estimate.

- *Is occupation a risk factor in vCJD?* Among occupational groups exposed to BSE, to date farmers are the only group to have an excess over the incidence of CJD for the population as a whole. Between 1990 and 1996 four cases of CJD occurred in farmers who were known to have had cases of BSE on their farms. In addition, two farmers' wives succumbed to CJD. The affected farmers were aged between 54 and 64 and had signs and symptoms typical of sporadic CJD. They did not have glycosylation patterns associated with vCJD. To date, no one has demonstrated a link between these cases and BSE.

---

[6] It is thought that domestic cats caught FSE and exotic ruminants a related TSE through the consumption of BSE-infected food

The story in a nutshell

What happened?

59 This is a summary of the more significant events in the BSE story. In responding to the emergence of BSE, the Ministry of Agriculture, Fisheries and Food (MAFF) and the Department of Health (DH) took the lead. For the most part, Wales, Scotland and Northern Ireland followed that lead. This summary will focus on the action taken by MAFF and DH.

60 A TSE known as scrapie has been endemic in the sheep population of the UK for nearly 200 years. In the later stages of the disease the fabric of the brain is attacked. The pathologist can diagnose the disease by the spongiform appearance of the diseased brain. At the end of 1986 pathologists at the Central Veterinary Laboratory (CVL) identified similar degenerative changes in the brain samples of diseased cattle from two different herds. These were early cases of BSE.

61 By May 1987 this novel disease had been confirmed in four herds. No publicity, even within the State Veterinary Service (SVS), had been given to these early cases and it is likely that others had gone unrecognised and unreported. From May, however, the fact of the existence of a novel disease was gradually disseminated and Mr John Wilesmith, head of the CVL's Epidemiology Department, was asked to investigate its cause.

62 Over the next six months, as he carried out his task, reported incidents of the disease proliferated. By 15 December 1987 there were 95 confirmed cases on 80 farms. Mr Wilesmith had formed the provisional view that the cause of the outbreak was contaminated meat and bone meal (MBM) that had been incorporated in cattle feed. His confidence in this theory grew stronger early in 1988, and he concluded that the likely contaminant was offal of scrapie-infected sheep, rendered down to make MBM. Enquiries of feed compounders tended to confirm this view.

63 On 18 May 1988 Mr John MacGregor, the Minister of Agriculture, on the advice of Mr William Rees, the Chief Veterinary Officer (CVO), decided on what proved to be the principal step taken to eradicate BSE. A prohibition on feeding ruminant protein to ruminants ('the ruminant feed ban') was introduced on 14 June 1988 to take effect on 18 July. This was, at the time, regarded as a measure to protect animal health. The risk that BSE posed to human health had not, however, been ignored.

64 Officials at MAFF had been concerned from the outset at the possibility that BSE might pose a risk to human health. Diseased cattle were going into the human food chain. Scrapie was not transmissible to humans, but there was no certainty that the same would be true of BSE. By 19 February 1988, 264 cases of BSE from 223 farms had been confirmed. On 24 February Mr Derek Andrews, the Permanent Secretary, forwarded a submission to Mr MacGregor. This recommended that BSE should be made a notifiable disease and that a policy of compulsory slaughter with compensation should be introduced. Mr MacGregor had reservations about such a policy and accepted the suggestion that the advice of Sir Donald Acheson, the Chief Medical Officer (CMO), should be sought on the implications that BSE had for human health.

Findings and Conclusions

65 Sir Donald, in turn, recommended that an expert working party should be set up to advise on the implications of BSE. This was done. The Working Party was chaired by Sir Richard Southwood.

66 Before the first meeting of the Southwood Working Party, and at the same time that the ruminant feed ban was introduced, Mr MacGregor, on the advice of his officials, introduced a requirement for compulsory notification of all cases of BSE.

67 On 21 June 1988 the Southwood Working Party made interim recommendations that included the compulsory slaughter of animals showing symptoms of BSE and the setting up of a committee to advise on research. The Government accepted these recommendations and, on 8 August 1988, an Order came into force making slaughter of BSE suspects compulsory. Compensation of 50 per cent of the sound value of the animal was paid if, on post-mortem, it was shown to have had BSE and 100 per cent if it did not. Although made under the Animal Health Act 1981, the primary object of this measure was to take sick animals out of the human food chain.

68 By 13 January 1989, 2,296 cases of BSE had been confirmed on 1,742 farms.

69 The *Southwood Report* was submitted to Ministers on 9 February 1989. This endorsed Mr Wilesmith's conclusion that the source of infection was probably scrapie-infected meat and bone meal. It concluded that it was 'most unlikely that BSE would have any implications for human health'. It recommended that the Health and Safety Executive (HSE) and the authorities responsible for human and veterinary medicines, which had already been alerted by the Working Party, should take appropriate measures to address possible risks posed by BSE, and advised manufacturers of baby foods not to include in their products ruminant offal including thymus, which, from what was known about scrapie, would be most likely to be infective. Sir Richard Southwood clarified later in February that this offal did not include liver or kidney.

70 The Working Party concluded that the risk posed by BSE-infected animals which had not yet developed clinical signs did not justify any further measures to protect human food. The Government accepted this, and on publication of the *Southwood Report* announced that secondary legislation would make it illegal to sell baby food containing the types of offal identified by the Report. MAFF Ministers, however, had concerns which, after discussion with officials and with DH and after wide consultation, led, on 13 November 1989, to the introduction of a ban on the use for human consumption of Specified Bovine Offals (SBO), namely those tissues in cattle considered most likely to be infective. This became known as 'the human SBO ban'. Tissues from cattle aged under six months were exempt from the ban on the basis that scrapie infectivity had not been found in lambs of this age.

71 Meanwhile, on 27 February 1989, the establishment of a committee chaired by Dr David Tyrrell was announced. The Tyrrell Committee was to advise on research in relation to BSE, thus implementing one of the first recommendations of the Southwood Working Party. This Committee met three times and delivered to the Minister of Agriculture and the Secretary of State for Health what they described as an 'Interim Report' on 13 June 1989. This identified the key research questions that needed to be answered and set in an order of priority the research studies needed to answer those questions.

INTRODUCTION

72 The Report was not published until 9 January 1990. By this time funding had been put in place which enabled the Food Minister, Mr David Maclean, to announce that all projects identified by the Tyrrell Committee as 'urgent' or of 'high priority' had either been put in train or would start as soon as possible. Experiments to check the belief that BSE was transmissible had been put in hand at an early stage. In September 1988 transmission to mice by intracerebral inoculation of brain tissue had been confirmed. By February 1990 transmission to cattle had been established by the same route and transmission to mice by oral ingestion had been achieved.

73 Meanwhile, on 28 July 1989, the EU banned the export of UK cattle born before 18 July 1988 and of offspring of affected or suspect females. This was the first of a number of restrictions placed by the EU on the export from the UK of live cattle and (from June 1990) of beef.

74 By the end of 1989, 10,091 cases of BSE had been confirmed in the UK.

75 Anxiety had been expressed in many quarters that 50 per cent compensation might be inadequate to procure full compliance with the requirement to notify BSE suspects and, on 14 February 1990, Mr John Gummer, who had succeeded Mr MacGregor as Minister of Agriculture, introduced entitlement to 100 per cent compensation.

76 On 1 March 1990 the EU restricted exports of live cattle to those aged less than six months. Importing Member States were required to ensure that these were slaughtered before they reached that age. Offspring of whatever age of affected or suspected females continued to be banned from export.

77 On 3 April it was announced that Dr Tyrrell was to chair a new expert committee – the Spongiform Encephalopathy Advisory Committee (SEAC). The Committee had a wider membership than the Tyrrell Committee and wider terms of reference:

> To advise the Ministry of Agriculture, Fisheries and Food and the Department of Health on matters relating to spongiform encephalopathies.

78 It was government policy in relation to BSE to act on 'the best scientific advice'. Thereafter the Government was to look to SEAC to provide that advice.

79 One of the recommendations of the Southwood Working Party had been the need for surveillance of CJD cases in order to detect whether there were any changes in their incidence that might be attributable to BSE. In May 1990 the CJD Surveillance Unit was set up under Dr Robert Will, a consultant neurologist at the Western General Hospital in Edinburgh.

80 On 10 May 1990 it was announced that a Siamese cat had died of a spongiform encephalopathy – the first known case of feline spongiform encephalopathy (FSE). This resulted in a rash of media comment, speculating that the cat had caught BSE and that humans might be next. Humberside Education Authority had already banned beef from school meals and a number of other Authorities threatened to follow this example. Public statements by the CMO and by Mr Gummer that beef was safe to eat failed wholly to reassure. The House of Commons Agriculture Committee announced an Inquiry into BSE. After receiving evidence from most of the key players in the BSE story, the Committee reported on 12 July 1990 that, while there were too many unknowns to say anything with absolute certainty, 'we heard no evidence of any sort to constrain those taking a more balanced view of the risks from eating beef'. The measures taken by the Government 'should reassure people that eating beef is safe'.

81 On 8 June 1990 the EU Council of Ministers agreed that bone-in beef exported from the UK must come from holdings where BSE had not been confirmed in the previous two years, while boneless beef was required to have obvious nervous and lymphatic tissue removed.

82 Meanwhile, there had been controversy as to whether the SBO that had been banned from human food should be permitted to be fed to animals. Pet food manufacturers had voluntarily ceased to incorporate it in their products. UKASTA, the feed producers' trade association, had pressed strongly for a ban on including SBO in the material rendered to make MBM for inclusion in pig and poultry feed, and advised their members to exclude it. MAFF officials and Ministers opposed a ban on the ground that it was without any scientific justification. SEAC was about to advise on this question when, early in September, a pig, which had been inoculated with BSE-infected brain tissue, succumbed to the disease. In an emergency meeting SEAC advised that, as a precautionary measure, SBO should not be fed to any animals. MAFF, which had anticipated this possibility, immediately banned the incorporation of SBO or its products in animal feed ('the animal SBO ban'). Export of feed containing SBO to the EU was also banned. This was followed in July 1991 by a ban on the export of material derived from SBO to third countries.

83 Among the many matters on which SEAC was asked to advise were slaughterhouse practices. There was concern that the removal of brain and spinal cord (both SBO) in slaughterhouses might contaminate meat going for human consumption. There was also concern about the practice of the mechanical recovery of remnants of meat and other tissues adhering to the vertebral column, in that these might include scraps of spinal cord not cleanly removed by slaughterhouse operators. SEAC advised that head meat should be removed before brain, but that no further measures were necessary provided that the rules were properly followed and supervised. This advice was implemented first by guidance and then, in March 1992, by statutory regulation.

84 By the end of 1990, 24,396 cases of BSE had been confirmed in the United Kingdom.

85 One of a number of recommendations of the House of Commons Agriculture Committee was that the Government should 'establish an expert committee to examine the whole range of animal feeds and advise on how industries which produce them should be regulated'. Some debate ensued as to how to implement this recommendation, but on 6 February 1991 MAFF announced the establishment of an Expert Group on Animal Feedingstuffs chaired by Professor Eric Lamming. It met on 14 occasions over the next year and reported on 15 June 1992. The Group considered the steps taken to prevent the BSE agent being transmitted to animals in feed and concluded that they were satisfactory and adequate. In particular the Group considered whether the practice of feeding animal protein to animals should be discontinued. It decided that there was no scientific justification for such a step. It did, however, recommend that:

> . . . an independent Animal Feedingstuffs Advisory Committee be established to take an overview of all feedingstuffs issues.

86 Although the Government initially accepted this recommendation, it subsequently decided not to proceed with it.

87 With compulsory slaughter of sick animals and the human SBO ban to deal with potentially infective tissues in apparently healthy animals incubating BSE, the Government considered that there were in place appropriate measures to deal with the risk that BSE might be transmissible to humans in food. Action was taken to see that medicinal products both for humans and for animals were not sourced from potentially infective bovine tissues. Ruminants were protected by the ruminant feed ban and other animals by the animal SBO ban. No further major measures were considered necessary to protect human or animal health in the period with which we are concerned. In March 1992 SEAC concluded 'that the measures at present in place provide adequate safeguards for human and animal health'. Several relatively uneventful years were to pass before it became apparent that the measures in place were not achieving all that had been expected of them.

88 Because of BSE's lengthy incubation period, it was appreciated when introducing the ruminant feed ban that years would pass before it would have a visible effect. What was not known was the rate at which cattle had been infected in the period up to 18 July 1988, when the ruminant feed ban came into force. At the time of the *Southwood Report* suspected cases of BSE were being reported at the rate of about 400 a month. It was considered that these had been infected with scrapie and that this source would have continued to infect cattle until the ban at about the same rate. Whether, or to what extent, recycling of BSE might have increased the rate of infection was not known.

89 It soon became apparent from the numbers of BSE cases reported[7] that the rate of infection had not reached a plateau, but had been increasing rapidly in the years leading up to the ruminant feed ban, and that the reason for this was the effect of recycling the BSE agent in MBM.

90 Thus the Government found it had to deal with many more cases infected before the ban than it had expected. But of even more concern were cases in cattle that had been born after the ban (BABs). The first of these was announced on 27 March 1991.

91 When exploring the possible sources of infection of the BABs, the CVL epidemiologists were able to rule out maternal transmission in most cases. The likely source of infection of the earlier BABs was thought to be ruminant feed in which ruminant protein had been incorporated before the ban and which was in the distribution pipeline, or still unused on farms when the ban came into force. This remained the view of MAFF officials at the beginning of 1994, by which time Mrs Gillian Shephard had succeeded Mr Gummer as Minister of Agriculture. Cross-contamination of ruminant feed by non-ruminant feed in the feedmills was

[7] For statistics, see vol. 16: *Reference Material*

considered, but discounted after September 1990, when the animal SBO ban should have prevented SBO from being incorporated in any animal feed.

92 In the course of 1994 opinions changed as to the source of infection of BABs. By August the CVL had reached the conclusion that the more recent BABs had been infected by feed which had been contaminated in the feedmill by feed containing ruminant protein. Two factors had led to this conclusion. First, there had been an increasing volume of evidence, some of it cogent, of widespread infringement of the animal SBO ban, so that SBO was contaminating non-ruminant feed. Second, interim results of an experiment, which started in 1992, indicated that a single quantity of as little as 1 gram of infective material – the size of two peppercorns – had sufficed to infect cattle to which this had been fed.

93 MAFF officials approached the problem of the cross-contamination of cattle feed on two fronts. Their primary emphasis was on tightening up the implementation of the animal SBO ban. This was facilitated by the transfer of enforcement functions in slaughterhouses to central government. What had been the responsibility of some hundreds of individual local authorities became the task of a new national Meat Hygiene Service (MHS) from 1 April 1995. A revised statutory scheme was introduced that required SBO to be identified by a distinctive blue dye and kept separate at all times from other material. At the same time plants rendering SBO were required to do so in separate facilities. The consultation process was thorough and lengthy, with the result that the introduction of the new Regulations was not completed until August 1995. Their introduction was combined with a campaign of more rigorous enforcement and monitoring of the Regulations by the MHS and the Veterinary Field Service (VFS).

94 At the same time as tightening up on the implementation of the animal SBO ban, MAFF officials took steps to address cross-contamination in feedmills. So far as these were concerned, effective monitoring of compliance with the ruminant feed ban had been initially impossible for want of any method of testing for the presence of ruminant protein in animal feed. It had been hoped that an 'ELISA test' would be perfected within about 12 months, capable of detecting this. In the event, it was not until 1994 that the test was ready for use, and even then its results were not sufficiently reliable to provide evidence that would support a prosecution for breach of the Regulations. The test was, however, employed on a voluntary basis, with cooperation from UKASTA, and resulted in at least some feedmills taking steps to reduce the possibility of cross-contamination.

95 Hindsight confirms that, between 1989 and 1994, the ruminant feed ban had resulted in a steady but substantial year-on-year reduction in the numbers of infections, and that the measures taken in 1994 and 1995 radically accelerated this decline (see Volume 16, Figures 3.2 and 3.34).

96 The years 1994 and 1995 also saw developments in relation to the risks posed by BSE to human health. An interim result of a pathogenesis experiment conducted by the CVL demonstrated infectivity in the distal ileum (small intestine) of a calf within six months of oral infection with BSE. This led MAFF, with the agreement of DH, to extend the human SBO ban to include the intestines and thymus of calves which had died aged over two months.

INTRODUCTION

97 On 27 July 1994 the European Commission decided that existing restrictions on the export of UK beef should be replaced with two measures. One was a ban on export of bone-in beef except from cattle which had not been on holdings where BSE had been confirmed in the previous six years. The other measure affected beef from cattle which had been on such a holding within that time. This could not be exported unless it was deboned with adherent tissues removed. In December 1994 the Commission amended this decision to exempt from these measures beef from cattle born after 1 January 1992. Subsequently in July 1995 this exemption was replaced with one that exempted beef from cattle less than 30 months of age at slaughter.

98 In July 1994 Mrs Shephard was succeeded by Mr William Waldegrave, who oversaw the introduction of the MHS. He in turn was succeeded by Mr Douglas Hogg in July 1995. At the direction of Mr Hogg, the MHS set about raising standards of meat inspection, a task that was to prove to require the employment of several hundred additional staff.

99 More rigorous monitoring of slaughterhouses in 1995 disclosed a number of occasions on which Meat Inspectors had applied the health stamp to a carcass to which fragments of spinal cord remained attached. This led SEAC to recommend a ban on the practice of extracting mechanically recovered meat (MRM) from the spinal column of cattle. MAFF accepted that advice and introduced the ban in December 1995.

100 In the course of 1995 a number of events served to increase public anxiety that it might be possible to contract CJD as a consequence of eating beef. Cases of CJD were reported in farmers whose herds had had BSE and in several young people – the latter being particularly significant because up until then the disease had almost invariably struck down its victims late in life. A distinguished scientist questioned the safety of beef offal. These events received wide media coverage. The CMO and the Secretary of State for Health each responded with public assurances that it was safe to eat beef.

101 The first two months of 1996 saw the CJD Surveillance Unit and SEAC concerned at an increasing number of young victims of CJD. On 16 March SEAC advised the Government that a new variant of CJD had been identified in young people and that the most likely explanation was that these were linked to exposure to BSE before the introduction of the SBO ban in 1989. A series of urgent meetings of Ministers and then of the Cabinet ensued, and SEAC's advice was sought as to further precautionary measures.

102 On 20 March 1996 the Government announced the likelihood that the recent cases of CJD in young people had resulted from exposure to BSE before 1989 and stated its intention to adopt further precautionary measures in accordance with SEAC's advice. These were that carcasses from cattle aged over 30 months must be deboned and that the use of MBM in feed for all farm animals would be banned. These measures proved inadequate to reassure the public and, within two weeks, were replaced with a total ban on cattle over the age of 30 months being used for human food or animal feed.

103 By 20 March 1996 approximately 160,000 cattle affected by BSE had been slaughtered. In addition about 30,000 cattle suspected of BSE, but not confirmed to have the disease, were slaughtered. These figures can be compared with over 3.3 million cattle slaughtered and destroyed under the Over Thirty Month Scheme in the period from March 1996 to the end of 1999.

104 This brief narrative has concentrated on events that have been most in the public eye. As we explained above, we shall also cover in later chapters of this volume precautionary measures taken in areas which, while important, did not come to the attention of the general public. These include medicines, cosmetics and occupational health.

Why did it happen?

105 The Report of an Inquiry such as this inevitably focuses on the areas where things went wrong. It is those areas that government and the public are most anxious to have thoroughly explored. For this reason we think it desirable to give at the outset an overview of why things happened in the way that they did.

106 Why initially a cow or cows developed BSE will probably never be known. Why the early case or cases began a chain of transmission that ended with hundreds of thousands of cattle becoming infected is now clear. It was because of the practice of rendering cattle offal, including brain and spinal cord, to produce animal protein in the form of meat and bone meal (MBM), and including MBM in compound cattle feed. This resulted in the recycling and wide distribution of the BSE agent.

107 Many have expressed the view that it was not surprising that a practice as unnatural as feeding ruminant protein to ruminants should result in a plague such as BSE. Had BSE emerged soon after this practice was introduced, there might have been force in this reaction. However, the practice of feeding MBM to animals in the UK dates back at least to 1926, when it was given statutory recognition in the Fertilisers and Feedingstuffs Act of that year. It is a practice which has also been followed in many other countries. It was recognised that it was important that the rendering process should inactivate conventional pathogens. Experience had not suggested that the practice involved any other risks. In these circumstances we can understand why no one foresaw that the practice of feeding ruminant protein to ruminants might give rise to a disaster such as the BSE epidemic. Accusations have been made both against the Government and against renderers of causing BSE by relaxing rendering standards. As we shall explain, changes in rendering practices and regulatory requirements are unlikely to have made any substantial difference.

108 There were a number of factors that made it inevitable that, whatever measures were taken in response to its emergence, BSE would be a tragic disaster:

- it had an incubation period of five years on average;
- it tended to strike a single cow in a herd;
- it had clinical signs which were similar to those of a number of other diseases in cattle;
- it was impossible to diagnose before clinical signs appeared; and
- it was transmissible to human beings, but with a much longer incubation period than that in cattle.

109 These factors had the following consequences:

- the emergence of the disease may well have gone undetected for ten years or more from the time of the first cases. A farmer would not be likely to send a single casualty for a post-mortem. It was only when, by chance, several cases were experienced on the same farm that the pathology was carried out that disclosed the new disease;

- by the time that BSE was identified as a new disease, as many as 50,000 cattle are likely to have been infected;[8]

- it is also likely that by this time some of the human victims had been infected;

- it was not until nearly ten years after BSE was identified as a new disease in cattle that the first human victims succumbed to the disease, thus showing that, contrary to expectation, it was transmissible to humans.

110 Given the practice of pooling and recycling cattle remains in animal feed, this sequence of events flowed inevitably from the first cases of BSE. It was inevitable that, whatever measures were taken, many thousands of cows would succumb to the disease in the years to come. It was inevitable that if humans were susceptible to the disease, some would be infected with it before its existence was even suspected.

111 The measures that were taken in response to the emergence of BSE greatly reduced the scale of the disaster. The MBM component of feed was diagnosed as the vector responsible for the disease with commendable speed, and the ruminant feed ban was a swift and appropriate response. That ban reduced the rate of infection by 80 per cent overnight and established a diminishing trend which would, ultimately, have resulted in the eradication of the disease. Unhappily, as the cases born after the ban were to demonstrate, there were shortcomings in formulating and carrying out both the ruminant feed ban and the animal SBO ban, which should have provided a second line of defence against infection of cattle feed. These shortcomings had serious consequences. Over 41,000 cattle that developed clinical signs of BSE in the years that followed were infected after the ruminant feed ban came into effect. Many more must have been infected but slaughtered before the signs developed. When the link between BSE and the new variant of CJD became apparent in March 1996, the Government was unable to demonstrate that the source of infection had been completely cut off. Had they been able to do so, some of the drastic measures that followed might have been avoided. The reasons for these shortcomings receive detailed consideration in our Report.

112 There is a popular misconception that the Government did nothing to protect the public against the risk BSE might pose to human health until the likelihood of transmissibility was demonstrated in 1996. It is important to emphasise that the most significant measures to protect human health were taken at a time when the likelihood of transmissibility to humans was considered to be remote. Those were the compulsory slaughter and destruction of sick animals introduced in August 1988 and later, in November 1989, the human SBO ban, which was intended to remove from the human food chain those parts of apparently healthy cattle most likely to be infective if the animals were incubating BSE. At the same time steps were taken to ensure that bovine ingredients of medicines came from BSE-free sources.

---

[8] S9 Anderson para. 1

113 These were vitally important measures. For a period of nearly ten years continuous consideration was given to addressing the possibility that BSE might be transmissible to humans, although few believed that there was any likelihood of it. This is a matter for commendation.

114 Yet again, however, there were shortcomings: shortcomings which led to delay in introduction of the precautionary measures, and shortcomings in formulating and carrying out the ban. Despite the SBO ban, some potentially infective bovine tissues continued to enter the human food chain. The reasons for these shortcomings also receive detailed consideration in our Report.

115 The other casualty of the BSE story has been the destruction of the credibility of government pronouncements. Those responsible for public pronouncements – or at least some of them – were aware of the possibility that humans might have become infected before the slaughter policy and the SBO ban were introduced. They saw no reason to draw attention to this. They believed that the measures taken had effectively removed the 'theoretical risk' of infection. They were concerned that the public should not be misled by scaremongers or the media into believing that it was dangerous to eat beef when this was not the case. Ministers and, on occasion, the Chief Medical Officers, made statements about the safety of beef which were intended to reassure the public. Insofar as these statements were believed, many clearly treated them as assurances that BSE posed no danger to human beings. In the case of some, there was a growing scepticism as the media reported cases of possible human victims of BSE which were then challenged by the Government. When on 20 March 1996 it was announced that cases of new variant CJD were probably attributable to contact with BSE before precautionary Regulations were introduced, the reaction of the public was that they had been misled, and deliberately misled, by the Government.

116 We have examined with care the public pronouncements that were made about the risks posed by BSE, and have concluded that allegations of a government 'cover-up' of the risks posed by BSE cannot be substantiated. There were, however, mistakes in the way risk was communicated to the public, and there are lessons to be learned from these.

117 As we go through the story we shall describe in greater detail what happened and how it came to happen in the way it did. We shall consider the response to BSE of the individuals principally concerned in the story. At the end of this volume we shall review what went right and what went wrong, before turning to the lessons to be learned from the BSE story.

2. Setting the context

118 In this chapter we provide some basic information about the context in which BSE emerged and in which people, both within government and without, had to respond. We do this in order to assist readers in understanding the significance of various parts of the narrative story which follows. We set out thumbnail sketches of the industries that were principally affected by BSE and some key features of how government works. More detailed descriptions of all these are to be found in the background volumes. We also explain some of the concepts involved in handling risk.

The cattle industry

119 At the time BSE emerged, beef and dairy farming was the largest sector of UK agriculture (see vol. 12: *Livestock Farming*). The output from milk, fattened cattle and calves totalled some £5 billion, nearly 38 per cent of the entire UK agricultural output. With a cattle population of some 12.7 million, the UK produced 97 per cent of the beef and veal required to supply the needs of the domestic market, and sufficient liquid milk to supply 100 per cent of domestic demand for milk and almost 70 per cent of domestic demand for butter and cheese.

120 This impressive degree of self-sufficiency was the result of the policies of successive governments which, in the period after the Second World War, had sought to increase domestic food production in order to reduce reliance on imported food and to foster rural communities. Incentives to increase production levels even further were provided in 1973, when the UK joined the European Economic Community. The possibility of increased exports to Member States, coupled with the support regimes of the Common Agricultural Policy (CAP), encouraged farmers to maximise their outputs, even if this led to surplus production.

121 The increase in output from the cattle industry was achieved in a number of ways. The most important of these was a combined breeding and feeding programme which produced cows with a genetic capability to give high milk yields if fed with high-protein feeds. Thus it became regular practice for farmers to supplement the forage-based diet of cattle with protein concentrates that they would buy from special animal feed manufacturers. The protein in these concentrates might come from animal sources in the form of meat and bone meal (MBM), bloodmeal, feather meal or fishmeal, or from non-animal sources, mainly in the form of soyabean meal.

122 Although soya-derived protein may seem the more 'natural' option to the layman, animal-derived protein produced as great or a greater increase in milk yield, and its use provided an outlet for animal waste that would otherwise have had to be disposed of in some other way. Small quantities of animal by-products had been used in animal feed since the beginning of the 20th century. Most farmers were well aware of the practice and had no problem with it.

Findings and Conclusions

123 Since the purpose of protein concentrates in feed was primarily to facilitate the high milk yield of dairy cows, these concentrates were used more in dairy herds than in beef herds. Dairy calves would have protein concentrates included in their feed from a week after birth, whereas calves used for beef production were unlikely to receive concentrates until they were at least 6 months old. However, since almost two-thirds of beef produced in the UK originated in dairy herds, we cannot conclude that the cattle whose flesh we were eating had been fed less protein concentrate than those whose milk we were drinking.

Slaughterhouses

124 Cattle that were destined for human consumption had to be slaughtered in a licensed slaughterhouse or abattoir (see vol. 13: *Industry Processes and Controls*). Sick cattle or those that had died on the farm would instead be taken to a knacker's yard or a hunt kennel and their meat and by-products would not enter the human food chain.

125 In the 1980s there were around 1,000 slaughterhouses in England, Wales and Scotland, although this number was steadily decreasing as economies of scale and higher health and environmental standards pushed the smaller premises out of business. This decline in the number of slaughterhouses meant that more cattle had to travel long distances between the farm and slaughterhouse, and it was not unusual for the largest slaughterhouses to receive cattle from all over Great Britain.

126 At this time the hygienic production of meat was governed in England and Wales by Regulations made under the Slaughterhouses Act 1974 and the Food Act 1984. There was in fact a two-tier system of regulation that differentiated between plants producing meat entirely for domestic consumption and those producing some or all of their meat for export to other EU Member States. The regulations for export slaughterhouses were more wide-ranging and required a more thorough system of inspections.

127 Slaughtering an animal, cutting it up and separating its constituent parts is a messy business however it is done. In the 1980s most large slaughterhouses had adopted a production-line type of procedure which enabled them to carry out the process as quickly as possible.

128 In a typical large slaughterhouse animals were unloaded from lorries into the holding area and then moved towards the slaughter hall in single file along special passageways. They were then fed one by one into a pen for stunning. There were two methods of stunning used for adult animals. The captive bolt method involved firing a metal bolt into the animal's brain, leaving a hole in its skull; the non-penetrative concussion method involved firing a mushroom-shaped bolt at the animal's head, thus rendering the animal unconscious without penetrating its brain or skull. It was common practice, following captive bolt stunning, to insert a pithing rod into the hole in the skull in order to cause further damage to the brain and spinal cord, and thus to prevent the animal from kicking due to reflex muscular action.

129 Once the animal was unconscious, its hind legs were shackled and it was hoisted to an overhead rail, known as the slaughter line. Hanging with its head closest to the floor, the animal could then be moved around the plant to the various stages of the slaughtering process. It would first be moved along until it was directly over the bleeding trough, where it would finally be killed by severing the large blood vessels in its neck. Blood would either be allowed to pour into the bleeding trough, or alternatively it would be sucked out through a hollow bleeding knife attached to a vacuum pump.

130 Once bled, the carcass was moved down the line to be dressed. First the forefeet, hind feet, udder or pizzle were removed with a knife, then the hide would be pulled off with a powered hide puller, and after that the head would be cut off. (Head meat would later be harvested either at the slaughterhouse or at special head-boning plants.) Then the abdominal wall would be cut open and the internal organs would tumble out onto the inspection table. Organs such as liver and kidneys which would go for human consumption were separated out and sent to the 'offal room' for sorting. The rest of the 'abdominal mass' was sent, either down chutes or in containers, to a different area known as the 'gut room'.

131 The final stage in the process involved splitting what was left of the carcass and removing the spinal cord. A cut would be made down the length of the spinal column using a mechanical saw.

132 Hygiene Regulations demanded that each carcass had to be inspected by a qualified inspector at various stages in the process in order to establish its fitness for human consumption. Only when parts unfit for human consumption had been removed from it could a 'health stamp' be applied to the carcass by the inspector.

133 Responsibility for the regulation of slaughterhouse practices was split between the Ministry of Agriculture, Fisheries and Food (MAFF) and the local authorities (see vol. 14: *Responsibilities for Human and Animal Health*). The Minister of Agriculture, Fisheries and Food was responsible for making Regulations under the Slaughterhouses Act 1974, and in particular had the power to make Regulations about the construction, layout and equipment in plants. The local District Councils or Unitary Authorities were responsible for the enforcement of these Regulations. They issued licences to slaughterhouses and to slaughtermen, they provided the meat inspectors, and they had the power to make byelaws (subject to confirmation by the Minister) to ensure that slaughterhouses were kept in sanitary conditions and were properly managed.

134 Meat and other animal by-products that were classified as unfit for human consumption had to be disposed of within 48 hours of slaughter. Complex Regulations prescribed how unfit meat was to be handled and much was sent direct to renderers for processing. Unprocessed blood could be sprayed on fields as a fertiliser, subject to the agreement of the local authority responsible for the slaughterhouse and the licensing of the recipient farm.

Renderers

135 The rendering process involved the crushing and heating of the raw material supplied from slaughterhouses (see vol. 13: *Industry Processes and Controls*). The process led to the evaporation of the moisture in the material, which then enabled the fat, known as 'tallow', to be separated from the remaining high-protein solids, known as 'greaves'. The greaves were further processed by pressing, centrifuging or by solvent extraction in order to remove more tallow. The resultant protein-rich material was then ground into meat and bone meal (MBM). In the 1980s both tallow and MBM had a good commercial value.

136 Rendering is not a new industry. It has existed in some form for centuries, producing tallow for candles and soap. However, it was only at the beginning of the 20th century that the production of MBM for animal feed became important. The production and use of MBM steadily increased throughout the first half of the century and, when national self-sufficiency became an important issue during the Second World War, Regulations actually prescribed its use in animal feed. The production of MBM and tallow continued to increase after the war.

137 From the 1960s onwards there was a change in technology from older-style 'batch-processing' systems to faster and more efficient high-volume 'continuous rendering' systems. By the 1980s most plants used a continuous rendering system, and the economies of scale forced older and smaller plants to close down, leaving fewer than 100 rendering plants in England, Wales and Scotland at this time. Two firms dominated the market, with Prosper De Mulder processing 64 per cent of the red meat waste in England and Wales by the early 1990s, and William Forrest & Son (Paisley) processing 74 per cent of the red meat waste in Scotland.

138 During the 1950s the process of solvent extraction became the preferred method of extracting tallow from greaves. The process involved pumping a benzene-based solvent through a heated vessel of greaves so that the tallow dissolved in the solvent. The tallow was then separated out from the solvent and the greaves were heated further so as to vaporise and remove any solvent that was still present. By the late 1970s this method was being phased out because of the increased price of solvents, the risk of fire and explosion entailed in their use, and because animal feed manufacturers wanted to buy MBM with a higher fat content.

139 Up until the 1980s the rendering industry was virtually unregulated in terms of quality control and production methods (see vol. 14: *Responsibilities for Human and Animal Health*). In 1981 Regulations came into force to ensure the microbiological safety of processed protein. In the context of increasing deregulation by government, it was decided that the best way to do this was by testing the microbiological safety of the finished MBM, rather than by prescribing set production procedures. In effect this gave renderers a lot of freedom in determining their preferred production processes and it allowed for a diversity of processes in different plants. Advice about new Regulations reached renderers through the UK Renderers' Association (UKRA), the primary trade association representing renderers' interests.

140 In the 1980s the end-products of the rendering process – MBM and tallow – were widely used in the manufacture of a diverse range of products. MBM was used as a protein source in animal feed, and in fertiliser. Tallow was used in the manufacture of many human foods, such as edible fats, and when further processed into glycerine it was used even more widely, for example in jellies and in baking. It was also used in animal feed and pet food, as well as in pharmaceuticals, cosmetics and in a range of industrial products. Meanwhile, gelatine, produced from the hide and bones of animals in a completely separate industry and process, was also used in a wide range of products including human food, the coatings of tablets, cosmetics, glue, bone china and photographic chemicals.

The animal feed industry

141 In the 1980s animal feed was made up of a mixture of various constituents, primarily cereals and cereal by-products, as well as oilseed meals, MBM and other protein concentrates, fats, molasses, vitamins, minerals and, in some cases, small amounts of medicinal additives. Feed manufacturers produced both ready-to-use compound feeds and protein concentrates which farmers could use if they preferred to mix their own feed on the farm.

142 In the early 1980s there were about 400 feed companies, although this number was in decline. The five largest companies dominated the market, producing 54 per cent of the UK feed output between them, while farmer co-operatives and smaller local and regional compounders produced the rest.

143 Feedmills produced many different kinds of feeds for different animals. The nutritional composition of the feeds was determined according to the specific requirements of each species, and then the particular ingredients that would meet these requirements were chosen on the basis of cost-efficiency. Medicinal additives and growth stimulants were added when appropriate on a species-specific basis. Some species-specific feeds were potentially dangerous to other species. Most feedmills produced these different feeds in the same equipment. There were several points in the manufacturing process where material could build up on or in machinery and cause cross-contamination in the next batch. The UK Agricultural Supply Trade Association (UKASTA) drew up a Code of Practice to try to minimise cross-contamination of feedstuffs during the production process.

The meat industry

144 Meat that had been 'health stamped' as fit for human consumption in the slaughterhouse was sent to butchers or meat processors to convert it into the forms in which it is purchased and eaten (see vol. 13: *Industry Processes and Controls*). In the post-war period processed meat products had become more popular than fresh carcass meat, and by the early 1990s there were over 700 meat processors in the UK. Some processed meat products contained mechanically recovered meat (MRM). This is residual matter left attached to the bones of carcasses after the cuts of meat have been removed. The bones are then put under high pressure so that what is left can be stripped from them in a slurry. In the early 1980s a major source of bovine MRM was the bovine spinal column.

145 In the fresh meat sector there had been a shift away from high street butchers towards supermarkets as the preferred place to buy meat, and in the 1980s Tesco, Sainsburys and ASDA between them accounted for nearly 50 per cent of retail beef sales in the UK. One reason why supermarkets had become more popular was that they had sought to improve the quality of their meat and meat products. They had done this primarily through the development of quality assurance schemes which provided an audit trail from farm to consumer and assurance about the origin, husbandry and health of the cattle (see vol. 12: *Livestock Farming*). These schemes had been actively encouraged by the Meat and Livestock Commission (MLC), a non-departmental public body whose role was to promote greater efficiency in the livestock industry.

The pharmaceutical industry

146 Bovine materials were, and are, also used in pharmaceutical, medical and veterinary medical products (see Annex 1 to Chapter 2 in vol. 7: *Medicines and Cosmetics*). The UK pharmaceuticals industry is one of the largest in the world. In 1997, for example, UK exports were worth over £5 billion and accounted for around 12 per cent of the world market. There were over 400 pharmaceutical manufacturers and research organisations in the UK, although the market was dominated by multinationals such as Glaxo Wellcome, SmithKline Beecham and Zeneca.[9]

147 Bovine materials from the slaughterhouse are used directly in pharmaceuticals. Several injectable medicines are derived directly from bovine sources. Hormones such as insulin and glucagon may be derived from bovine pancreases, and protein products such as aprotonin and heparin are derived from bovine lungs and intestinal mucous respectively. Sutures and some medical devices such as heart valves and pericardium patches are also derived directly from bovine materials, in this case the intestines, heart and serous membranes.

148 Bovine materials are also used indirectly in the manufacture of certain types of vaccine. Cells which are used to grow these vaccines are nourished in nutrient-rich cultures that contain serum from the blood of foetal or new-born calves, or bovine serum albumin, which derives from the blood of older cattle. Bacterial cells are grown in nutrient-rich broths containing peptone derived from bovine meat, and some allergens are produced in special culture media which contain digests of calf brain and ox liver. In all these cases the bovine materials are not a constituent of the final product, but they are used in an ancillary way in the manufacturing process.

149 Tallow and gelatine are also used in several pharmaceutical and medical products. Gelatine is widely used as a pill coating and tallow is a constituent of most creams and ointments.

Other uses of bovine products

150 Bovine materials are used in a wide range of processes and products in many different industries. They are used in toothpaste, chewing gum and pet food; in fertilisers and cosmetics; and in such varied products as fire extinguisher foam, buttons, handles, lubricants and racquet strings. Bovine materials are used in the manufacture of paint. Cattle skins are used for hides, and other bovine materials are included in cleaning agents used in leather processing.

---

[9] *Britain 1999: The Official Yearbook of the United Kingdom, London,* The Stationery Office, 1998, p. 475

Government and BSE

151 MAFF had lead responsibility on most BSE matters and was the 'sponsor department' for those industries which found themselves implicated in the generation and spread of the disease. This raises a question of conflict of interest which we shall discuss later in this volume. MAFF officials took the lead on research into the disease. Its veterinarians and scientists had particularly important advisory roles about its causes and nature and negotiated with their counterparts abroad about measures to control it. They had considerable national and international stature. On a number of occasions the Chief Veterinary Officer (CVO), or an Assistant CVO, acted as the authoritative government voice.

152 The risk from BSE to human health took matters beyond MAFF's departmental borders. Acting as the authoritative public voice on the safety of beef was a role undertaken by the Chief Medical Officer (CMO) at the Department of Health (DH), and it was the CMO who had oversight of the response within his Department. He and his colleagues were closely involved in considering and agreeing with MAFF measures to reduce risks to human health via food, pharmaceuticals, occupational exposure and other pathways. They mainly relied on advice from outside experts and committees.

153 Measures affecting most aspects of agriculture and health in Wales, Scotland and Northern Ireland were the responsibility of Departments overseen by the Welsh, Scottish and Northern Ireland Offices. Others directly concerned with the response to BSE included the Health and Safety Executive (HSE), because of risks through occupational exposure; the Department of Trade and Industry (DTI) as sponsor Department for the cosmetics and toiletries industries; the Department of the Environment (DoE) in respect of the effects of various methods of waste disposal such as carcass burial and incineration; and the Department of Education and Science (DES), both in handling funds for the Research Councils sponsoring much of the BSE research, and in giving advice about dissecting bovine eyeballs.

154 Three general features of the arrangement of legislative powers and duties described in vol. 14: *Responsibilities for Human and Animal Health* bore directly on how BSE was handled:

- Although Departments in Wales, Scotland and Northern Ireland had responsibility for many agricultural and health matters, the guiding principle was that issues affecting the safety of food, medicines and other consumer products, and the prevention and control of infectious animal and human disease, should be dealt with consistently on a UK-wide basis.

- The main Acts of Parliament governing the different areas in which BSE impacted were a heterogeneous collection of legislation. Each of those covering animal health, food safety, wholesomeness of feedstuffs, control of pollution, medicines safety, consumer protection, and occupational risk had its own set of basic concepts, preferred approach and basic machinery on matters requiring public intervention. Associated with each major Act or EU instrument was a shoal of subordinate legislation reflecting the differing powers, duties, sanctions and enforcing agencies. There could be no uniform approach to the response to BSE.

- Although central government was largely responsible for the Regulations made about BSE, it usually fell to local government to enforce them.

155 Volume 15: *Government and Public Administration* explains how policy is developed and implemented within Departments, the main terminology and procedures that crop up throughout the other volumes, the relationship between Ministers and officials, and how accountability operates.

156 The volume also describes conventions for consultation and cooperation within and between Departments. The need for 'joined-up government' is not new. It reflects a basic characteristic of institutions. Policy matters rarely have neat boundaries or single solutions. Each Department, division or agency reasonably enough has its own agenda, reflecting its particular set of statutory responsibilities. It is necessary to secure agreement about the efforts of different agencies with different responsibilities, priorities and especially budgets, in order to achieve common objectives.

157 During the 1980s and 1990s decision-making was affected by legislative and financial control pressures, and by administrative developments:

- **the existing legislation.** Departments generally had to make do with existing primary legislation, although it was often not ideally suited to addressing the problems of BSE. New secondary legislation could be introduced, but this required clearance, consultation, and time to introduce;

- **resource planning.** Money to run Departments and finance their operations had to be voted by Parliament under itemised heads. The justification for bids was rigorously scrutinised by the Treasury as part of the control of government spending. Voted money could not be switched at will to different purposes, nor could Departments overspend. This system involved an annual cycle of bids and negotiations for resources for the next three years. The cost of any proposed new action was therefore a major consideration;

- **cuts in resources.** The heavy squeeze on public spending on administration year on year throughout the period, both in Whitehall Departments and in local government, required MAFF to make significant cuts in running costs; it reduced its staff numbers by 12 per cent between 1986 and 1996. Research budgets were being slashed. Making room for BSE work involved jettisoning something else. Strict staff ceilings were in operation. Unclear prospects made recruitment for many types of post difficult, and staff in post were overloaded;

- **value for money and charging.** There was increased emphasis on business efficiency, charging for services or certificates, and measured performance targets. Setting up Executive Agencies took considerable management time, including that involved in setting up systems for charges and fees; and

- **deregulation.** A key aim of the Government was to lift the burden of state regulation from industry, especially small businesses. Instructions and government papers were issued urging this on Departments. Proposals for new measures had to be tested against their cost to industry. Enforcement was expected to be done with a light touch.

Handling risk

158 In a primitive society, the major hazards are those posed by nature. In a complex modern society the acts of individuals or corporate bodies may also involve serious hazards to other members of society. All governments intervene in many different ways to reduce the exposure of their citizens to hazards created by nature or by human acts. Dealing with such hazards is one of the most important functions of government.

159 Every action taken to reduce exposure to hazard has its price. Many administrative actions taken for this purpose involve government expenditure, to be recovered in one way or another from the citizen. Statutory measures which prohibit or regulate potentially hazardous activities impose costs on those to whom the measures apply and may stifle innovation. Where the activities are commercial, these costs are likely to be passed on to the customer or consumer. Restriction of freedom of choice for the individual will usually be part of the cost of a safety restriction – sometimes the most significant part.

Risk evaluation

160 When considering whether to impose a safety measure the Government has to balance the benefits that will be achieved from reducing or eliminating exposure to a hazard against the costs that the measure will involve. This process involves what is sometimes described as 'risk evaluation'.

161 A risk is not the same as a hazard. A hazard is an intrinsic propensity to cause harm. Natural phenomena, physical substances, human activities will be hazardous if they have an intrinsic propensity to cause harm. A risk is the likelihood that a hazard will result in harm. A risk can usually be evaluated once the nature of the hazard and the degree of exposure to it are identified. Risk evaluation involves considering both the likelihood that a hazard will cause harm and the severity of the harm that is threatened.

Risk management

162 Action to reduce or eliminate a risk may involve destruction of a substance or prohibition or regulation of an activity that gives rise to a hazard. Alternatively it may involve eliminating or reducing the exposure to the hazard. Risk management involves identifying the options for reducing or eliminating the risk and their likely efficacy, estimating the costs involved in each option, deciding which, if any, of the available options to exercise, implementing the chosen options and monitoring the results.

163 In some circumstances past experience enables the statistician or the epidemiologist to calculate with some precision the effect that an option will have on reducing risk. Management of the risks associated with road traffic is such an example. It is often possible to calculate the number of lives that a particular road safety measure is likely to save. In such circumstances one can decide on principles or guidelines that will govern risk management, such as the maximum expenditure that can be justified per life saved.

BSE and risk

164 BSE was not like that. Attempts could be made to evaluate the risk to cattle. So far as other animals, and humans, were concerned, however, nobody knew whether BSE was a hazard or not. In such a situation the Government has to decide what precautionary measures to adopt against the possibility that the risk exists. One technique that can be adopted is known by the acronym ALARP. This calls for weighing the efficacy that any particular measure will have in reducing the notional risk against the cost and other consequences of introducing the measure. The aim is to reduce the possible risk so that it is As Low As Reasonably Practicable. It involves an exercise in proportionality that often calls for nice judgement.

3. The early years, 1986–88

165 This is the first of a number of chapters which tell, in summary form, the story detailed in Volumes 3 to 9 and 11.

Identification of a new disease in cattle

166 The epidemic of BSE may have started with a single diseased cow. Why should that cow have developed BSE? It is possible that the disease developed spontaneously as a consequence of a genetic mutation. It is possible (though we believe less likely) that a mutant strain of the scrapie agent transmitted to one or more cows. There are other possibilities. No one will ever know.

167 When was the first case? The epidemiologists, with their skills in back calculation, suggest in the 1970s. Where was it? Again no one can say, though epidemiologists would point to the early concentration of cases in the West Country as suggesting that BSE may well have come from there.

168 Did the first case get ill on the farm and end up in the knacker's yard, or was it sent to be slaughtered for human food – perhaps before the signs of the disease were even showing? We cannot know. What we can deduce is that, by one route or another, the animal's head, together with other unwanted offal, was sent to the renderers. The parts carrying the BSE infection contaminated the batch of meat and bone meal (MBM) produced from the rendering. That MBM was sold to a food compounder and mixed into cattle feed, contaminating that feed. That feed may have infected many cows and some of these, by a similar series of events, infected many more. Thus, like a chain letter, the spread of the disease was almost exponential.

169 The disease spread wide, and it spread at first unnoticed. It spread wide because MBM may travel long distances from renderers to the feedmill and the cattle feed produced by the mill may be widely distributed. The calves which eat the feed may end their lives far from the farms on which they were born.

170 It spread at first unnoticed because most infected cattle were slaughtered before showing clinical signs of the disease. When clinical signs did appear, they were similar to those of some other diseases of cattle. Only histopathology of the brain could reveal the existence of the new disease. Before that could happen the carcass had to be sent by a vet to one of the regional State Veterinary Investigation Centres (VICs), and from there the brain had to be sent to the Pathology Department of MAFF's Central Veterinary Laboratory (CVL) at Weybridge. Most cattle infected with BSE went for slaughter before the clinical signs developed ('subclinical cases'). Where a single cow fell ill, the farmer was unlikely to want a post-mortem examination and, for some reason, not yet clear, BSE tended to strike down single cows in a herd.

171 The first brain from a cow with what we now know as BSE reached the CVL in September 1985. It came from a herd in Pitsham Farm in Sussex where unusually a number of cattle had been struck down with symptoms that we now recognise as typical of BSE. The CVL pathologist identified the condition of the brain as spongiform encephalopathy, but concluded that this, and a kidney condition from which the animal had also suffered, was probably caused by toxicity of some description.[10]

172 At the end of 1986 pathologists at the CVL identified the possibility that cattle had developed a spongiform encephalopathy that was transmissible in the same way as scrapie was in sheep. This followed the submission of brain samples from a herd in Kent and another from near Bristol. Mr Raymond Bradley, head of the Pathology Department, remarked in a note to colleagues:

> If the disease turned out to be bovine scrapie it would have severe repercussions to the export trade and possibly also for humans.

173 One witness described meeting Mr William Rees, the Chief Veterinary Officer (CVO), who had just heard the news, with 'steam coming out of his ears'.[11]

174 The CVL pathologists identified the emergence of a new disease, which they considered might be a bovine form of scrapie, as soon as could reasonably have been expected. They are to be congratulated – particularly Mr Gerald Wells and Dr Martin Jeffrey, who carried out the initial histopathology.

Restraints on information

175 CVL staff thought that they might have identified a bovine form of scrapie, but they were not sure. The experts in this field were the members of the Neuropathogenesis Unit (NPU) in Edinburgh. If the CVL had consulted them at this stage, the NPU would have confirmed that there were very strong indications that this was indeed a new Transmissible Spongiform Encephalopathy (TSE). In the event the CVL did not seek the collaboration of the NPU until June 1987, and Mr Wells did not get confirmation from the NPU of his diagnosis until the end of July. Having regard to the importance of this matter, we think that Dr William Watson, the Director of the CVL, should have sought the assistance of the NPU from the outset.

176 It was important that MAFF should discover not merely the nature of the problem, but also its scale. If private vets and members of the VI (Veterinary Investigation) Service around the country were told of what the CVL had found and asked to look out for cattle with similar signs, reporting of cases, which might otherwise go unremarked, would be encouraged. Unfortunately, in the first half of 1987 there was a policy that one Senior Veterinary Investigation Officer described as 'a total suppression of all information on the subject'. This was encouraged by an understandable anxiety on the part of Mr Wells that MAFF should not go public until the CVL was sufficiently sure of its ground to advance a scientifically responsible claim to have discovered a new disease. In March 1987 a proposed publication about BSE in *Vision*, a VI Service newsletter, did not proceed. The

---

[10] Vol. 3, paras 1.7–1.33
[11] Vol. 3, paras 1.34–1.40

decision was Dr Watson's, who should not have permitted Mr Wells's concern to prevail over the desirability of effective surveillance.

177 Events after March 1987 demonstrated a policy of restricting dissemination of information about BSE. The principal reason for this was concern about 'the possible effect on exports and the political implications' should news get out that a possible TSE in cattle had been discovered in Britain. Publication to the VI Service of information about BSE eventually took place in June. This was not in *Vision*, which was circulated to Veterinary Investigation Officers (VIOs) not only in England and Wales, but also in Scotland. Instead a circular letter was sent to Senior VIOs in England and Wales, describing the clinical signs and the pathology and calling for notification of similar cases to a Senior Veterinary Officer at the State Veterinary Service headquarters at Tolworth, Surrey. It directed that VI staff should not consult Research Institutes or University Departments, or publish anything about BSE or discuss it at meetings without clearance. A proposed letter by a VIO to the *Veterinary Record* describing the clinical signs and the pathology of BSE was refused permission for submission to the journal.

178 Primary responsibility for this policy lay with Mr Rees, the CVO, but it received support from his subordinates, Dr Watson and Dr Bernard Williams, the head of the VI Service. We can see why there were concerns that reports of a possible TSE in cattle might harm the industry and, in particular, the export market. But this did not justify suppression of information needed if disease surveillance was to operate effectively. Dr Watson and Dr Williams should have urged the merits of publication and Mr Rees should have permitted it.

179 An article by Mr Wells for the *Veterinary Record*, which compared the pathology of BSE and scrapie, was embargoed and it was made plain that comparisons with scrapie were not acceptable. This line was taken at the instigation of Mr Rees. He should have permitted publication of the article and he should have permitted comparisons with scrapie.

180 Had there been a policy of openness rather than secrecy, this would have resulted in a higher rate of referral of cases to MAFF in the earlier part of 1987. This, in turn, might have led to a better appreciation of the growing scale of the problem and hence to remedial measures being taken sooner than they were.

181 In the second half of 1987, restraints on publication of information about BSE were progressively relaxed. Articles about BSE were submitted to the *Veterinary Record* and the disease was the subject of discussion at a number of agricultural trade meetings. In October articles about the disease appeared in the farming and national press. The number of cases reported increased rapidly. At the end of May there had been 6 identified cases and 13 suspected cases. By the beginning of September there were 66 suspect cases, of which 8 were histopathologically confirmed. By the end of October the figures were 120 and 29, and by the end of the year 370 suspects, of which 132 were confirmed.

What was the cause of BSE?

182 The CVL had only one qualified epidemiologist in 1987, Mr John Wilesmith, who headed a small Epidemiology Department. He knew nothing of BSE until late in May, when he was asked by Dr Watson to investigate its epidemiology. There were then 6 confirmed cases on 4 farms, but as we have seen the numbers were about to escalate.

183 Mr Wilesmith prepared a questionnaire, rolled up his sleeves and set off in person to visit farms on which BSE suspects had been reported. Soon Mr M Cranwell had to be seconded from Starcross VIC in Exeter to assist him. By this time, unknown to Mr Wilesmith, thousands of cattle had been infected by recycling of earlier cases and were incubating the disease. Mr Wilesmith assumed, quite naturally, that each new case was an index case (that is, arising as a fresh incident) and that there was some common factor causing all of them. The search was on for that common factor. Vaccines, hormones and organophosphates were considered but ruled out: the disease had been found in cattle exposed to none of these.

184 From the outset feed was a runner. In August Mr Wilesmith noted that lamb MBM was used in commercial dairy rations, but added that it was not a recent introduction. This was a major conundrum. If feed was the cause, what novel ingredient or feature had suddenly started to make the feed infective?

185 Mr Wilesmith carried out calculations which indicated that the exposure of the cattle population to the BSE agent was likely to have begun in the winter of 1981–82. Had anything occurred at about this time to explain the disease?

186 Further investigations were put in hand to explore, with the help of the feed and rendering industries, why it might be that cattle feed had suddenly started infecting cattle. By the end of April 1988 Mr Wilesmith had reached no conclusion on this. He had, however, concluded that feed was the source of infection and that the source of infection in the feed was MBM made from sheep affected by scrapie. He set out these conclusions in a report, recommending a temporary ban on the inclusion of MBM in cattle and sheep feedstuffs, while further enquiries were made.

187 Mr Wilesmith and his colleagues are to be congratulated on the rapid identification of cattle feed as the cause of the cases of BSE that were being reported, and on the advice of a ban on feeding MBM to cattle and sheep. As we shall see, this advice was promptly implemented and cut off most of the source of infection, turning an escalating disease into one that would peak and decline.

188 Mr Wilesmith had, however, made some tentative conclusions which were to prove erroneous. He concluded that the cases being reported were all index cases. He concluded that the common source of infection was scrapie-infected feed which would result in the incidence of BSE rising sharply over a short period of time before maintaining a constant incidence. In a paper published at the end of 1988 he identified a number of factors which might explain why cattle feed had become infective around 1981–82. These included an increase in the amount of scrapie-infected sheep going for rendering and changes in the rendering process which had reduced the temperature applied. In the following year he refined these ideas and decided that particular significance attached to one specific change in the rendering process. The use of solvent to extract tallow had been widely abandoned at just about the right time to explain the outbreak of the disease. This process might well have played an essential role in inactivating the scrapie agent. When Mr Wilesmith learned of this change he commented that it was 'too good to be true'. In that, he was correct.

189 Mr Wilesmith's tentative conclusions were reasonable on the data available to him at the time, but they were wide of the mark, as he was in due course to acknowledge. The cause of infection of the cases being reported was not the scrapie agent in the feed, but the BSE agent itself. The cases were not first generation cases, but the consequence of recycling of BSE. Far from the incidence of BSE infection being likely to prove constant, it had been escalating year on year and was, in 1988, infecting cattle at a rate that probably exceeded 10,000 cases a month.

190 Changes in rendering processes may have had some effect on inactivation of the BSE agent, but they were not decisive or even significant.

191 Mr Wilesmith's tentative conclusions were widely accepted. They led to misconceptions, some of which have survived to the present day. We will deal with them shortly. They receive detailed consideration in Volumes 2 and 3.

The scrapie theory

192 The conclusion that BSE had been transmitted from scrapie-infected sheep was generally accepted. It was a reassuring conclusion. Sheep affected by scrapie had been eaten by humans for 200 years or more, without apparent ill effect. It was likely that scrapie in cattle would prove similarly innocuous. Although, as the years passed, evidence mounted that discredited the scrapie theory, this was never made clear to the public and most people are still under the impression that cattle caught BSE from scrapie-infected feed.

193 The conclusion that rendering changes had permitted the BSE agent to survive unscathed, whereas previously it had been inactivated, is also still widely accepted. There are two variations on this theme:

  i. Some accuse the Government of having recklessly relaxed the Regulations governing rendering, or of having failed to impose a sufficiently rigorous regulatory regime.

  ii. Some accuse the rendering industry of having put the safety of their product at risk by cutting corners in order to cut costs.

194 Neither of these accusations is valid. There was no relaxation by the Government of rendering standards. Up to 1981 the rendering industry was largely unregulated. In 1981 Regulations were introduced that set minimum standards for the product of renderers, to be checked by regular sampling. The Regulations were strengthened in 1989.[12] A more complex alternative involving the licensing of rendering plants was not pursued, but this would not have addressed the problem of BSE and the proposed criteria for the grant of licences would not have prevented it. That problem was not foreseen, nor was it reasonably foreseeable.

[12] See Volumes 13 and 14

195 By the same token the changes made by the rendering industry to their processes did not, overall, make them more vulnerable to BSE. Neither the old nor the new processes would have inactivated the BSE agent. No rendering process has yet been devised which can guarantee to do so, though infectivity is reduced.

196 The theory that the rate of infection would have reached a plateau led to the conclusion in 1989 that the scale of the problem could be related to the rate at which cases were being reported. The Southwood Working Party on Bovine Spongiform Encephalopathy reported that year on the basis that the effect of recycling could be 'minimal and undetectable', in which case 350 to 400 cases a month could be expected. In early 1993 cases were being reported at a rate of around 1,000 a week.[13]

197 These misconceptions involve no criticism of Mr Wilesmith. They demonstrate that in 1987 and 1988 lack of data made it impossible to appreciate the nature and extent of the disaster that had already occurred.

The ruminant feed ban

198 While Mr Wilesmith was exploring why cattle were succumbing to BSE, consideration was also being given to the implications that the disease might have for humans. Before turning to that part of the story, let us follow the reaction to Mr Wilesmith's advice that the practice of including animal protein in cattle feed should be subjected to a temporary ban.

199 If Mr Wilesmith's conclusions were tentative, Mr Rees, the CVO, thought that the picture was clear. In a submission to Mr John MacGregor, the Minister of Agriculture, Fisheries and Food, he advised that he was:

> . . . satisfied from the information produced by the investigating teams that the source of the transmissible agent which has caused BSE is through meat and bone meal derived from sheep material in which the rendering process has failed to inactivate the scrapie agent. Affected sheep material is continuing to be processed and it must be assumed therefore that cattle continue to be exposed to infection.[14]

200 He advised that the feed industry should be asked to agree a voluntary withdrawal of MBM from ruminant feed, but that if they refused, a mandatory ban should be imposed.

201 Mr MacGregor was even more decisive. On 19 May 1988 he determined that there should be a 'speedy and compulsory ban on sheep meat material in feed for ruminants'. It fell to Mr Alan Lawrence, a Grade 7 official in MAFF's Animal Health Division, to implement this decision in consultation with departmental lawyers and with the benefit of advice from his administrative and veterinary colleagues. It was decided that the ban should extend to the feeding of ruminant protein to ruminants. In effect the ban was subsequently operated as if it encompassed all animal protein, for no renderers attempted to segregate their raw materials in order to produce non-ruminant MBM. The ban was achieved by an

---

[13] See vol. 4: *The Southwood Working Party, 1988–89*
[14] YB88/5.6/11.3

Order[15] signed by Mr MacGregor and Welsh and Scottish Office Ministers on 10–14 June. This made it an offence to sell, supply or use for feeding to ruminating animals any feedstuff in which the offender 'knew or had reason to suspect' that any animal protein had been incorporated. The ban was initially only up to the end of 1988, but it was subsequently to be extended, and finally made permanent.

202 This simple Order has been described by one distinguished epidemiologist as:

> A spectacularly successful control measure . . . one of the notable success stories of global disease control.

203 It has, today, come close to eradicating an epidemic that, at its height, was of gigantic proportions. Primary credit for this goes to Mr Wilesmith and his Department for their diagnosis of the source of infection, but credit also is due to Mr Rees and Mr MacGregor for their prompt and decisive response. Unhappily, though, the measure was not a total success. There were shortcomings in its implementation. We turn to consider why this was.

204 The question arose in the course of consultation as to when the ban should come into effect. After consulting its members, the UK Agricultural Supply Trade Association (UKASTA) asked for a three-month period of grace to enable the industry to clear from the distribution channels all stocks of ruminant feed that had already been compounded. After taking advice from the veterinarians in MAFF, Mr Lawrence proposed a two-month period of grace. MAFF's press office advised that a delay as long as this would lead to accusations of risking the further spread of the disease simply to make life easy for the industry. Mr MacGregor, on the advice of Mr Alistair Cruickshank,[16] compromised and decided that the ban should come into effect on 18 July – five weeks from the date of the Order.

205 We initially questioned the grant of this period of grace, but concluded that our reservations were the result of being wise after the event. Mr Kevin Taylor, one of the MAFF veterinarians involved in the preparation of the ruminant feed ban, explained to us his reasons for viewing a period of grace of as long as two months as perfectly acceptable from a veterinary point of view. On the basis of the information then available it did not seem to him that such a delay was going to make very much difference. The industry had been exposed to infected feed for 380 weeks. A few weeks more would not make a great deal of difference.

206 In June 1988 MAFF officials reasonably expected, on the basis of Mr Wilesmith's advice, that the rate of infection was likely to have stabilised at about 60 cases a month. Mr Taylor considered that if no period of grace had been granted, farmers and the industry would initially have disregarded the ban. We found force in these points and reached the conclusion that the compromise period of grace decided upon by Mr MacGregor could not be criticised. Had it been appreciated that cattle were being infected at the rate of thousands of cases a week, we have no doubt that a very different approach would have been adopted.

207 Much later it became apparent that infected feed had continued to be fed to cattle on a substantial scale after 18 July. Nearly 12,000 cattle born after the ban (BABs) in 1988 and over 12,000 born in 1989 subsequently developed clinical signs

---
[15] The Bovine Spongiform Encephalopathy Order 1988
[16] MAFF Under Secretary (Grade 3) responsible for the Animal Health Group

of BSE. A far larger number must have been infected, but slaughtered before signs became apparent. Some of these cases will have resulted from accidental contamination of feed. Some will have resulted from farmers, who had little or no means of knowing whether their feed contained ruminant protein, continuing to use the feed they had in stock. But we are satisfied that some feedmills and feed merchants deliberately continued to sell cattle feed containing animal protein after the ban come into effect.

208 Had the only source of contaminated feed been existing stocks of cattle feed made up before the ban came into effect, the BABs would have come to an end once this had been consumed. In the event, over 5,600 cattle born in 1990, 4,500 born in 1991, 3,000 born in 1992, 2,200 born in 1993 and 1,000 born in 1994 were to go down with the disease. With hindsight, it is clear that most of these infections resulted from cross-contamination of cattle feed with pig and poultry feed, containing infective MBM, in the feedmills. The risk, indeed the certainty, of a degree of cross-contamination when the same production lines are used to produce different batches of feed is, and was in 1988, well established. One reason that has enabled us to conclude that cross-contamination did indeed result in infection of cattle is knowledge that we now have as to the quantity of infectious material that suffices to transmit BSE orally in cattle.

209 An experiment carried out by the NPU has demonstrated that ½ gram of homogenised brain from BSE-infected cattle is sufficient to transmit the disease orally across the species barrier to sheep and goats. Another experiment carried out by the CVL has demonstrated that 1 gram of such material can transmit the disease orally to cattle.[17]

210 The results of these experiments were not available when the ruminant feed ban was introduced. What consideration was given at that stage to the amount of material that might infect? What consideration was given to the question of whether cross-contamination might pose a risk of infection? UKASTA witnesses spoke of receiving repeated reassurances from MAFF right up to 1994 that a large amount of contaminated feed would be necessary to infect a cow.

211 We found no specific evidence of when or by whom such assurances were given. A number of MAFF administrators spoke of their understanding that a large amount of infective material was needed to infect. Some of the professionals – Dr Watson, Mr Kevin Taylor, Mr Bradley – told us that they had no idea what the minimum quantity would be. There was general surprise, when the result of the attack rate experiment was made known, that as little as 1 gram had sufficed to infect. Although there is no record of Mr Keith Meldrum[18] reassuring UKASTA that there was no need to worry about cross-contamination, he is recorded as telling representatives of the cattle industry in June 1988 that feedmills presented at worst a low contamination risk and would not be investigated. He advised at the same meeting that MBM could safely be used as fertiliser because the dose that might be received by grazing cattle would almost certainly be too low to cause disease.

212 Was there any valid basis upon which Mr Meldrum could have concluded in 1988 that cross-contamination in the feedmill would not involve sufficient quantities of infective material to give rise to transmission? We have concluded that

---

[17] See vol. 2: *Science*
[18] Mr Meldrum succeeded Mr Rees as Chief Veterinary Officer in June 1988

there was not. Mr Wilesmith told us that he had concluded that a very small amount of infective material would suffice to infect. This he deduced from the small inclusion rate of MBM in calf rations. He believed that his view should have been widely shared by administrators at MAFF. Those who designed the experiments at the NPU and CVL, to which we have referred above, envisaged the possibility that ½ or 1 gram would suffice to infect. Had the question of the amount of material needed to infect been explored at the time of the imposition of the ruminant feed ban with those best placed to advise, the conclusion should have been reached that this amount might be very small.

213 Mr Meldrum told us that if he or any other MAFF or industry representative had known at the time that the infective dose was so low as to lead to cross-contamination problems, the issue would have been pursued. As it was, the existence of a danger from cross-contamination was not considered to exist at the time.

214 We have concluded that at the time that the ruminant feed ban was imposed, there was a lack of rigorous thought about its implementation. One person who should have given more thought to this was Mr Meldrum. He had knowledge of how feedmills operated, and of the problem of cross-contamination between batches. He assumed this would not matter but did not have adequate grounds for that assumption. A failure to attach significance to the possibility of infection through cross-contamination in feed was understandable when the apparent rate of infection was only about 60 cases a month. However, in the course of September 1988, 435 cases of BSE were reported in Great Britain. Once this was apparent, Mr Meldrum should have ensured that proper consideration was given to this matter. This should have led to guidance being given to both the feedmills and to those farmers who mixed their own feed, on the need to take precautions to minimise cross-contamination.[19]

215 Mr Meldrum is a man of great energy and industry. He had only just taken up the reins of the CVO. His national and international duties were onerous. These are considerations which should temper any criticism of his oversight on this occasion.

216 Failure to appreciate that cross-contamination mattered carried with it a failure to appreciate the importance of a test that would detect cross-contamination. When the ruminant feed ban was introduced, there was no test which would detect animal protein in compound feed, let alone ruminant protein. Without such a test the Order was unenforceable. Steps were put in hand to develop, in-house, the ELISA technique so as to produce a test that would identify ruminant protein in feed. This was not treated as a matter of priority. Deliberate breach of the ban was not considered likely and accidental cross-contamination was not considered to be cause for concern. Development of the ELISA test followed a leisurely course and did not approach achievement until the end of the period with which our Inquiry is concerned.[20]

Exports

217 The United Kingdom exported very little compound feed, but did export significant quantities of MBM. This was exported initially to Europe to

---

[19] For detailed discussion see vol. 3: *The Early Years, 1986–88*, paras 4.117–4.171
[20] See further: vol. 2: *Science* and vol. 5: *Animal Health, 1989–96*

manufacturers of concentrates who re-exported their products to the Middle East or North Africa. Some have suggested that the United Kingdom should have imposed a ban on the export of MBM when the ruminant feed ban was introduced to try to make sure that foreign countries did not infect their cattle with BSE. This would have been difficult. Renderers were still permitted to sell MBM to British purchasers for incorporation in pig and poultry feed. Most MBM that was exported was used for the same purpose. An attempt to prohibit exports would have been likely to be challenged in the Courts. It could be argued convincingly that foreign importers could be adequately protected by warnings that MBM should not be fed to cattle.

218 Were adequate warnings given? Mr John Gummer urged before the ruminant feed ban was introduced, when he was the junior Agriculture Minister, that we had a moral duty to warn our neighbours of the danger of feeding MBM to cattle. Under European law this country was obliged to give notice of the ruminant feed ban to all EU members and did so. What of the countries that were not members of the EU? Mr Meldrum told us that he relied on the customary means of communicating with them on the subject of animal diseases. He notified the Office International des Epizooties, which passed the information on to all members in a report of its annual General Session in May 1989. In February 1990 Mr Gummer, by now the Minister of Agriculture, Fisheries and Food, insisted that Mr Meldrum take the further step of writing a letter of warning to Chief Veterinary Officers of all countries which imported MBM from the UK. There is scope for arguing that Mr Meldrum should have done this earlier. We think the argument is academic. The only country outside the EU where it is suspected that cattle were infected with BSE as a result of importing MBM is Switzerland, and it seems that the MBM in question reached Switzerland via Belgium. If this occurred after the ruminant feed ban, both Belgium and Switzerland were aware that ruminant protein was suspected to be the cause of BSE. Accordingly we have seen no need to pursue this issue further.

Human health implications

219 BSE had implications for human health in many different ways. The one of which the public was most aware was the possibility that BSE posed a risk through food. Responsibility for addressing this risk was shared by MAFF and the Department of Health (DH). Mr Meldrum emphasised to us that DH was responsible for assessing risk to human health. He told us that he did his best to avoid making public comments on this matter. He saw MAFF's role as being risk management, together with the provision of advice to DH on matters that fell within the expertise of the veterinarians.

220 We have not found it easy to draw a distinction between risk evaluation and risk management. Throughout the BSE story, MAFF officials and Ministers appear to us to have proceeded on the footing that it was their responsibility to see that whatever left the slaughterhouse to go into the human food chain was safe to eat. MAFF made the running in considering both what was and what was not safe to go into the food chain, and how what was not safe should be kept out of it. Problems arising over the safety of animal feed, which were unquestionably MAFF's responsibility, tended to mirror problems of the safety of human food. In relation to the latter, DH was consulted, but not often actively involved in the initial formulation of policy. Whether DH should have been more involved is a matter that we shall consider.

221 BSE also posed a potential risk to human health as a result of the use of bovine products or by-products in the making of pharmaceuticals and cosmetics. So far as the former were concerned, DH had responsibility for human medicines and MAFF for veterinary medicines. Responsibility for the safety of cosmetics fell to the Department of Trade and Industry (DTI). These areas, and the occupational risks posed by BSE to those who handled cattle, or their products, we consider in separate chapters of this volume.

222 MAFF Ministers were first informed about BSE after the General Election in June 1987. Mr MacGregor was appointed Minister of Agriculture, Fisheries and Food, and Mr Gummer his Minister of State. Mr Donald Thompson retained his post as MAFF Parliamentary Secretary. In a note to him about the disease, Mr Rees commented, 'There is no evidence that the bovine disease is transmissible to humans,' a statement that was to be frequently repeated. Mr Thompson met officials on 22 July. The Permanent Secretary, Sir Michael Franklin, observed that the establishment of any risk to human health was the highest priority, and Mr Thompson said that he was particularly concerned about this. In a paper for him, which was subsequently seen by the Minister, Dr Watson advised that there was no reason at all to believe that any risk to human health existed.

223 By the end of July, 46 probable cases of BSE had been identified involving 18 herds. Both Mr Thompson and Sir Michael Franklin had raised concerns about human health. Mr Rees did not share those concerns. He viewed BSE as an animal health not a human health problem. Dr Watson thought it very unlikely that BSE posed a risk to human health.

224 In mid-September Mr Rees prepared a progress report for Ministers. This included a statement that DHSS was aware of the problem.[21] Dr Watson had told Mr Rees that Dr Thomas Little, the Deputy Director of the CVL with responsibility for veterinary medicines, had discussed BSE with DH colleagues at a meeting of a subcommittee of the Committee on Safety of Medicines. Regrettably Mr Rees did not explain to Ministers the limited nature of the communication that had occurred. There had been an informal discussion in the margins of that meeting, but news of BSE had gone no further within DH.

225 By this time there were 73 suspected cases in 36 herds across 11 counties. In a Q&A briefing for the media in October, Mr John Suich, who headed the Animal Health Division, included the following:

> Q : Can it be transmitted to humans?
>
> A : There is no evidence that it is transmissible to humans.

226 On 11 November 1987 he repeated this comment in a briefing for Mr Thompson, adding the suggestion that reassurance could be drawn from an analogy with scrapie.

---

[21] The Department of Health and Social Security (DHSS) split into two separate Departments – DH and DSS – during 1988

227 On 4 December Lord Montagu of Beaulieu wrote to Mr MacGregor expressing concern at the fact that cattle with BSE were being slaughtered for human consumption. He suggested that:

> Perhaps this is an area where the Ministry should make the disease notifiable and pay compensation at the full value for animals infected.

228 It seems that this letter served as a catalyst for formal consideration by MAFF officials of whether action should be taken to address the possibility that BSE might be transmissible to humans, though other letters from the public were received to similar effect. Mr Rees chaired a meeting of MAFF officials on 15 December. It was agreed that a paper should be prepared for Ministers setting out the options. On 29 December an article in *The Times*, headed 'Mystery Disease Strikes at Cattle', observed that there was no indication of whether the disease was transmissible to humans. By the end of the year, 370 suspect cases had been reported and 132 had been confirmed.

229 The options to be submitted to Ministers were discussed by, among others, Mr Rees, Mr Cruickshank, Dr Watson, Mr Meldrum, Mr Wilesmith and Mr Lawrence. The submission was perfected by 16 February 1988 and forwarded by Mr Cruickshank to Mr Edward Smith, the Deputy Secretary at MAFF. In his covering minute, Mr Cruickshank remarked:

> We do not know where this disease came from, we do not know how it is spread and we do not know whether it can be passed to humans. The last point seems to me the most worrying aspect of the problem. There is no evidence that people can be infected but we cannot say there is no risk.

This was an acute analysis of the position so far as humans were concerned. Mr Cruickshank's analysis of this aspect of BSE was not to be bettered, or even significantly augmented, by the scientists who were to consider the problem in the months to come.

230 The submission itself observed that it was uncertain whether the disease was transmissible to humans, and continued:

> We could therefore be criticised for allowing affected animals to be sold for human consumption. MAFF are already being asked to advise on whether there is any risk to humans.

231 The option recommended was a policy of slaughter of affected animals with payment of compensation, the principal advantage of which was to enable the Government to answer criticism about human health implications. The submission took some pains to emphasise that payment of compensation was appropriate as the measure would be taken mainly for public health reasons, not in order to eradicate the disease.

232 Mr Smith forwarded the submission to the Permanent Secretary, now Mr Derek Andrews, adding that as the policy was in the interests of public health, it would not be appropriate to look to the industry to fund it.

233 It is remarkable that MAFF officials had prepared this submission, whose recommendation was based essentially on an evaluation of risk to human health, without involving anyone at DH. The expressions of concern in the summer of 1987 by Sir Michael Franklin and Mr Thompson, coupled with the growth of the epidemic, called for joint consideration by MAFF and DH, with assistance from experts in TSEs, as to whether BSE might be transmissible to humans. Had this course been followed, we have little doubt that a joint submission would have been made to both MAFF and DH Ministers to the same effect as that which went forward to Mr MacGregor, but backed by conclusions as to the uncertainty about risk to humans that would have carried more weight than those of MAFF officials alone. It might moreover, as we shall see, have brought together those licensing veterinary and human medicines to consider their shared problems.

234 We sought explanations for the failure to involve DH from Dr Watson, Mr Cruickshank and Mr Rees. We have summarised their explanations in Volume 3.[22] We find that the true reasons were (i) a belief on the part of some that BSE was an animal and not a human health problem and (ii) a degree of interdepartmental reserve which led Dr Watson, Mr Rees and Mr Cruickshank to conclude that BSE was their problem to be resolved without the need for outside assistance – or interference – from DH. In this, each of them was at fault. The consequence, as we shall show, was a lengthy delay in reaching a decision as to the precautionary action to be taken.

Mr MacGregor's reaction

235 Mr MacGregor's previous office had been Chief Secretary to the Treasury. We believe that MAFF officials anticipated that he would have reservations about a policy that involved paying compensation out of public funds to farmers for the slaughter of sick animals. In this they were correct. Mr MacGregor's initial reaction to the submission was to be 'very cautious'. He expressed concern that if compensation were paid for slaughtering cattle with BSE, there would be a 'read across' to situations where the destruction of diseased crops had been ordered without payment of compensation. Rhizomania, a disease of sugar beet, was an example.

236 Mr Cruickshank told us that he and his colleagues considered Mr MacGregor's reaction to the submission to be a peremptory rejection. Sir Derek Andrews demurred at this description, and so would we. Mr MacGregor's initial reaction to a policy that involved payment of compensation was unfavourable, but he nonetheless agreed that the advice of the Chief Medical Officer (CMO) should be sought. His reaction affected, however, the manner in which the CMO, Sir Donald Acheson, was approached. The intention had been to tell him that MAFF wished to introduce a slaughter and compensation policy and to ask him to advise whether or not BSE posed a risk to humans. Had that approach been adopted, we think it likely that Sir Donald would have endorsed MAFF's proposed policy. As it was things took a different turn.

237 It was unfortunate that Mr MacGregor did not share his officials' view of the merits of the slaughter and compensation policy. It would not, however, be fair to criticise him for his reservations, for they did not lead him to reject the policy. His

[22] Vol. 3: *The Early Years, 1986–88*, paras 5.125ff

decision to consult the CMO before reaching a final decision fell well within the range of responses that were reasonably open to him.

Sir Donald Acheson's advice

238 Mr Andrews wrote to Sir Donald Acheson on 3 March 1988. He described the nature of BSE. This was the first that Sir Donald had heard of the disease. Mr Andrews then raised the question of whether BSE might be transmissible to humans. He wrote:

> It would be very helpful therefore to have your advice on the view we should take of the possible human health implications and how we should handle questions about the risks to human health.

239 This put the ball of recommending what action should be taken into Sir Donald's court, and with no warning at all. Sir Donald's reaction was to call an interdepartmental meeting to consider the matter.

240 Those present at this meeting were not able to form a firm view as to whether or not BSE posed a risk to human health. It was agreed to recommend to Health Ministers that a small group of experts be set up to advise on the human health risks and possible preventive measures. Sir Donald commented that he thought it highly likely that the advice would be that carcasses of affected animals should not go for human consumption.

241 We found this decision disappointing. MAFF officials had formed the view that unless one could be confident that they posed no risk to humans, sick animals should not be permitted to be slaughtered for food. The Southwood Working Party, set up on Sir Donald's recommendation, was to take the same view immediately it met. This was, we feel, no more than common sense. Referring the matter to an expert Working Party was bound to result in significant delay. A better and more robust response would have been to recommend that the practice of eating diseased cattle should cease at once. We have concluded, however, that it would not be fair to criticise Sir Donald for the course that he took. He was put in an invidious position, being asked for advice without notice on policy that had significant consequences. Those whom he summoned to help him decide on what to do expressed uncertainty. In these circumstances, we find that the decision to recommend that the matter be referred to an expert group fell within the range of reasonable responses open to Sir Donald.

242 Delay did indeed result, however. Over three months were to elapse before the Southwood Working Party was constituted and met for the first time. During this period MAFF came under increasing pressure to take action. On 22 April 1988 a front page article in *Farming News* accused MAFF of seriously underestimating the extent of BSE and referred to disquiet about whether the disease posed a danger to humans. By then there had been 421 cases confirmed in 352 herds.

243 Mr MacGregor continued to set his face against any suggestion that the Government should fund a compulsory slaughter and compensation scheme. He accepted a recommendation that BSE should be made a notifiable disease – a measure designed to give MAFF a better picture of the incidence of the disease and the power, if necessary, to impose movement controls on animals. BSE was made notifiable in June 1988 by the same Order that introduced the ruminant feed ban. The rate of reporting leapt almost overnight from 60 cases a month to 60 cases a week. The Order required that the heads of all these cases be surrendered to MAFF; the brains were then removed and examined by the CVL. So far as the proposal for compulsory slaughter was concerned, discussions were carried on with the farming industry to explore the possibility of an industry-funded scheme. Industry was told that there was no question of government funding being provided. Industry's response was that it was for the Government to fund compensation if compulsory slaughter were to be introduced.

244 On 4 June 1988 an article in the *British Medical Journal*, co-authored by a doctor and a dietician, pointed out that if BSE were transmissible to humans it might be years before infected individuals succumbed. The authors wrongly assumed that animals showing signs of sickness would not enter the food chain, but went on to say that it was 'naïve, uninformed and potentially disastrous' to assume that animals incubating the disease but not yet showing signs posed no risk to humans.

245 On 20 June the Southwood Working Party met for the first time.[23] They were horrified to learn that animals sick with BSE were being slaughtered for food. The next day Sir Richard Southwood wrote to Mr Andrews recommending that carcasses of BSE-affected animals should be condemned and destroyed. Mr MacGregor's officials advised him that compulsory slaughter should be introduced and that the Government would have to pay compensation under the Animal Health Act 1981 – they recommended that this should be fixed at 50 per cent of the value of a sound animal. Mr MacGregor wrote to Mr John Major at the Treasury urging, though with reluctance, that payment of compensation at this level be approved.

246 At the same time, Sir Donald Acheson informed Mr David Mellor, the Health Minister, that destruction of the carcasses of clinically affected animals was essential on the grounds of risk to humans. It was on this basis that the consent of the Treasury was given to the payment of compensation. Mr MacGregor had suggested that the cost of this measure would be around £250,000 a year on the basis that cases would continue to be reported at a rate of about 60 a month. He cannot yet have been aware of the increase of the reporting rate consequent upon the notification requirement.

247 The Order providing for compulsory slaughter and destruction of cattle suffering from BSE came into force on 8 August. Nearly six months had gone by since MAFF officials had first recommended this course.

---

[23] See vol. 4: *The Southwood Working Party, 1988–89*

4. The Southwood Working Party and other scientific advisory committees

The Southwood Working Party

248 The Southwood Working Party[24] consisted of Sir Richard Southwood, Professor of Zoology at Oxford University; Professor Anthony Epstein FRS,[25] a virologist; Professor Sir John Walton,[26] a neurologist; and Dr William Martin, a veterinarian who had just retired from the Directorship of the Moredun Research Institute in Edinburgh. Sir Richard emphasised to us that they were not experts in the narrow sense of having particular expertise in TSEs. Each was, however, a scientist of the highest standing in his field and together they were well placed to consider the available data and to give a considered view as to what implications these suggested that BSE might have for human health.

249 This was precisely the task that Sir Donald Acheson wanted the Working Party to perform. When writing to Sir Richard on 8 April 1988, he suggested a first meeting of the group as soon as possible, a small number of additional meetings at the end of the summer and 'a very brief note with recommendations'. In the event a substantial report was delivered in February 1989. The Working Party's wide terms of reference were:

> To advise on the implications of Bovine Spongiform Encephalopathy and matters relating thereto.

250 The Report addressed both human and animal health. The original reason for this had been to 'play down the human health issue'. Sir Richard had, however, been anxious from the outset to have broad terms of reference and he had also been determined that the Report should be published. Happily the breadth of the terms of reference did not inhibit MAFF officials from recommending, before the Working Party had been fully constituted, that a ruminant feed ban should be introduced.

251 The Working Party were served by a joint secretariat, consisting of Mr Alan Lawrence, an official in MAFF's Animal Health Division who was given special responsibility for BSE, and Dr Hilary Pickles, a Principal Medical Officer whom Sir Donald Acheson appointed to take the lead in DH in relation to BSE.

252 Although the Working Party took longer than had been hoped to produce a Report, they lost no time in making important interim recommendations. They had asked what happened to material from affected animals and been told that these animals would usually go to be slaughtered for human food, in the same way as healthy animals. They told us that they were horrified by this and felt it was their

---
[24] Who are the subject of Volume 4 of this Report
[25] Now Sir Anthony Epstein
[26] Now Lord Walton

job to stop it happening immediately. In consequence, after their first meeting on 20 June 1988, Sir Richard wrote to Mr Andrews recommending that carcasses of clinically affected animals be destroyed by incineration or a comparable method. The removal of the head was not an adequate safeguard as that was not the only source of infection. This recommendation was accepted and implemented. The measure proved of crucial importance in protecting humans, and also animals, from the risk of infection with BSE. The Working Party are to be commended for their prompt and decisive action.

253 The Working Party made further immediate recommendations: that an expert working party should be set up to advise on the research in hand and the research required in relation to BSE; that priority should be given to a study to see whether BSE transmitted from cow to calf; and that tests be carried out to see whether scrapie could be transmitted to cattle. This was further wise advice promptly given. It led to the setting up of the Tyrrell Committee on research.

254 The Working Party were not to meet again until November. In the meantime, the two secretaries, and Mr Wilesmith, who had been asked to act as expert adviser to the Working Party, set about drafting sections of the Report.

255 The second meeting of the Southwood Working Party on 10 November 1988 led to interim recommendations that the ruminant feed ban, which was due to expire at the end of the year, be extended indefinitely, and that milk from cows affected with BSE be destroyed. Dr Richard Kimberlin, who had retired from being Acting Director of the Neuropathogenesis Unit (NPU), Edinburgh, to run his own consultancy in TSEs, attended this meeting. Experiments at the NPU had recently demonstrated that BSE could be transmitted to mice, and there was discussion about the likelihood of transmission from cow to calf. There was also discussion about whether it was safe to eat ox brain. The Working Party decided that it would not be appropriate to ban the eating of UK ox brain but that it was worth consideration whether products containing brain should be required to be labelled, leaving the consumer to make his or her own choice. The Working Party subsequently dropped the idea of labelling as they were informed that this would involve complications under European law.

256 It was agreed at the second meeting that those responsible for occupational health and for the safety of medicines should have their attention drawn to the need to address potential risks posed by BSE. Again, the Working Party are to be commended for taking action to safeguard human health in advance of delivering their Report. We shall consider the response to their action when we come to consider the topics in question.

257 The Working Party met again on 16 December and had a final meeting on 3 February 1989. The contents of their Report were considered in detail on both occasions, and we shall now consider these.

Epidemiology

258 The first eight pages of the Report consisted of a history of BSE and an account of what was known about TSEs. These were largely written by Mr Wilesmith, Mr Lawrence and Dr Pickles, the latter topic being a summary of a substantial number of published papers, with which members of the Working Party would have made themselves familiar. There then followed a chapter on 'the cause of BSE: the epidemiological evidence'. This had been written by Mr Wilesmith. It set out the tentative conclusions that we have detailed in the previous chapter, including the following:

- the epidemiology was typical of an extended common source epidemic;
- all affected animals appeared to be index cases;
- the common feature was the use of commercial concentrates in feed;
- a possible explanation for the emergence of BSE was a change in the exposure of cattle to ovine-derived protein and the scrapie agent due to
  i. more scrapie-infected material going to be rendered;
  ii. changes in the rendering processes.

259 A subsequent chapter, also written by Mr Wilesmith, dealt with 'the Future Course of the Disease'. This stated that the effect of recycling of BSE was impossible to quantify and possibly minimal and undetectable, in which case a constant incidence of 350–400 cases a month could be expected. The possibility of maternal transmission was recognised, but it was observed that this would be unlikely to sustain BSE in the national cattle population.

260 The Working Party did not see it as their role to conduct a critical review of Mr Wilesmith's conclusions. We do not suggest that they should have done. The Report did nothing, however, to dispel the impression that the conclusions in question had been reached, or endorsed, by the members of the Working Party. In a covering letter to Ministers, published with the Report, the Working Party thanked Mr Wilesmith and others for their assistance and added, 'The Report, however, remains our own.' We think that the Working Party should have made it plain that the section of the Report dealing with epidemiology had been provided by Mr Wilesmith and was based on data which they had not been able to review. In the event their Report added weight to a number of epidemiological conclusions which subsequently proved to be fallacious, the most significant being that the cases of BSE were index cases of cattle infected with scrapie. It was this theory which gave so many the false reassurance that it was very unlikely that BSE was transmissible to humans.

Risk to humans

261 In the most important part of their Report, the Working Party set out their views on the possibility that BSE might be transmissible to humans. These were, in summary:

- Humans were susceptible to spongiform encephalopathies.
- Neural and, to a lesser extent, lymphoid tissue carried the infection, while the risk was far less with other tissues.
- Parenteral inoculation was more efficient in transmitting disease than oral or topical exposure.

- The greatest risk in theory would be from parenteral injection of material derived from bovine brain or lymphoid tissue.

- Medicinal products for injection or surgical implantation using bovine tissues might be capable of transmitting infectious agents.

- Direct inoculation of bovine tissue could arise accidentally in certain occupations.

- In these and in other circumstances the risk of transmission of BSE to humans appeared remote.

262 The Working Party commented that because the risk of transmission of BSE to humans could not be entirely ruled out, action had been taken to remove known affected cattle from the human food chain. The Medicines Licensing Authority had been alerted to potential concern about BSE in medicinal products and would ensure that scrutiny of source materials and manufacturing processes now took account of the BSE agent. The Health and Safety Executive had also been alerted to potential concern about BSE.

263 The Working Party had this to say about possible risks from eating animals incubating BSE but not yet showing clinical signs:

> It has been suggested, although clinically affected animals are being slaughtered and destroyed, that consideration should be given to products containing brain and spleen being so labelled, to enable the consumer to make an informed choice. The Working Party believes that risks as at present perceived would not justify this measure.

264 They went on to state, however:

> We consider that manufacturers of baby foods should avoid the use of ruminant offal and thymus.

We shall from now on describe this piece of advice as 'the baby food recommendation'.

265 There were a number of matters which the Working Party did not explain in their Report:

- What did they mean when they said that the risk of transmission of BSE to humans appeared 'remote'?

- Why did they consider that the risk appeared remote?

- Why did they recommend that affected cattle should be slaughtered and destroyed?

- Why did they make the baby food recommendation?

- Why did they not recommend any other precautions to protect human food from subclinically infected animals?

266 All these matters we raised with the members of the Working Party.

267 They explained that they intended the word 'remote' to bear the meaning that this word has when used to describe a risk in a medical context. In that context a remote risk is one that is highly unlikely to prove significant, but which it is unreasonable to ignore. Reasonable precautions should be taken to try to prevent a remote risk. The Working Party set out to advise what those precautions should be. They told us that in doing so:

> Our approach to risk was in accord with the then developing application of analysis to public risk which involved the balancing of the perceived magnitude of the risk against the practicability or achievability of successive steps for its reduction. The magnitude of a risk comprises both its likelihood and the scale of the danger.

268 This approach is sometimes known as ALARP (As Low As Reasonably Practicable). It requires an exercise in proportionality. When deciding whether a precaution is 'reasonably practicable' it is necessary to weigh the cost and consequences of introducing the precaution against the risk which the precaution is intended to obviate.

269 Why was the risk considered remote? Our reading of the Report led us to conclude that the Working Party had drawn comfort from the way that scrapie behaves. Sheep infected with scrapie have been slaughtered for human food for hundreds of years, without doing any harm. If BSE was the scrapie agent in cattle, it was likely that it would behave in the same way.

270 The Working Party confirmed to us that this was indeed their reasoning. But they emphasised that they did not assume that BSE would behave like scrapie. They recognised the possibility that, whether or not scrapie was the source of the infection, BSE in cattle might behave more virulently than scrapie in sheep. Because of this possibility, reasonable precautions needed to be taken against the possible risk from eating BSE-infected meat.

271 The Working Party concluded that reasonable precautions against the risk from eating BSE-infected meat involved taking sick animals out of the food chain, but that no precautions were needed in respect of subclinically infected animals, other than the baby food recommendation.

272 We have a number of criticisms to make of this part of the Working Party's Report. In the first place they did not make it clear that, in describing the risk as remote, they were intending to indicate that steps should be taken to reduce the risk as low as reasonably practicable. We think that they should have done.

273 In the second place, we do not consider that the Working Party correctly applied the ALARP principle. Animals with BSE that had developed clinical signs of the disease were to be slaughtered and destroyed. No steps were to be taken, however, to protect anyone other than babies from the risk of eating potentially infective parts of animals infected with BSE but not yet showing signs. It is true that infectivity of the most infective tissues – the brain and spinal cord – rises significantly shortly before clinical signs begin to show. It is also true that there were reasons to think that babies might be more susceptible to infection than adults. But we do not consider that these differences justified an approach that treated the

risk from eating brain or spinal cord from an animal incubating BSE as one in respect of which there were no reasonably practical precautions that need be taken.

274 We believe that part of the Working Party's problem was that they were in no position to reach an informed view of how the ALARP principle should apply. They were not aware of the practice of mechanical recovery of meat, which sucked from the spinal column the residue left attached after removal of meat – a residue likely to include portions of spinal cord. Nor, so we believe, did they have in mind that it was reasonably practicable to identify and remove the potentially infective tissues in the course of the slaughterhouse processes.

275 In these circumstances, we do not criticise the Working Party for failing to recommend the precautionary measure that MAFF was subsequently to put in place – the SBO ban. What we feel they should have done was to point out that cattle subclinically infected with BSE were entering the human food chain, that some tissues of such cattle were potentially infective, and that consideration should be given to identifying such steps as were reasonably practicable to prevent their being eaten, not just by babies, but by everyone.

276 There is a further aspect of the way the *Southwood Report* dealt with risk that caused us concern. The Working Party said of the risk of transmission of BSE through the use of medicinal products:

> Although the risks appear remote the Working Party recommended that the attention of the Licensing Authority, the Committee on Safety of Medicines, the Committee on Dental and Surgical Materials and the Veterinary Products Committee be drawn to the emergence of BSE so that they can take appropriate action.

277 The Working Party told us that they had described these risks as remote only because of the action that they had been assured was being taken to address them. They had initially considered that some medicinal products sourced from bovine materials, which were injected, might carry a relatively high risk of transmission. With the assistance of Dr Pickles they had taken all proper steps to get those responsible for the safety of medicines to start taking action to address this risk. They had intended to include in their Report details of some of the steps that could be considered to prevent the BSE agent entering into pharmaceutical manufacture. However, as we describe in paragraphs 901–906 below, in response to concerns expressed by officials responsible for medicines licensing, they had been persuaded to tone down their Report and make no mention of these by the assurance that action was being taken.

278 The action taken by the Working Party, assisted by Dr Pickles, to galvanise those responsible for the safety of medicines was praiseworthy. The Working Party told us that they were anxious to avoid raising, by their Report, concerns about the safety of vaccines that would lead to a vaccine scare which could result in children being exposed to much greater risk than that posed by BSE. We sympathise with their anxiety. It led, however, to their Report giving the reader a false impression of their assessment of the risk relating to medicinal products. The Working Party should not have allowed this. They could have avoided doing so, without creating a vaccine scare, simply by saying that they had had concerns about the implications that BSE might have for certain medicinal products and had referred those concerns to the Committee on Safety of Medicines and the Veterinary Products Committee, which had undertaken to address them. Unfortunately, the wording of the Report was to give some who were responsible for dealing with medicinal products, both human and veterinary, the impression that these would involve no more than a remote risk, even if no remedial measures were taken.

279 Similarly, the sections of the Report that dealt with occupational safety gave the impression that occupational risks were remote whether or not steps were taken to address them. The Working Party had commendably taken steps before publication of their Report to ensure that occupational risks were addressed. Dr Pickles had written to, and met with, the Health and Safety Executive (HSE) on their behalf. Their Report recommended that the HSE consider whether further guidance should be given. However, it seemed to us that the effect of this recommendation was likely to be uncertain, given the indications in the body of the Report that the risk was remote and that no specific additional guidance on BSE was thought necessary. As with medicines, we consider that the Working Party should not have used words that conveyed the impression that the risk were, even in the absence of precautionary measures, remote.

280 By the time that the Working Party came to finalise their Report, their interim recommendation that an expert committee be set up to advise on research had been implemented. The Tyrrell Committee had been established. In their Report the Working Party drew attention to a number of areas where research was needed for further consideration by that Committee. They also recommended the monitoring of CJD cases, since any human cases of BSE would probably present as CJD. The achievement of the CJD Surveillance Unit in identifying in 1996 the emergence of variant CJD demonstrated the wisdom of this recommendation.

281 The draft of the Working Party's Report had a sting in its tail. It referred to the fact that BSE had resulted from the practice of feeding animal protein to herbivores, and noted that this practice opened up new pathways for infection. It continued:

> We believe that the inevitable risks are such that it would be prudent to change agricultural practice so as to eliminate these novel pathways for pathogens.

282 When MAFF officials learned that this was to be included in the Report they were horrified, as they read it as an attack on the practice of incorporating MBM in animal feed. Animal Health Division commented to the Permanent Secretary that the rendering industry processed over 100,000 tonnes of raw material every month, thus providing a source of animal feed and industrial raw material, and also a 'waste disposal' service for the slaughtering industry. A paper setting out those implications was quickly prepared and sent to the Working Party. Dr Martin also wrote to Sir Richard, urging restraint on this topic. Restraint there was, for an amendment was made to the draft which was intended to make clear something that Sir Richard later confirmed. The Working Party was not recommending that the practice of rendering animal protein should cease, but that its continuance should depend upon finding a rendering process capable of destroying all pathogens.

283 We have criticised some aspects of the *Southwood Report*, but those criticisms should not obscure the vital benefit that the Working Party provided in putting an immediate stop to the practice of eating BSE-diseased animals, in bringing immediate pressure to bear on those responsible for the safety of human medicines and occupational health to address the risks posed by BSE, and in giving wise advice about research. When the Report was published, it was generally well received by those who were expert in the field. Nonetheless a number of experts raised, at the time, the question of the risk posed by subclinical animals, and many more, when giving evidence to us, claimed to have identified the need to address this problem at the time. Pressure to do so was soon to build up and lead to the decision to introduce the SBO ban.

284 The Working Party's risk assessment had, necessarily, been based on very limited data. In August 1988 Sir Richard, replying to a medical correspondent, wrote:

> My colleagues and I have made various recommendations based, I have to admit, largely on guesswork and drawing parallels from the existing knowledge of scrapie and CJD.

In a summary section of their Report, the Working Party wrote:

> Our deliberations have been limited by the paucity of the available evidence. Further research work in this area is essential.

In their General Conclusions, after observing that it was most unlikely that BSE would have any implications for human health, the Working Party added this warning:

> Nevertheless, if our assessment of these likelihoods are incorrect, the implications would be extremely serious.

285 Unfortunately, this warning and the tentative nature of the Working Party's conclusions were not appreciated or were lost sight of. Right up to 1996 the *Southwood Report* was cited as if it demonstrated as a matter of scientific certainty, rather than provisional opinion, that any risk to humans from BSE was remote.

Other scientific advisory committees

The Consultative Committee on Research into SEs (The Tyrrell Committee)[27]

286 One of the first recommendations to be made by the Southwood Working Party in June 1988 was that an expert Consultative Committee on research should be set up. In February 1989 it was announced that, following this recommendation, a Consultative Committee had been set up, chaired by Dr David Tyrrell.[28] The other members were Dr Watson,[29] Professor John Bourne,[30] Dr Robert Will,[31] and Dr Richard Kimberlin.[32] The terms of reference were:

---
[27] Detailed consideration of the work of The Tyrrell Committee appears in vol. 11: *Scientists After Southwood*
[28] A microbiologist who was Director of the MRC Common Cold Unit
[29] Director of the Central Veterinary Laboratory
[30] Director of the Institute for Animal Health
[31] Consultant Neurologist at the Western General Hospital Edinburgh
[32] Ex-Acting Director of the NPU, who had retired to set up an independent consultancy, advising on TSEs

> To advise the Ministry of Agriculture, Fisheries and Food and Department of Health on research on transmissible spongiform encephalopathies including:
>
> > (a)   work already in progress or proposed;
> >
> > (b)   any additional work required;
> >
> > (c)   priorities for future relevant research.
>
> In the context of these terms of reference, transmissible spongiform encephalopathies include those affecting both domestic and wild ruminants and man.

287 The Committee moved fast. After three meetings it presented an 'Interim Report' to the Government on 10 June 1989. This identified a number of research questions that needed to be answered about BSE under the headings: epidemiology, pathology and molecular studies. Research studies needed to answer these questions were identified and graded with three stars for highest priority, two stars for medium priority and one star for low priority. We consider the adequacy of the research carried out into BSE in Chapter 12 below.

288 In commenting on the research questions, the Committee observed:

> We need to be sure that the disease really came from sheep and to know whether it is likely to establish itself long-term in bovines.

289 In their conclusions the Committee stressed that more research was needed:

> If the preliminary studies and arguments-by-analogy used to determine our present control policies turn out to be incorrect, it will be essential to have well-documented facts available so that current policies can be effectively revised.

290 The Report was produced in haste as an interim one because the Committee was anxious that there should be no delay in seeking provision of resources for essential research and getting the projects under way. The Committee emphasised the importance of having the projects peer-reviewed and suggested that:

> A standard mechanism may be needed to oversee this co-operation and co-ordination beyond the lifetime of our Committee.

291 The Committee asked for guidance as to whether they were expected to have a continuing role in peer review and project coordination.

292 Mr Gummer decided that the Government should respond to the *Tyrrell Report* by initiating all research projects falling within the top two of the Tyrrell Committee's priority categories, and Mr Roger Freeman, Parliamentary Under-Secretary at DH, conveyed to him his Department's agreement with this response. Delay then occurred in ensuring that the necessary funding was in place. This was not achieved until January 1990, when the Government published the Report and announced that work was in hand to implement the projects recommended by the

Tyrrell Committee as urgent and of high priority, some of which were already in progress. It is creditworthy that Mr Gummer, in accordance with the advice of his officials, proffered by Mr Andrews, and with the support of DH, decided that all these projects should be pursued.

The Spongiform Encephalopathy Advisory Committee (SEAC)

293 No further assistance was sought from the Tyrrell Committee. Officials at MAFF and DH agreed that it was desirable that a new expert standing committee should be formed to meet from time to time to advise on questions about BSE, but that this new committee should not publish reports. Its role would include having a general overview of research. Dr Tyrrell was invited, and agreed, to chair this new committee. Mr Gummer announced the setting up of SEAC on 3 April 1990. Its terms of reference were:

> To advise the Ministry of Agriculture, Fisheries and Food and the Department of Health on matters relating to spongiform encephalopathies.

294 A detailed account of the setting up, membership and activities of SEAC appears in Volume 11, together with discussion on its role. In this volume we shall refer from time to time in the course of the narrative to questions asked of, and advice given by, SEAC. Contrary to the expectation, and to some extent the wishes, of its members, SEAC found itself given the role of providing policy advice on almost every decision that the Government was faced with in handling BSE.

295 We should record our respect for the dedication of the members of both the Tyrrell Committee and SEAC. Members of the latter found themselves called upon to provide much more assistance than they had been led to believe would be the case. Independent scientists in this country have an admirable tradition of agreeing to serve on committees performing functions in the public interest. Members of SEAC, who exemplified this tradition, found that it involved a considerable burden.

5. The animal health story

296 By the time that the *Southwood Report* was published, the two major measures that the Working Party had recommended were in place. The ruminant feed ban had been extended – not indefinitely as the Working Party had recommended, but for a further year. An indefinite extension was to come later. If feed were the only means of infecting with BSE, the ban should in due course eradicate the disease. So far as the risk to humans was concerned, the Working Party considered that slaughter and destruction of animals showing clinical signs sufficed to protect against the remote risk of transmission as a result of eating infective tissue. So far as occupational risks and risks in relation to medicinal products were concerned, the Working Party had alerted those responsible for addressing these.

297 Substantial further measures were, however, to be taken to address food risks, for both humans and animals. These were, first, the ban on using Specified Bovine Offal (SBO) for human food ('the human SBO ban'), followed by a ban on incorporating SBO in animal feed ('the animal SBO ban'). Our task of reviewing the action taken in response to BSE up to 20 March 1996 requires us to examine the circumstances in which these measures were introduced. It also requires us to review the various measures that were taken in response to BSE and how they were enforced and monitored. That is a complex, but important, part of the BSE story. It is important because there were significant shortcomings in both the human health and the animal health measures, and in their enforcement and monitoring. Had we attempted to cover all of this in simple chronological order in our Report, the result would have been to confuse. Accordingly we decided at this stage to divide our coverage into two. In Volume 5 we have traced the story of measures taken to protect animal health. In Volume 6 we have followed the story that relates to the protection of human health.

298 We propose to follow the same course in this volume. In this chapter we shall cover that part of the story which is told in detail in Volume 5. We shall moreover subdivide the topics in the same way as we have in that volume. This means that we shall give separate treatment to the ruminant feed ban and the animal SBO ban. The former was the measure designed to protect cattle and other ruminants. The latter was designed to protect non-ruminant animals, but provided fortuitously an additional line of defence for cattle, which proved of great importance.

299 It may be thought that we have got our priorities wrong in considering animal health before human health. The reality is that, although introduced in the interests of animal health, the action taken to eradicate BSE was of critical importance in protecting humans should BSE prove, as indeed it did, to be a zoonosis. It is for this reason that we considered it logical to look first of all at that part of the BSE story which was motivated by the immediate demands of animal health.

Ruminant feed ban

300 Mr Kevin Taylor[33] became responsible for providing veterinary advice on all aspects of the control of BSE from the time that it became a notifiable disease in 1988. He told us that there was no practical way in which the ruminant feed ban could be enforced, as there was no test which could identify rendered ruminant protein in animal feed. Effectiveness depended on voluntary compliance with the ban. Because of the long incubation period, years would elapse before it would become apparent whether there had been strict compliance with the ruminant feed ban. We consider that it was reasonable to expect that neither feedmills nor farmers would deliberately incorporate MBM in cattle feed. Other sources of protein were available that were only marginally more expensive.

301 No guidance was given to the County Councils and Unitary Authorities, whose duty it was to enforce the ban. We had evidence which suggested that some local authorities made attempts to check on compliance with the ruminant feed ban by sampling, but found this impossible. It is possible that others may have checked the records of feedmills to ensure that MBM was not a component of cattle feed, although strictly they had no statutory right to demand to see these.[34] In general we do not believe that any steps were taken by local authorities to enforce the ban during any part of the period with which we are concerned.

302 Mr MacGregor proposed that the introduction of the ruminant feed ban should be handled in a low-key way on the assumption that MAFF had a system for notifying all those who were affected, and in particular farmers. In the event MAFF officials made no attempt to contact renderers, the feed trade or farmers directly, but relied upon meetings with trade associations, or farmers' unions, together with a press release, in order to publicise the introduction of the ban.[35]

303 Representatives of the feed industry told us that when the feed ban was introduced, a number of factors combined to detract from any impression of urgency about its implementation:

- the grant of a period of grace in which to use up current stocks of feed;
- the absence of any feed recall;
- the fact that neither import nor export of MBM was to be prohibited;
- uncertainty as to whether MBM was indeed the vector of BSE and, if it was, as to which rendering systems were unable to inactivate it; and finally,
- the belief that a very large amount of infective feed would have to be consumed to transmit the disease.

304 Farmers who gave evidence told us that they did not appreciate the gravity of the situation at the time. It was only the occasional farmer who had experience of BSE and that experience was normally of no more than a single case. They continued to use up any stocks of cattle feed remaining at the time that the ban came into force.[36]

---

[33] Veterinary Head of Notifiable Diseases Section, 1986–91; Assistant Chief Veterinary Officer, Animal Health and Welfare Veterinary Section, 1991–97
[34] Vol. 5: *Animal Health, 1989–1996*, Chapter 2, paras 2.10 and 2.43
[35] Vol. 5: *Animal Health, 1989–1996*, Chapter 2, paras 2.31–2.35
[36] Vol. 3: *Early Years, 1986–88*, paras 4.86–4.113

FINDINGS AND CONCLUSIONS

305 A relatively relaxed attitude to enforcement of the ban was illustrated by the decision of Mr Meldrum in February 1989 that the development of an ELISA test, to detect the presence of ruminant protein in animal feed, should be carried out 'in house' by a senior scientific officer at Worcester VIC, Mr Mike Ansfield. This course had a number of attractions, not least that MAFF would retain the intellectual property in the test, which might prove commercially valuable. It was estimated, however, that it would take between 12 and 18 months to develop the test. The more costly alternative of seeking external collaboration in producing a test would have been likely to produce swifter results.[37] As we have commented above, this attitude was a consequence of a failure to appreciate the need to guard against cross-contamination of cattle feed.

306 Although no anxieties were expressed about the adequacy of the action taken by MAFF to eradicate BSE, there were concerns about the risk that BSE might pose, in the interim, to humans and to non-ruminant animals. The scale of infection of cattle during the period before the ruminant feed ban was introduced proved to have been greatly underestimated. By the end of 1988 cases were being reported and confirmed at a rate of over 100 cases a week. The Southwood Working Party had envisaged cases remaining on a plateau at about that rate, but by the end of April 1989 the rate had increased to about 150 cases reported each week. In June 1989 the Government announced its intention to ban SBO (brain, spinal cord, tonsils, thymus, spleen and intestines) from all human food ('the human SBO ban'). This led a large part of the feed industry to impose a voluntary ban on including those categories of offal in animal feed, a ban that MAFF made statutory in September 1990 ('the animal SBO ban') after a number of cases of Feline Spongiform Encephalopathy (FSE) had been identified and BSE had been experimentally transmitted, by inoculation, to a pig. Those events are dealt with later in this chapter.

307 The first case of FSE in May 1990 led to considerable public concern about its implications for human health and to an Inquiry into BSE by the Agriculture Committee of the House of Commons. The primary concern of the Committee was the implications of BSE for human health. So far as animal health was concerned, the Committee observed that the ruminant feed ban, if strictly applied, should arrest BSE. They recommended, however:

> That the Government establish an expert committee to examine the whole range of animal feeds and advise on how the industries that produce these should be regulated.

308 This recommendation was accepted by the Government, which set up the Lamming Committee (the Expert Group on Animal Feedingstuffs) in 1991.

309 By the end of 1990 MAFF officials and Mr Gummer, now the Minister of Agriculture, Fisheries and Food, had no reason to doubt the efficacy of the ruminant feed ban. The rate of reported cases had soared until in some weeks these exceeded 400, but they were all cases of cattle born before the ruminant feed ban came into force. Mr Ansfield appeared to have made substantial progress in the development of an ELISA test. His test could detect both ovine and bovine protein in meat and bone meal (MBM). It remained to test it on compound feed.

[37] Vol. 5: *Animal Health, 1989–1996*, Chapter 2 paras 2.52–2.54

The Animal Health Story

310 The development of the ELISA test resulted in some concern on the part of UKASTA. It feared that the test would identify small quantities of ruminant protein in cattle feed resulting from cross-contamination with pig and poultry rations in the feedmills, or from traces in tallow incorporated in cattle feed. Its concern was not that this would be sufficient to infect cattle, but that it might result in prosecution of its members for breach of the Regulations. This concern was conveyed by Dr Danny Matthews[38] to Mr Meldrum. He told us that at this point he did not recall any concerns at MAFF that cross-contamination of feed might be taking place on a scale sufficient to undermine the effectiveness of the ruminant feed ban.

The first BAB

311 On 22 March 1991 the first BAB[39] was reported to Mr Gummer. This was made public by a news release five days later. It caused considerable excitement within MAFF as urgent consideration was given to whether it was a case of maternal transmission or whether it might have been infected by feed. It was, however, only the first of what was to become first a trickle, then a stream and finally a flood. By the end of the year 300 BABs had been reported, of which only 11 had been confirmed. Investigations by Dr Matthews and his colleagues suggested that at least the majority of these cases were caused by feed containing MBM that was still in the feed chain when the ban came into force on 18 July 1988.

312 The Lamming Committee[40] met for the first time on 15 February 1991. At their second meeting on 13 March they heard evidence from Mr Meldrum. He told them that he was not totally content with the current controls, as at

> . . . present there was no test for ruminant protein in feed. However, an ELISA method was currently being evaluated for use in the field.

313 He said that he was fairly confident that on-farm mixers would observe the controls, despite the absence of a test.

314 Unfortunately, hopes that the ELISA test was almost ready for use were dashed when it was found that most compound feeds produced a positive result even when they included no MBM.

315 When the Lamming Committee reported in June 1992, they commented about BSE that the evidence suggested that in the majority of cases the controls were working, despite the fact that the ruminant feed ban and the SBO ban were to a considerable extent dependent on self-regulation by the industry. They welcomed the development of the ELISA test.

316 As the number of BABs increased, so did MAFF officials' conviction that feed containing ruminant protein had been fed to cattle for a significant period after the ban came into force. In September 1992 Dr Matthews minuted Mr Meldrum commenting that it was clear that the major compounders had needed at least three months to clear stocks, in some cases longer. He added that smaller compounders, who were disproportionately represented among suppliers to owners of BAB cases,

---

[38] The Senior Veterinary Officer at Tolworth responsible for BSE
[39] BSE victim Born After the ruminant feed Ban came into force
[40] Professor G E Lamming, Professor of Animal Physiology, Nottingham University; Professor P C Thomas, Principal and Chief Executive, Scottish Agricultural College; Mr C Maclean, Technical Director, Meat and Livestock Commission; and Dr E M Cooke, Deputy Director, Public Health Laboratory Service

having not been party to discussions prior to the introduction of the ban, might be expected to have taken longer to clear their stocks.

UKASTA's information about breaches of the ban

317 By this time 220 BABs had been confirmed. Mr Meldrum wrote to Mr James Reed, the Director-General of UKASTA, suggesting that there had been a time lag of between three and six months before the ban became fully effective. In response to Mr Meldrum's request for information, UKASTA asked all companies represented on its Executive Committee to answer a questionnaire. At a meeting on 10 November, they gave Mr Meldrum the results of this survey, on condition that the information would be treated with the utmost confidence. The survey showed that most compounders had continued to manufacture cattle feed containing ruminant protein into July 1988 and did not clear stocks from their premises until August or September, or even, in a few instances, October. When giving evidence to us, UKASTA representatives suggested that the stocks of cattle feed may have been cleared by incorporation in feed for non-ruminants, so that the survey may not have disclosed deliberate breach of the ban by UKASTA members. We reject this suggestion, as did Mr Meldrum. The contemporary evidence of the meeting on 10 November is unequivocal. That evidence is reinforced by the fact that over 11,000 cattle born in the last five months of 1988 contracted BSE, as did a further 12,600-odd that were born in 1989.

318 These figures will, of course, reflect the use by farmers after 18 July 1988 of feed purchased before that date, but we are satisfied that they also reflect deliberate breaches of the ban by some compounders and others in the supply chain. As a whole the animal feed industry does not emerge from the BSE story with credit.

319 MAFF officials seized eagerly on evidence of breaches of the ruminant feed ban, for the alternative explanation that maternal transmission was occurring was less palatable. In November Ministers were told that there was clear evidence that ruminant feed containing MBM would have been available for six months after the ban came into force. At the year end a MAFF progress report expressed continued confidence that the ban would bring the epidemic to an end.

320 The first half of 1993 saw MAFF officials frustrated in their desire to start testing feed for the presence of animal protein by continuing difficulties in developing the ELISA test – accentuated by suspension of work on the project while the Worcester VIC was relocated to Luddington. MAFF's difficulties were compounded by the fact that they had no legal power to carry out random sampling. Samples could only be taken when there were reasonable grounds for suspecting that the Regulations were being broken. The lengthy incubation period made it difficult to demonstrate such grounds.

321 In September a briefing paper prepared for Mrs Gillian Shephard, who had succeeded Mr Gummer in July as the MAFF Minister, and Mr Nicholas Soames, the Parliamentary Secretary at MAFF, stated that there had been 4,010 confirmed BABs, the great majority of which had had access to ruminant protein in their feed. The paper went on to make the point that the animal SBO ban introduced in 1990 had had the effect of reinforcing the ruminant feed ban.

322 Problems in relation to sampling continued in the first half of 1994. The ELISA test was ready for field testing, but sampling capacity at Luddington was limited and there was no hope of embarking on large-scale monitoring at feedmills. Furthermore, the lawyers were having difficulty finding a path through the maze of different Regulations relating to animal feed that would enable mandatory sampling to be introduced.

323 Towards the end of 1993 Mr Wilesmith had begun to feel concerned that cross-contamination might be taking place at feedmills manufacturing multi-species rations. This concern was taken up by Mr Bradley early in the following year. In a minute to Mr Kevin Taylor, he commented that they had both believed that the animal SBO ban would have stopped any infected ruminant protein getting through into the animal feed chain, but if the SBO ban was being abused there was a weakness in this argument. It was at this time that concern was growing about reports of non-compliance with the animal SBO ban.[41]

324 By the middle of 1994 MAFF officials had worked out a sampling procedure which they recommended in a submission to Mr Soames. Sampling of cattle feed should initially be carried out on farms on a voluntary basis. Any positive results would lead to mandatory sampling at the feedmill which had supplied the feed. MAFF would carry out the testing themselves rather than entrusting the ELISA test to the local authorities, which had statutory responsibility for enforcing the ban. There were a number of reasons why MAFF officials wished to keep the testing 'in house' – one being apprehension that some local authorities might prove over-assiduous in enforcing the ban. It seems to us that the test was not sufficiently robust at this stage to be used in statutory enforcement of the ban. The decision of MAFF officials that MAFF should use the test on a voluntary basis under a uniform scheme to operate across the country was reasonable.

Cross-contamination in feedmills

325 In June 1994 the possibility that cross-contamination in feedmills was a cause of some of the BABs was discussed with UKASTA's Scientific Committee. They commented that equipment used in feedmills was being updated 'as and when required'. This was the start of a series of meetings between MAFF officials and UKASTA in which each had a similar hidden agenda. MAFF was concerned not to do anything that would lead UKASTA members to cease using animal protein as an ingredient of feed for non-ruminant animals. UKASTA, for its part, was anxious that its members should be able to continue to do this without incurring risk of prosecution should it result, on occasion, in cross-contamination of ruminant feed. UKASTA was to threaten repeatedly that it might have to advise its members to cease using animal protein, while MAFF officials sought to allay UKASTA's anxieties by reassuring its members that sampling was not being used as a precursor to prosecution. In reality, the limitations of the ELISA test, coupled with the requirement under the Order to prove knowing incorporation of ruminant protein, meant that MAFF officials were in no position to contemplate enforcing the ruminant feed ban by criminal proceedings.

[41] See paras 441ff

FINDINGS AND CONCLUSIONS

326 In July 1994 Mr William Waldegrave succeeded Mrs Shephard as Minister of Agriculture, Fisheries and Food. The following month he was informed of the first four BABs to be reported that had been born in 1991.

327 By September 1994 a number of factors had combined to indicate that cross-contamination in feedmills was a serious problem:

- Inactivation studies had shown that the three systems which provided most of the UK rendering capacity were not capable of inactivating the BSE agent.

- Epidemiological investigations had revealed a correlation between the incidence of BABs and the ratio of cattle to pigs in the different counties. The incidence of BABs was highest in those counties where mills were producing both pig and cattle feed in large quantities. Mr Wilesmith concluded that cross-contamination was likely to be occurring at the mills rather than on the farms, although he recognised that cross-contamination on the farm was possible.

- There was clear evidence of failures to comply with the requirements of the animal SBO ban.

- Four BABs born in 1991 had been confirmed.

- Voluntary on-farm ELISA testing had produced the first positive result.

- Interim results of MAFF's attack rate experiment had shown that 1 gram of infective material was sufficient to transmit BSE when administered orally.

328 This last factor produced a radical change of attitude on the part of both MAFF and UKASTA to the dangers of cross-contamination of feed. In reporting to Mr Waldegrave on 21 November 1994, Mr Richard Packer, the Permanent Secretary at MAFF, stated:

> The trade's protestations that cross-contamination never occurred have been reversed; they are now more or less telling us that where the same mill is used for ruminant and non-ruminant feed, some cross-contamination is inevitable, although this is usually at low levels.

329 Mr Packer had plainly been misinformed. UKASTA had expressed concern about cross-contamination at the outset, but had been led to believe that this would not matter because a large quantity of infective material had to be eaten in order to result in infection.

330 At this point UKASTA appeared to come closest to advising its members to cease using MBM in feed. It attempted to elicit from MAFF an assurance that the rendering processes would produce MBM that was 'safe'. Mr Packer was not prepared to provide this. However, he did produce for UKASTA a statement summarising the steps MAFF had taken to prevent transmission of BSE to cattle. This emphasised that the controls over the implementation of the animal SBO ban were being strengthened and that more effective rendering processes were being adopted. The statement ended:

> The Ministry considers there to be no reason in principle why [ruminant protein] should not continue to be used in non-ruminant feed, even in

> premises preparing feed for ruminant and non-ruminant species, provided
> that steps are taken to prevent accidental inclusion in ruminant rations.

UKASTA accepted this statement as satisfactory reassurance and the use by its members of MBM in non-ruminant feed continued.

331 We had evidence from some of the major feed compounders that once they had been made aware that cross-contamination was a cause for concern, they took steps to identify the critical control points and to modify their production lines so as to reduce the risk of contamination occurring. UKASTA and MAFF reached agreement under which compounders were permitted, under a quota system, to submit samples to Luddington for ELISA testing in order to check that their production was free of contamination. This sampling was carried out in parallel with sampling by MAFF of feed on farms on a voluntary basis. Mr Meldrum told us of at least one occasion on which this led to the identification of a mill where cross-contamination was occurring, and to the mill in question taking steps to remedy the problem. Problems were, however, still being experienced with the ELISA test and it was apparent to both MAFF and UKASTA that it was capable of giving false positives and false negatives.

332 As at 23 January 1995, the number of confirmed BABs had risen to 15,771, of which 812 had been born in 1990 and 9 in 1991. In the following month it was confirmed that the attack rate study had demonstrated that 1 gram of material was sufficient to produce oral transmission. When this was reported to Mr Waldegrave, he asked whether further steps needed to be taken to ensure that compounders' feedlines were clean. Mr Meldrum replied that the short answer was 'No'. The important thing was to prevent infected material entering the feedlines. As to this, the only action that he could recommend was to continue to intensify controls on the disposal of SBO. Mr Waldegrave accepted this advice.

333 In May 1995 MAFF officials were giving consideration to arranging advisory visits to feedmills in order to give guidance on how to avoid cross-contamination and, at the same time, to replacing voluntary sampling on farms with unannounced sampling visits to mills. Our impression is that UKASTA was less than enthusiastic about these proposals. Its first duty was to protect its members' interests and it showed a continued awareness of the need to protect its members from the risk of prosecution. However, consideration of voluntary visits and sampling was overtaken by a Decision of the European Commission[42] adopted on 18 July 1995. This required routine monitoring of feedmills, and in particular of mills which produced both ruminant and non-ruminant feed, to include official ELISA tests for the presence of animal protein.

334 Discussions with UKASTA about implementing this Decision did not receive an enthusiastic response. UKASTA did, however, cooperate in the drafting of a letter from MAFF to all manufacturers and mixers of feedstuffs, drawing attention to the need to avoid cross-contamination and giving guidance on how to do so. We found this a bland document. In particular it made no mention of the fact that experiments had demonstrated that as little as 1 gram of infective material could result in oral transmission of BSE. A revised Advisory Note directed specifically to farmers was drafted by MAFF in November 1995. This was an admirable document giving detailed advice on all the different ways in which feed might become

[42] Commission Decision 95/287/EC

contaminated on the farm or in the course of farm mixing. Unfortunately, this draft got bogged down in the course of the consultative process, involving input from the Spongiform Encephalopathy Advisory Committee (SEAC) and the Parliamentary Secretary, and had not been sent out when it was overtaken by events in March 1996. This was one of a number of examples in the BSE story of the best being the enemy of the good.

335 By 24 August 1995 the number of confirmed BABs had risen to 21,475, of which three had been born in 1992. Although it was not initially appreciated, the effect of the Commission Decision requiring mandatory sampling of feed was to give MAFF officials the right to enter mills and carry out the sampling. An Animal Health Circular was drafted instructing State Veterinary Service staff on measures to implement the mandatory sampling regime, which was initiated early in 1996. Although the ELISA test was still not perfected – we understand that it remains imperfect to this day – the first round of tests produced four positive results from 25 mills tested.

336 On 6 July 1995 Mr Douglas Hogg succeeded Mr Waldegrave as Minister of Agriculture, Fisheries and Food. One of his first acts was to introduce the Specified Bovine Offal Order 1995 which, as we explain later in this chapter, dramatically improved the regime for enforcing the animal SBO ban. Later in the year, Mr Hogg discussed with Mr Meldrum whether further measures should be taken in the feedmills to address the risk of cross-contamination. Mr Meldrum explained that mandatory sampling was to be introduced and advised that it would not be practical to require feedmills to set up separate production lines for ruminant and non-ruminant feed. Mr Hogg accepted this advice.

337 SEAC reviewed from time to time the implications of the BABs and the action that MAFF officials were taking to address the cause of infection.[43] The Committee urged the importance of the development of the ELISA test, but in general endorsed the action that MAFF was taking. On the identification of the probable link between BSE and the new variant cases of CJD, SEAC's attitude changed. Members considered that it was of paramount importance to bring the BSE epidemic to a close as swiftly as possible, thereby protecting both animal and human health. To achieve this SEAC proposed a ban on the use of *all* meat and bone meal of mammalian origin in farm animal feed. This would remove all possibility of the contamination of ruminant feed. The Government accepted this advice and gave effect to it on 29 March 1996.[44]

338 As at end-June 2000 the number of confirmed BABs stood at 41,538. Of those 179 were born in 1995 and 2 in 1996. For each confirmed case, several will have been slaughtered before developing clinical symptoms. Almost all of these cases will have resulted from eating MBM derived from apparently healthy animals, because animals showing signs of BSE were being slaughtered and destroyed.

What went wrong?

339 When looking back with the benefit of hindsight, we have identified a number of things that went wrong in the history of the ruminant feed ban.

---

[43] See Volume 11
[44] The Bovine Spongiform Encephalopathy (Amendment) Order 1996

340 At the time that the ban was introduced, it was thought that all that the Regulations needed to do was to prevent the deliberate inclusion of ruminant protein in cattle feed. The Regulations were not designed to make unlawful the accidental contamination of cattle feed with small quantities of feed containing ruminant protein. Nor did they confer adequate powers of entry, inspection of records and sampling.

341 For the same reason, the development of a test to detect the presence of small amounts of ruminant protein in cattle feed was not treated as a matter of high priority. Five years were to elapse before the ELISA test was developed to a point at which some practical use could be made of it.

342 These shortcomings were symptomatic of a lack of rigorous thought about the implementation of the ruminant feed ban and the risk of cross-contamination at the time that it was introduced, which we have discussed in Chapter 3 above.

343 The risk of cross-contamination was then masked by the introduction of the animal SBO ban. We believe that it was because of the second line of protection apparently afforded by this ban that the Lamming Committee had no concerns about the possibility of contamination of ruminant feed. The Committee expressed concerns about the lack of control of on-farm mixing, but not in the context of BSE.

344 No sense of urgency attended the introduction of the ruminant feed ban. This was because of a fundamental misunderstanding of the scale of infection that was taking place. It was believed that infection had probably been occurring at a uniform rate of perhaps no more than 60 cases a month. In fact, the latent snowballing effect of recycling had boosted the rate of infection to 10,000 cases a month or more.[45] No one is to be criticised for failure to appreciate the scale of the problem. We do, however, censure (although we do not have the means to identify) those in the feed industry who deliberately breached the ruminant feed ban by continuing to supply ruminant feed that contained animal protein after 18 July 1988, when the ban came into force.

345 For some years MAFF officials proceeded on the basis that all necessary steps had been taken to eradicate BSE. As Mr Thomas Eddy[46] was to remark to Mr Waldegrave in February 1995, the long incubation period meant that five years had to elapse before it could become apparent whether precautionary arrangements and compliance by the industry were adequate. As the numbers of BABs increased, and their dates of birth grew later and later, MAFF officials progressively extended the period of carry-over of cattle feed containing ruminant protein that they assumed must have occurred. To an extent they were correct and we do not feel that they can be criticised for not appreciating until 1994 that a significant cause of infection of BABs was cross-contamination of cattle feed.

346 At that stage there were a number of alternative options to address the problem. The most radical was to prohibit the use of MBM in all animal feed. As Mr Meldrum remarked to Mr Hogg, the economic consequences of this would be 'devastating' and a serious waste disposal problem would be created. In the absence of evidence that BSE was transmissible to humans, we do not consider that this extreme measure was called for. To have adopted it, simply to prevent cross-

---

[45] We base this figure on the rate of confirmation of cases five years later, assuming that for every cow that developed clinical signs there would have been several infected cattle slaughtered before signs developed
[46] Head of Animal Health (Disease Control) Division, MAFF

contamination of feed in feedmills and on farms, would have been an admission of defeat. Other, less drastic, viable options were open.

347 At one time we were attracted by the view that feedmills should have been required to process feed for ruminants and feed for non-ruminants in separate production lines. We were, however, persuaded that to have insisted on the heavy expenditure necessary to achieve this would also have been disproportionate. MAFF's approach was to concentrate on procuring proper implementation of the SBO ban. This included requiring renderers to process SBO in dedicated plant. We consider that it was reasonable for MAFF officials and Ministers to conclude that it was not necessary to require feedmills to undertake, in parallel with renderers, the expense of installing duplicate lines. Instead MAFF sought to encourage feedmills to take voluntary steps to reduce the potential for cross-contamination.

348 With hindsight, we can deduce that the measures that MAFF had already taken had had a dramatic cumulative effect in reducing infection year on year. Looking back five years from end-June 2000, we see only 232 BABs which were born in 1995, and only 2 born in 1996. But for the events of March 1996 MBM would have remained part of the diet of pigs and poultry and MAFF would have been able to claim that, by a combination of the ruminant feed ban and the animal SBO ban, they had virtually eradicated infection of cattle with BSE.

349 It is this consideration which has led us, at the end of the day, to conclude that no criticism need be made of the somewhat muted attempts by both MAFF officials and UKASTA to get feedmills and farmers to take steps to tackle cross-contamination. When it was appreciated that this was occurring, and that a quantity as small as 1 gram of infective material would suffice to transmit the disease orally, one might have expected UKASTA urgently to draw these facts to the attention of its members and MAFF to do the same in relation to cattle farmers and to feedmills that were not members of UKASTA.

350 We suspect that the more measured approach that was adopted was explained by a shared reluctance on the part of MAFF and UKASTA to adopt a course that might lead to feed compounders ceasing to use animal protein as a feed ingredient. For the reasons that we have given, we do not feel that this was an unreasonable attitude to adopt.

Introduction of the animal SBO ban

351 In Chapter 4 we examined the consideration given by the Southwood Working Party to the risk that attached to eating beef or offal from animals infected with BSE but not yet showing clinical signs (subclinical animals). We saw that the Working Party did not consider that the risk posed to humans (other than babies) justified any precautions. The same was true in relation to the risk involved in feeding such matter to animals, although the Working Party expressed some general reservations about the practice of disposing of animal waste in this way.

352 As we have pointed out, the virulence of the infectivity of subclinical animals is indicated by the fact that, despite the ruminant feed ban and the animal SBO ban, over 41,000 cattle born after 18 July 1988 developed clinical signs of BSE. Most of these would have been infected by MBM derived from apparently healthy cattle, since clinically affected animals were removed from the human and animal food chains.

353 In June 1989 the Government announced that it had decided to go beyond the precautions recommended by the Southwood Working Party and to ban Specified Bovine Offal (SBO) from human food.[47] MAFF officials had reservations about imposing the human SBO ban. These included apprehension that it might lead to public pressure for further precautionary measures. These concerns were soon to prove well founded.

354 Even before the human SBO ban was introduced, the pet food industry had been considering whether to stop incorporating in pet food those bovine tissues most likely to be infectious if they came from an animal incubating BSE. The major pet food manufacturers have a guiding principle, which is that nothing should be incorporated in pet food which is not fit for human consumption. No sooner had the Government announced that it intended to introduce a ban on including certain types of bovine offal in human food than the Pet Food Manufacturers' Association advised its members to exclude this offal from their products.

The voluntary animal SBO ban

355 At this time farmers began to express concern about purchasing pig and poultry feed that contained animal protein – particularly protein derived from those parts of cattle which had been banned from human consumption. Some supermarkets were also showing a reluctance to purchase meat from animals that had been reared on such feed. In order to restore customer confidence UKASTA decided in July 1989 to advise its members to insist that any MBM which they purchased for incorporation in animal feed should be SBO-free. This led the UK Renderers' Association (UKRA) to threaten that its members would be forced to refuse to accept SBO for rendering if there ceased to be any custom for the end product. Mr Meldrum persuaded UKASTA to defer introducing its voluntary ban until the human SBO ban came into force.[48]

356 It was at this time that Mr Gummer succeeded Mr MacGregor as Minister of Agriculture. In September he received a submission from his officials about UKASTA's proposed voluntary ban. They had considered, in the light of the *Southwood Report*, whether any restrictions should be placed on feeding animal protein to non-ruminants and decided that there was no scientific justification for this. In their submission to Mr Gummer, they warned of 'serious implications' if UKASTA went ahead with its proposed ban. Renderers would be likely to refuse to accept 1,500 tonnes of SBO per week. Slaughterhouses left with SBO on their hands might be forced to close. Public pressure might grow for a complete ban on animal protein in animal feed.

357 At a meeting with UKASTA on 2 October 1989, Mr Gummer sought to dissuade the Association from its proposed ban, arguing that there was no scientific justification for this. He said that the human SBO ban was only being introduced for 'administrative convenience'.[49] UKASTA remained unmoved by this and by

[47] See Chapter 6 below
[48] This proved to be 13 November 1989
[49] See paras 564ff as to the basis for this statement

continued pressure from MAFF officials to drop its ban. Later in the month Mr Lawrence wrote:

> Despite all our efforts UKASTA seem hell bent on pursuing their potentially damaging course . . . I am concerned and aggrieved that UKASTA seem blind to the consequences of their actions.

358 On 9 November, four days before the human SBO ban came into force, UKASTA issued a circular to its members recommending that their contracts for the purchase of MBM should stipulate that this must be SBO-free.

359 Not all feed compounders refused to accept MBM derived from SBO. A limited market developed for this, at a lower price than SBO-free meal. Renderers sought to satisfy the demands of those customers seeking meal that was free of SBO by insisting that slaughterhouses separate the SBO from other offal. Renderers collected the SBO in separate containers for processing as waste, but charged for doing so. Renderers had, however, no means of ensuring that slaughterhouses complied strictly with this requirement. MAFF officials continued to protest that there was no justification for the ban.

360 In introducing a voluntary SBO ban, UKASTA and UKRA were doing no more than responding to customer demand. They were not concerned with the question whether or not their customers' perceptions were scientifically sound; or with the adverse financial consequences that the ban had for slaughterhouses. These were matters of legitimate concern to MAFF. There were good grounds for believing that pigs and poultry had shown themselves impervious to TSEs – a point on which Mr Meldrum sought and obtained confirmation from Professor Southwood. We have no criticism to make of MAFF officials' and Ministers' vigorous opposition to the voluntary SBO ban at this stage of the story.

361 In the months that followed, MAFF came under increasing pressure to introduce a statutory ban on the incorporation of SBO in animal feed, and some pressure to introduce a total ban on feeding animal protein to animals. Ministers sought reassurance that there was no merit in these proposals. Their officials assured them that they had no scientific justification. This reassurance Mr Gummer conveyed to the Prime Minister, Mrs Margaret Thatcher, at the end of January 1990, when she in her turn queried whether it was desirable to continue feeding animal protein to pigs and poultry. Over the months that followed, MAFF officials continued to insist that there was no scientific justification for an animal SBO ban. Then came the cat.

The cat

362 On 9 May 1990 Mr Gummer was informed that a Siamese cat had died of a spongiform encephalopathy. This was the first known case of Feline Spongiform Encephalopathy (FSE). The public reaction was predictable. Had the cat caught BSE? If BSE could be transmitted to a cat, why not to humans? The media had a field day. We deal with the human health implications of the cat later. Here we are concerned with the implications it had in relation to animal feed.

363 It was not clear at the time whether there was any connection between BSE and the cat. It was possible that cases of FSE had occurred in the past, but had never been diagnosed. Mr Gummer understood from Mr Meldrum that there was no likely connection between the cat and BSE. Mr Meldrum should not have given this reassurance, for it put the matter too high.

364 The cat led to renewed public concern about the practice of feeding SBO to pigs and poultry. In a meeting with Mr Gummer, Sir Simon Gourlay, the President of the National Farmers' Union (NFU), suggested that MAFF should introduce a statutory SBO ban for pig and poultry feed, thereby regaining the initiative and restoring public confidence. Mr Gummer's response was that there was no scientific justification for such action, which would be unlikely to allay public concern but would merely move the debate to another vulnerable area. The NFU was not convinced. In June 1990 it issued advice to farmers recommending that they should not use animal feed that included SBO.

365 That there was no scientific justification for an SBO ban remained MAFF's public position. The cat had changed nothing. SEAC had, however, been asked to give urgent consideration to the implications of the cat. SEAC then indicated that it wanted to give consideration to pig and poultry feed. This led Mr Gummer, who previously had seen no need to refer this matter to SEAC, to ask the Committee to consider the whole question of feeding animal protein to animals. Neither he nor his officials thought it appropriate to inform the feed industry or others that he had done so. Mr Andrews, the Permanent Secretary, remarked that 'the issue would have to be very carefully handled'.

366 The issue was carefully handled. Over a period of several months a paper was prepared for SEAC on the inclusion of SBO in feed for non-ruminants. This set out MAFF's reasons for concluding that there was no justification for preventing this practice and invited SEAC to endorse that conclusion. In August 1990 the paper was submitted to Mr Gummer for his approval, which it received. But before the paper could be considered by SEAC, it was overtaken by events (see paragraph 368 below).

367 The furore that greeted the announcement of the first case of FSE led MAFF to adopt an unnecessarily defensive approach to pressure for an animal SBO ban. Public pronouncements suggesting that the cat was no cause for concern did not carry conviction. MAFF witnesses emphasised to us that if any doubt had been expressed, this would have been treated as being of major significance, indicating a possible change of policy. We do not criticise MAFF officials for the cautious stance that they took, but we feel that it was ill-judged in that it harmed their credibility. They would have done better to state openly that, while MAFF did not consider that the cat called for any change of policy, SEAC had nonetheless been asked to advise whether it had any implications in relation to the composition of animal feed.

The pig

368 In August 1990 the whole picture was changed by the experimental transmission of BSE to a pig by injection of infectious material into the brain. This experiment had started 15 months earlier. In July 1990, in a note to Mr Gummer,

Mr David Maclean, the Minister for Food Safety, suggested that some contingency planning should be put in hand against the possibility that this experiment might produce a positive result. He expressed the view that in that event:

> We would have no option but to ban specified offals from pig and poultry feeds also. No-one should imagine that we could do anything else. It would be pie in the sky to believe that we could hold the line on this or somehow distinguish poultry feed from pig feed.

369 MAFF officials did not agree. In a note to Mr Andrews, which he told us he had cleared with his veterinary colleagues, Mr Robert Lowson[50] said that there was not much that they could do to prepare for the possibility that offal would have to be banned from pig and poultry feed, but that this would only become necessary if it was shown that transmission could be effected by the feed route. Transmission by inoculation would not justify a ban. Mr Andrews endorsed this view. It proved to be wrong.

370 When, on 20 August, news was received that BSE had been transmitted experimentally to a pig, Mr Meldrum and Mr Gummer agreed that this should be kept confidential until SEAC's advice had been obtained. An emergency meeting of SEAC was held on 7 September. A paper was prepared for this meeting by Mr Meldrum which put forward three options:

- Do nothing
- Ban MBM derived from SBO from animal feed
- Ban all MBM from animal feed.

The paper stated that the second option:

> ... would, in practice, simply add the weight of legislation to an arrangement which is already operating de facto on a voluntary basis. This is the option that holds most attraction for the Ministry's veterinary advisers.

371 By the time of SEAC's meeting, FSE had been reported in nine cats. SEAC concluded that the result of the pig experiment indicated that it would be prudent to exclude SBO from pig diet, and that the cases of FSE suggested that a cautious view should be taken of those species which might be susceptible to BSE. Accordingly SBO should be excluded from the feed of all species. Mr Meldrum's second option had found favour.

The statutory animal SBO ban

372 When Mr Gummer was informed of this advice, he accepted it. This accorded with an approach to BSE that he had decided to adopt as a matter of principle: decisions on what action should be taken in the face of any development should always be referred to SEAC, and SEAC's advice should be followed.

373 On this occasion Mr Gummer was determined that news of the result of the pig experiment should not leak out until MAFF was in a position to announce its response to it. The task of drafting appropriate Regulations was tackled by the

[50] Head of Animal Health Division, MAFF

MAFF lawyers based on instructions from officials in what had become the Animal Health and Veterinary Group, but without any wider consultation. Although implementation of an animal SBO ban would involve, as a matter of critical importance, practices in the slaughterhouse, Mr Keith Baker, the Assistant Chief Veterinary Officer responsible for meat hygiene, was not consulted. Instructions were given that the Territorial Departments in Wales, Scotland and Northern Ireland were to be informed 'at the latest possible moment and in such a way that as few as possible people were in the picture'.

374 SEAC confirmed its advice on 20 September. The draft Order implementing it was submitted for signature on 21 September. MAFF announced the making of the Order[51] in a news release on 24 September and the Order came into force on the following day. Mr John Maslin of the Animal Health Division was to describe the Order as made 'in haste and secrecy'. That was a fair description.

375 The new Order amended the Order that had introduced the ruminant feed ban. It prohibited the sale, supply and use of SBO, feedstuffs containing SBO, or animal protein derived from SBO for feeding to animals and poultry. It also prohibited the export to EU Member States of feedstuffs containing SBO or animal protein derived from SBO.

The operation of the statutory animal SBO ban

376 We noted earlier in this volume that the ruminant feed ban was not fully effective. One reason was that ruminant feed was contaminated by feed for pigs and poultry which contained bovine MBM from cattle incubating BSE. After September 1990, when the animal SBO ban came into force, this cross-contamination should not have mattered. Pig and poultry feed should not have contained any MBM derived from SBO. If a little of this got mixed with feed for cattle it should have caused no harm. This was not the reason for bringing in the animal SBO ban, but it should have been one of its effects.

377 The large number of BABs born after September 1990 shows that something went very wrong. Over 12,000 of these animals developed signs of BSE. A much greater number must have been infected with BSE, but were slaughtered and eaten before any signs developed. How were all those cattle infected? For the vast majority it was because their feed had been contaminated by pig and poultry feed infected with BSE. How was it that, despite the animal SBO ban, BSE was getting into pig and poultry feed? There is more than one answer. In Chapter 4 of vol 5: *Animal Health, 1989–96* we identify two reasons which probably played a minor part:

- The Order excluded from animal feed the SBO that was banned from human consumption, but did not identify all the potentially infective tissues and products which might go into animal feed.

- SBO was not always cleanly removed from the parts of the carcass that went to be rendered for animal feed.

---

[51] The Bovine Spongiform Encephalopathy (No. 2) Amendment Order 1990

378 Each of these sources of potential infectivity is, we believe, of insignificance compared with the primary source of the infectivity that resulted in BABs. This was that SBO was mixed, both deliberately and by accident, with carcass remains that were rendered for animal feed.

379 There were always going to be problems with enforcing the animal SBO ban. The financial temptation to pass off SBO as offal fit for incorporation in animal feed was considerable. There were ample opportunities, in the slaughterhouse, in the collection centres and at the renderers to give way to this temptation. Admixture of SBO with other offal was hard to detect. Those practical problems were compounded by the form of the Regulations that were put in place. They were, quite simply, unenforceable. To explain why this was, we shall have to lead the reader through a complex regulatory maze.

Before the ban

380 In order to understand the working of the animal SBO ban, it is necessary to appreciate the scheme that operated for dealing with meat unfit for human consumption, including SBO, before the ban was introduced. This is a topic of complexity, dealt with in detail in Chapter 4 of Volume 5. Here we shall give a greatly simplified account.

381 Animals killed for human consumption had to be slaughtered in a licensed slaughterhouse. The parts of the animal which were not wanted or were not fit for human consumption would normally be removed to a renderer to produce tallow and MBM – the latter being used as an ingredient of animal feed.

382 Fallen stock or animals put down on the farm would normally be collected by a knacker's yard or hunt kennel. Although they could not be used for human food, a variety of other uses were made of these carcasses. Remnants, including heads and spinal columns, would commonly go to be rendered to produce tallow and MBM used for animal feed. This waste from knackers and hunt kennels provided about 10 per cent of all rendered material.

383 The Meat (Sterilisation and Staining) Regulations 1982 (MSSR) were complex provisions designed to ensure that unfit meat was not used for human food. In a slaughterhouse, Meat Inspectors had to identify unfit meat and ensure that it was separated from the meat that was to go for human consumption. They applied a health stamp on the meat that was going for human consumption. The unfit meat, if not sterilised on the premises, normally had to be stained black. It could only leave the slaughterhouse after the issue of a permit authorising its removal to an approved destination, which would normally be a renderer. A copy of the permit would have to accompany the unfit meat to its destination, before being returned to the local authority which had issued it, so that a check could be made that the unfit meat had not gone astray.

384 At the knacker's yard and hunt kennel, the MSSR provided that all meat had to be treated as unfit for human consumption. Any remnants sent off to be rendered had to be stained black and accompanied by a movement permit.

385 Limited exceptions were made to requirements to stain and to obtain movement permits in respect of some categories of unfit meat when they were placed in a container of green offal. Green offal consisted of the intestine and stomach of the cow, together with their contents. Green offal was unfit for human consumption and was readily identifiable, and so was not required to be stained. It acted as a passport for the unfit material that it cloaked.

The human SBO Regulations

386 The human SBO Regulations[52] followed the scheme of the MSSR. Their broad effect was to add a parallel regime so that SBO had to be handled in a similar way to unfit meat under the MSSR. The Regulations applied only to slaughterhouses, as in knacker's yards and hunt kennels the whole of the carcass was already treated as unfit for humans and subject to the MSSR. SBO had to be stained in the same way as other unfit meat, stored separately from meat fit for human consumption and removed under cover of a movement permit. But there was no requirement that SBO should be kept separate from other unfit meat. On the contrary, the Regulations permitted SBO to go down the same chute as other unfit meat into the same container to be stained by a common stain and removed to the renderers as a single consignment.

387 There were one or two complications. Bovine intestine was an SBO. Under the 1982 MSSR, intestine and its contents, being a constituent of green offal, did not have to be stained even if found unfit for human consumption. Like other green offal, it could act as a passport for unfit meat in the same container, but that was not the case under the human SBO Regulations, under which bovine intestines were subject to the requirements of staining and movement permits. Nor did the Regulations explain how the system of movement permits should operate in respect of a mixed consignment of SBO and other unfit meat.

Enforcement

388 Regulations made under the Food Act 1984 and its successor the Food Safety Act 1990 fell to be enforced by the District Councils, of which there were 275, and by the unitary authorities in the Metropolitan and London Boroughs.[53] Thus the 1982 MSSR and the human SBO Regulations fell to be enforced by this tier of local authorities, which were also responsible for enforcing the Meat Inspection and Meat Hygiene Regulations in slaughterhouses. Their Environmental Health Departments employed Authorised Meat Inspectors (AMIs) and Environmental Health Officers (EHOs), who were trained in meat inspection, to enforce all these Regulations. Some slaughterhouses were approved to produce meat for export. In these, Official Veterinary Surgeons (OVSs) engaged by the local authorities were responsible for overseeing the implementation of all hygiene and meat inspection Regulations. This became the rule in all slaughterhouses – domestic and export – following the introduction of the European Single Market on 1 January 1993.

389 Hygiene standards varied enormously in British slaughterhouses from the lamentable to the good, with the majority tending towards the former rather than the

---

[52] The Bovine Offal (Prohibition) Regulations 1989
[53] Slightly different arrangements applied in Scotland, though again the main task fell to local authorities, and in Northern Ireland, where the State Veterinary Service was directly involved. For simplicity we focus here on England and Wales, but the features and failings we describe apply elsewhere

FINDINGS AND CONCLUSIONS

latter. This meant that the United Kingdom was unable to satisfy European standards and led, in 1995, to the transfer of meat hygiene enforcement functions in slaughterhouses from local authorities to central government, and to the establishment of a national Meat Hygiene Service, responsible to MAFF, to carry out those functions. Until that occurred there was a wide disparity, not merely in hygiene standards, but in the manner in which, and rigour with which, individual local authorities organised the fulfilment of their slaughterhouse enforcement obligations. In many slaughterhouses, staffing levels were such that Meat Inspectors had little time for anything except the vital function of ensuring that unfit meat did not go for human consumption. This was one reason why hygiene standards were so poor.

390 The MSSR 1982 were designed to ensure that unfit meat was not diverted into the human food chain. By 1989 in most slaughterhouses a routine had become well established under which the unfit material would be regularly collected by a local renderer to be turned into tallow and MBM for animal feed. Some was supposed to be stained and removed under a movement permit. Some travelled cloaked in green offal. Where such a routine was established, local authorities were permitted to authorise slaughterhouses to make out their own movement permits, and did so. In such circumstances, the Meat Inspectors and EHOs in many slaughterhouses devoted little time or effort to enforcing what seemed no more than formalities of movement permits and requirements as to staining and carrying unfit material in sealed and marked containers.

391 Once the container of offal left the slaughterhouse for the renderer, all supervision ceased. Often the container did not go direct to the renderer. Lorries would collect containers from a number of slaughterhouses, and sometimes they would be taken to collection centres, where offal from different slaughterhouses would be combined into larger consignments for onward carriage to the renderer. The MSSR 1982 envisaged that checks would be made on containers of unfit meat when in transit. So far as we are aware no such checks were ever made. The only checks carried out by District Councils were the reconciliation of movement permits once these were ultimately returned from the renderers, and the evidence was that this formality was, in practice, not an effective check.

392 If Meat Inspectors and EHOs had little time for enforcement of what may have seemed over-bureaucratic Regulations, the Government's deregulation initiative tended to convey, whether rightly or wrongly, the message that it was not desirable to be over-fastidious in insisting on compliance with the letter of the Regulations when there was no concern of substance that their object was not being achieved.

393 The evidence that has led us to these conclusions is set out in detail in Volumes 5 and 6. Some of it came to light when Mr Lawrence of MAFF was leading a team to investigate how enforcement of the Regulations worked in practice as part of the task of introducing the Meat Hygiene Service. Some of it came to light in 1994 and 1995, when very significant shortfalls were discovered in the quantities of SBO that were going for rendering. Some represents the testimony of individual witnesses given to the Inquiry.

394 One piece of evidence, which we found particularly significant, merits specific mention here. When the human SBO ban was introduced, it focused the attention of the Environmental Health Departments of the local authorities on the practical problems of the scheme established under the MSSR 1982. On 1 February 1990 Mr Mike Corbally of the Institution of Environmental Health Officers wrote to the Animal Health Division of MAFF with no less than 11 pages of enquiries and comments about the human SBO ban that the Institution had received. In particular, the requirements of the Regulations as to the containers in which unfit material was stored and transported and the formalities in relation to movement permits were proving difficult or impossible to comply with in practice. In 1994 MAFF was again to receive information that the system of movement permits was not working and 'had to rely on trust'.

The voluntary animal SBO ban

395 The MSSR and human SBO Regulations provided two parallel systems for handling all unfit meat. Renderers and the animal feed industry lost little time in introducing a practice of greater complexity. Under the voluntary animal SBO ban, described earlier in this chapter, feed merchants required renderers to supply MBM free of SBO, for incorporation into animal feed. The renderers, in their turn, required the slaughterhouses to segregate SBO from other offal. Other offal the renderers would pay for, as the raw material of MBM which they could sell on. SBO was unwanted waste. The renderers made a substantial charge for disposing of this.

396 The voluntary animal SBO ban was not complied with by all. Renderers found a market, albeit diminishing, of feed compounders who were happy to purchase, at a reduced price, MBM derived from SBO. Nor was there confidence that those who were purporting to comply with the voluntary ban were being scrupulous in doing so. It was difficult, if not impossible, to tell whether a container of decomposing offal contained an element of SBO. The financial temptation for slaughterhouses to pass off SBO as non-SBO material was considerable. Forced to trust the slaughterhouses, but with reservations about doing so, the renderers contracted with the feed merchants, not that they would supply MBM that was SBO-free, but that they would do their best to do so. Renderers, also, were under a temptation not to look too closely at the material that they were rendering to sell at a profit lest they should have to treat it as SBO to be disposed of at cost to themselves.

397 This, then, was the regime prevailing when the animal SBO ban was introduced.

The statutory animal SBO ban

398 The provisions in the Order bringing in the animal SBO ban were very short and simple. They made it an offence knowingly to sell or supply for feeding to animals or poultry, or to feed to any animals or poultry, any SBO. The same applied to any animal feedstuff known to contain SBO or where there was reason to suspect this. There was a fundamental problem with these provisions. Neither the feed compounder nor the farmer had any means of knowing whether animal protein incorporated in the feed had been derived from SBO. They were reliant on renderers to ensure that the MBM that they supplied was not derived from SBO. But the Order did not expressly make it an offence for renderers to manufacture MBM from SBO. It was arguable whether, on a proper construction of the Order, supplying such MBM to feed compounders was an offence. The renderer in his turn relied on the slaughterhouse, the knacker's yard and the hunt kennel to ensure that material supplied was separated into SBO and other offal. Yet the Order did not require this.

399 If those whose duty it was to comply with the animal SBO ban had no means of knowing whether ruminant protein incorporated into animal feed was derived from SBO, those responsible for enforcing the ban were in an even worse position. They had no means of proving that animal feed contained protein derived from SBO, let alone that those supplying the feed, or feeding it to animals, knew that it contained SBO. The Order was unenforceable.

400 In England and Wales, enforcement of Orders made under the Animal Health Act 1981 was the statutory responsibility of the County Councils and the Unitary Authorities in the Metropolitan and London Boroughs. Thus outside the Metropolitan and London Boroughs it was not the District Councils (responsible for the human SBO ban) but the County Councils that were responsible for the enforcement of the animal SBO ban. The County Councils sought to discharge that responsibility through the Trading Standards Officers employed by their Trading Standards Departments.

401 We had little evidence to suggest that Trading Standards Officers made any attempt to enforce the animal SBO ban, which is hardly surprising having regard to the practical problems of enforcement that we have described above. We did receive evidence of consideration being given by the Trading Standards Officers of one county, in conjunction with the State Veterinary Service and the local District Council, to taking action to address the practices of a particular renderer who allowed SBO to become mixed with offal that was going to be rendered to produce MBM for sale to feed compounders. It was concluded that there was no action that could be taken because:

   i. There was no provision in the animal SBO Regulations which made it an offence for a renderer to mix SBO material with non-SBO material.

   ii. It was impossible to demonstrate that MBM which was being sold for incorporation into animal feed was derived in part from SBO materials.

402 Much later, in 1995, after defects in the Regulations had been identified, new provisions were introduced which were enforceable.[54] They included the following requirements:

- On removal from the carcass, whether in the slaughterhouse, the knacker's yard or elsewhere, SBO had to be kept separate from all other material.

- SBO had to be dyed with a distinctive blue stain.

- SBO had to be removed to approved premises for disposal.

- SBO had to be kept separate from all other material at all stages of its progress from the slaughterhouse to final disposal.

- Records had to be kept of receipt and onward despatch of SBO at each stage of its journey from the slaughterhouse to final disposal.

[54] The Specified Bovine Offal Order 1995

403 Why were the shortcomings in the animal SBO Regulations not identified at the time that those Regulations were introduced? Why did the Regulations not include requirements such as those introduced in 1995? Broadly, witnesses from MAFF gave two answers to these questions:

> i. The Regulations were merely giving statutory force to the animal SBO ban that was already in place on a voluntary basis. This ban was being taken seriously and appeared to be operating satisfactorily.
>
> ii. A detailed statutory code for the handling of SBO already existed under the human SBO ban. Enforcement of this ban would have the additional benefit of ensuring that the animal SBO ban was complied with.

404 These views were implicit in this observation made by Mr Maslin in his submission inviting Mr Gummer to approve the draft Order:

> Enforcement is the responsibility of the Local Authorities. They are already monitoring and enforcing the Bovine Offal (Prohibition) Regulations 1989. In practice, the specified offal is being separated from other material at the abattoir. It is collected and processed separately by renderers. As with the existing ruminant feed prohibition, the ban on its sale, supply and feeding will, to a large extent, be self-policing. In these circumstances there would be little or no resource implications for Local Authorities.[55]

Reliance on the voluntary animal SBO ban

405 We have already drawn attention to the fact that the voluntary animal SBO ban was not universally applied. We have also drawn attention to the financial consequences of that ban, the motive that these gave for evasion and the doubts as to compliance – particularly in relation to the slaughterhouses. On the evidence that we received, there were no reasonable grounds for concluding that there was or would be satisfactory compliance with the animal SBO ban on a 'self-policing' basis. The voluntary animal SBO ban was not a satisfactory alternative to a statutory scheme that was capable of enforcement. We identify below the MAFF officials who should have appreciated this.

Reliance on the human SBO ban

406 Reliance on enforcement of the human SBO ban as a means of enforcing the animal SBO ban was misplaced for a number of reasons:

- For the reasons given above, Meat Inspectors, EHOs and OVSs were unlikely to devote much energy to enforcement of the technical requirements of staining and movement permits under the MSSR and the human SBO Regulations.

- Strict compliance with those Regulations was not practical and was not being insisted upon, as Mr Corbally's letter had demonstrated.

- It was of critical importance from the viewpoint of the animal SBO ban that SBO should be kept separate in the slaughterhouse and not mixed, whether

[55] YB90/09.21/14.3

FINDINGS AND CONCLUSIONS

- by accident or design, with carcass remains that were going to be supplied to renderers as fit for incorporation in animal feed. There was, however, no statutory requirement in the human SBO Regulations that such separation should take place. There was thus no relevant Regulation for the District Council officials to enforce.

- Witnesses suggested that the AMIs and the EHOs employed by the District Councils would have been ready to help out their colleagues on the County Councils by ensuring that SBO was in fact handled separately from other unfit material. Although we have no doubt that many District Councils and County Councils cooperate closely, we were not persuaded that District Council officials, whose responsibilities were to protect human health, would be enthusiastic about enforcing practices that had relevance only to animal health, the more so when those practices were not required by any Regulations. In 1994 a MAFF official was to report of the District Councils:

    It is clear that some Local Authorities see the legislation merely as an exercise in removal of SBO from carcasses and preventing its use for human foodstuffs.

    We did not find that attitude surprising. It reflected precisely the area of legitimate concern for District Council officials.

- If the two tiers of councils were to cooperate in trying to make the animal SBO ban work, it was desirable that they should have been given some guidance by MAFF officials as to what was expected of them. No such guidance was given. They were simply sent a copy of the 1990 Order and asked to arrange a meeting if they wished to discuss its enforcement. No such request was received.

Knacker's yards and hunt kennels

407 So far we have been concentrating on slaughterhouses, for they were the major suppliers of raw material to the renderers. Turning to knacker's yards and hunt kennels, we find a particularly unsatisfactory state of affairs.

408 The definition of SBO in the animal SBO Order followed that of the human SBO Regulations. This defined SBO by reference to offal from animals 'slaughtered' in the UK. The ban thus did not apply to any offal from fallen stock – the major source of knacker meat. It is not clear to what extent knacker's yards and hunt kennels took advantage of this lacuna and continued to use SBO as a source of animal feed, for MAFF made it plain that the Order was intended to apply to these premises. This error in the Order was remedied by amendment in 1991.[56]

409 Although the 1991 amendment of the animal SBO Order made it illegal to feed to animals SBO from fallen stock, or protein derived from this, there were no Regulations which required a renderer to separate SBO from other material. The handling of knacker meat was governed by the MSSR 1982, which treated all of it as unfit for human consumption. There was no statutory basis for insisting that knacker's yards or hunt kennels separate SBO from other material being sent to renderers. We are not aware of either County Councils or District Councils making

[56] By the Bovine Spongiform Encephalopathy Order 1991

any attempts to enforce the separation of SBO from other matter at knacker's yards or hunt kennels. Those renderers that were prepared to receive material from knackers for production of MBM – and they were a minority – insisted that it should be SBO-free. We are sceptical as to how rigorously the knacker's yards complied with that requirement.

SBO in transit

410 No Regulations required SBO to be kept separate from other offal when in transit to the renderers. There was scope for admixture, deliberate or accidental, when containers of SBO and non-SBO material were carried together on the back of the same vehicle and, more particularly, when stored together in collection centres. Neither District Councils nor County Councils considered it any part of their duties to check what was happening to SBO in transit.

Responsibility

411 We turn to the question of who should bear responsibility for the shortcomings in the animal SBO Order. Part of the problem was that the Regulations were introduced 'in haste and secrecy' and without consultation. Had there been consultation with those who would have to enforce the Order or those who were knowledgeable about problems in slaughterhouses – and we have particularly in mind Mr Keith Baker, the Assistant Chief Veterinary Office (ACVO), Meat Hygiene – mistakes might have been avoided.

412 Are Mr Meldrum and Mr Gummer to be criticised for their decision to keep the transmission to a pig and the measures being planned in response to it 'under wraps'? We do not think so. They were reacting to the furore that had been generated when the news of FSE in a cat was announced before MAFF had been advised by SEAC on its implications and whether any action was called for. With hindsight, however, we believe that it would have been better if MAFF had published SEAC's advice of 7 September and stated that the voluntary ban that was in place would be replaced with a mandatory ban after consultation. The fact that the voluntary ban was already in place, albeit that it was not universally observed, meant that MAFF could have justified taking a reasonable length of time to prepare the Regulations for a compulsory ban.

413 The fact that the ban was introduced in haste need not have prevented those who were responsible for its terms from giving rigorous thought to the question of how it was to work. We have not found it easy to identify the parts played by individual team members responsible for the Order, for instructions were given by telephone and memories are hazy as to precisely what took place. Those involved included Mr Maslin, Mr Lawrence, Mr Lowson and Mrs Elizabeth Attridge on the administrative side, Dr Danny Matthews, Mr Kevin Taylor and Mr Meldrum on the veterinary side, and Miss Gillian Richmond and Mr Ayyildiz Yavash from MAFF's Legal Department.

414 We do not consider that the lawyers are to be criticised for the contents of the Order. It was primarily for those instructing them to consider how the Order would work in practice. Furthermore, we note that Miss Richmond 'flagged up' a warning that officials might be criticised for including provisions which were unenforceable.[57]

415 We have concluded that, as head of the Animal Health Division, although he delegated the detailed discussions about the Regulations, Mr Lowson had ultimate responsibility on the administrative side of the team for ensuring that the terms of the Order were satisfactory. Mr Meldrum had lead responsibility for providing veterinary advice on the practicalities of the Order.

416 We do not consider that either Mr Lowson or Mr Meldrum gave rigorous consideration to the requirements of the animal SBO ban. They should have appreciated that the working of the voluntary animal SBO ban did not demonstrate that there would be satisfactory compliance with the statutory animal SBO ban on a 'self-policing' basis. And they should have appreciated that in the form in which the Order was drafted, it was obviously unenforceable. We do not say that they should have identified all the answers to the considerable problems posed by the ban. They should, however, have identified that the problems existed.

417 We would exclude from this criticism the lacuna in relation to fallen stock. This drafting error was not an obvious one, though it was quickly picked up. It was the kind of drafting point which can slip through the net when Regulations are drafted under pressure, and not one that we would necessarily have expected either Mr Lowson or Mr Meldrum to identify.

418 We have drawn attention to the fact that the Regulations did not require SBO to be kept separate or treated differently from other unfit material in slaughterhouses or elsewhere, although such separation was required under the contractual arrangements between slaughterhouses and renderers. We are satisfied that this separation requirement was not properly implemented and that, both by accident and by design, substantial quantities of SBO were supplied by slaughterhouses to renderers as material that was fit to be rendered for animal feed. Although the State Veterinary Service undertook the task of monitoring performance of the Regulations, four or five years were to pass before MAFF discovered that the ban was not being properly implemented. We turn to examine this part of the story.

Monitoring

419 Ministers looked to the State Veterinary Service (SVS) to monitor and alert them to any problems arising over the enforcement by local authorities of Regulations introduced by MAFF. The Food Act 1984 gave MAFF's veterinary inspectors the right to enter premises for this purpose. In 1989 Veterinary Officers (VOs) of the Veterinary Field Service (VFS) would make an annual visit to domestic slaughterhouses and a monthly visit to export-approved slaughterhouses to check that the various Regulations introduced by MAFF were being observed. These included the Meat Hygiene Regulations, the Meat Inspection Regulations, the MSSR and, after their introduction, the human SBO Regulations. Visits were also made by VOs to knacker's yards on an annual basis to check on observation of Regulations which applied there.

[57] YB90/9.00/7.1

420 Reports of visits had to be submitted to MAFF on a prescribed Meat Hygiene Inspection (MH1) form, which had space for entries in respect of each of the applicable Regulations. When the 1990 Food Safety Act replaced the Food Act 1984, the right of entry of MAFF inspectors was not preserved. This did not inhibit them from making their regular visits. They would normally, however, arrange to visit in the company of the district council EHO responsible for supervising the enforcement of the Regulations in the slaughterhouse in question.

421 The Animal Health Act 1981 gave MAFF's veterinary inspectors the right to enter premises on suspicion that the Regulations under that Act were not being complied with. They also made regular visits to renderers to check for the presence of salmonella in accordance with the Protein Processing Order of 1981.

422 If VOs found that Regulations were not being complied with, it was their duty to inform the relevant local authority of this, giving guidance where necessary. The breach would be recorded on the MH1 form and would thus be drawn to the attention of MAFF officials at headquarters.

423 When the human SBO ban was introduced in November 1989, no specific instructions were given to the VFS as to monitoring compliance with its requirements. The human SBO Regulations were simply added to the list of those that had to be checked on the MH1 form. Nor, initially, were any special steps taken to monitor compliance with the animal SBO ban upon its introduction in September 1990. In October 1990, however, Mr Andrews, the MAFF Permanent Secretary, suggested that the Ministry should carry out checks at slaughterhouses and renderers so that Ministers could be assured that no SBO was getting into animal feed. Mr Meldrum was quick to take up the suggestion. He asked Mr Keith Baker to make arrangements for VOs of the Field Service to make special visits to slaughterhouses and to renderers to check on the handling of SBO. Mr Baker was the ACVO at the head of the Meat Hygiene Veterinary Section. It was strange, on the face of it, that he should be charged with the checking of an animal health measure. We would have expected this duty to fall to Mr Kevin Taylor, who was at the time head of the Notifiable Diseases Section and subsequently became ACVO responsible for animal health and welfare. The explanation was that the only Regulations that made express provision for the handling of SBO were the human SBO Regulations, and monitoring of these fell logically within the province of the VOs with special training in meat hygiene.

424 The system that was set up required the Divisional Veterinary Officers to submit monthly returns of visits made by the VOs in their divisions. These were collated on a regional basis and sent to an officer of the Meat Hygiene Veterinary Section at Tolworth – initially Mr Stephen Hutchins, followed in 1991 by Mr Alick Simmons, who was himself succeeded in 1995 by Mr Andrew Fleetwood (an officer in the Notifiable Diseases Section). That official prepared a summary giving the national picture. Although Mr Baker set up this system, he told us that responsibility for it subsequently shifted to Mr Iain Crawford, as head of the VFS; thereafter he continued to receive the summaries of the returns, but only as 'a matter of politeness' because they were prepared by one of his staff. Mr Crawford told us that he had responsibility for advising on the practical problems of implementing policy in the field, but no responsibility for making policy decisions or advising Ministers on policy.

425 Mrs Attridge, head of the Animal Health and Veterinary Group, explained to us that while administrators in her group had responsibility for monitoring and keeping the Regulations in relation to the animal SBO ban under review, her 'eyes and ears' on the ground were the field vets.

426 We found this a confusing picture. No one person appears to have been responsible for keeping the adequacy of the monitoring of the animal SBO ban under review. As the story progressed the initiatives for tightening the system tended to come from Mr Meldrum.

Renderers

427 Initial returns on visits to renderers in January 1991 led Mr Meldrum to direct that there should be a further round of visits the following month and that thereafter renderers should be visited every two months. These early returns indicated that SBO was being kept separate from other material and that renderers were making sure that it did not get mixed with the material being processed for animal feed. Unofficial reports were nonetheless received by MAFF of wrongdoing in some rendering plants. As Mr Lawrence remarked, 'short of catching them in the act it is a pretty hopeless task'.

428 The reports disclosed one area of particular concern. Renderers used common plant for processing offal which produced MBM for incorporation in animal feed, and for processing SBO, whose product had to be kept out of animal feed and disposed of as waste. Methods of cleaning or purging the production plant between one batch and the next varied widely, with some plants doing nothing at all. Once again Mr Meldrum intervened. In September 1991 he asked Mr Simmons to draw up procedures which would ensure that no cross-contamination occurred at rendering plants. In consultation with UKRA, Mr Simmons prepared a 'Code of Practice for the Handling of SBO at Rendering Plants', which was distributed in August 1992. This provided for precautions to prevent 'comminglement' of SBO with other material. MAFF officials had expressed concern at the use of the description 'cross-contamination'. The precautions included cleansing or purging of plants between batches.

429 Had it been appreciated when the animal SBO ban was introduced that a very small quantity of infective material might suffice to transmit BSE orally,[58] we have no doubt that more urgent steps would have been taken to address the risk of cross-contamination in the course of rendering. As it was, the Code of Practice was a significant, if tardy, step in the right direction. But as we shall see, events in 1994 were to demonstrate that it was not enough.

430 Meanwhile, early in 1991, concerns had been raised about the disposal of protein produced from the rendering of SBO. SEAC had been consulted and advised that it was not satisfactory that it should be spread on fields as fertiliser. This led to the introduction of a statutory requirement that disposal of protein derived from SBO would have to be effected under licence, to be granted by MAFF.[59] The licensing scheme required data to be kept of weights of SBO received by renderers and of the protein derived from it, which enabled a rudimentary check to be made that SBO had not gone astray at the renderers.

[58] The NPU BSE-to-sheep experiment was to show that ½ gram was enough
[59] The Bovine Spongiform Encephalopathy Order 1991, article 9

431 The VFS continued to make regular visits to renderers to monitor the practice of keeping SBO separate from other material. These gave no indication that the position was other than satisfactory.

Slaughterhouses

432 Mr Baker's instructions to Divisional Veterinary Officers about monitoring in November 1990 had focused on information to be provided about the handling of SBO at renderers. They included, however, a request that visits should be made to slaughterhouses in order to discover 'how slaughterhouses are handling specified offal'. The response to this request varied so much in format and detail that a further round of reports was called for. For these a pro forma was used which called for information about brain removal, staining and movement permits. No mention was made of ensuring separation between SBO and other material. Not until August 1992 were field staff expressly instructed that the 'essential feature in effective control' was ensuring that SBO was kept separate from other material in the slaughterhouse and during transportation to rendering plants.

433 Both before and after these instructions, the returns received in respect of visits to slaughterhouses gave a satisfactory picture of practices observed. There were occasional reports of failures to observe the requirements of the human SBO Regulations in relation to staining or movement permits, but not to an extent that was significant.

434 This picture contrasted with a series of unofficial reports to MAFF of evasion of the animal SBO ban. In November 1990 Mr Lawrence was invited by Mr Peter Carrigan, who had a substantial business operating the gut rooms of slaughterhouses under contract, to visit a gut room to see the operations involved and to consider whether existing controls were sufficient. Mr Carrigan was in no doubt that they were not, and that there was widespread evasion of the animal SBO ban. This visit led Mr Lawrence to question the adequacy of the monitoring that MAFF was providing and to suggest that weight checks should be carried out at slaughterhouses and renderers to verify that the weight of SBO reflected the number of animals slaughtered. This suggestion was considered to be impractical by the members of MAFF's Meat Hygiene Veterinary Section.

435 Mr Lawrence also suggested that a distinctive marker might be added to SBO, thus facilitating its identification in and after it had left the slaughterhouse. We had evidence that enquiries were made as to whether a cheap marker or stain could be developed for this purpose. They did not lead to a successful outcome, but we were not able to establish why this was.

436 Reports continued to come in from 'trade sources', some considered reliable, that the Regulations were being disregarded and that SBO was being consigned by slaughterhouses to renderers unstained. One source of such information was Prosper De Mulder, the major UK renderer. This was typical of the cooperation provided by this company to MAFF throughout the BSE story. The company operated to high standards and showed a consistent concern that the Regulations should be effectively implemented.

437 This concern was shared by Mr Meldrum. His reaction to reports of disregard of the Regulations was to seek to improve the rigour of the monitoring by the VFS. In August 1991 a Circular was issued to the field staff, informing them of reports of non-compliance with the Regulations and instructing them to carry out the occasional unannounced visit to slaughterhouses. Notwithstanding this, the reports from the VFS continued to portray a satisfactory picture so far as slaughterhouse practices were concerned.

Knacker's yards and hunt kennels

438 Officers of the VFS were instructed to make monthly visits to knacker's yards and hunt kennels; the frequency reflected the fact that these premises were subject to significantly less local authority supervision than slaughterhouses. The instructions given were that on these visits staff should review 'the procedures for the disposal of waste material generally and the specified offals in particular'. We are not, however, aware of any returns which dealt with the manner in which SBO was handled at knacker's yards and hunt kennels. The only relevant Regulations were the MSSR 1982. These required all knacker meat to be treated as unfit for human consumption. They provided for staining and for movement permits in relation to this, but made no specific provisions in relation to SBO. We believe that this is why the returns from the VFS made reference, on occasion, to non-compliance with the MSSR but no reference to the handling of SBO.

'Cradle to grave' reviews

439 A significant improvement in the monitoring of the handling of SBO was introduced on the initiative of Mr Simmons in April 1993. He recognised that individual reports on slaughterhouses and on renderers did not give headquarters a complete picture of SBO disposal. He issued a new pro forma, Form MH6. This extended the scope of the return to cover all aspects of SBO handling from the slaughterhouse, to the collection centre, to the renderer and to the final disposal of the protein derived from the rendering of the SBO. Confirmation was sought that SBO was separated from other material at all stages of its journey.

440 In 1993 three sets of the new 'cradle to grave' returns were summarised by Mr Simmons. They indicated that practices were almost universally satisfactory. Occasional infringements 'of a minor nature', such as failing to stain all SBO or failing to identify SBO lines, were drawn to the attention of the local authorities, which took remedial measures.

The truth emerges

441 Despite the rosy picture painted by the returns from the VFS, unofficial reports of disregard by slaughterhouses of the Regulations were becoming more frequent. These reports led Mr Crawford to issue instructions on 1 February 1994 that all renderers processing SBO should be visited during the month of February unannounced. Full and detailed reports were to be provided of what was found. In summarising these reports on 25 March, Mr Simmons observed that both at collection centres and at renderers the constituents of stored material awaiting processing had to be taken on trust. His conclusions were that a small but significant amount of the total SBO processed, as a result of being inadequately identified and separated from other material at the slaughterhouse, in transit or at the rendering plant, was finding its way into processed protein that was being incorporated in animal feed. It was to become apparent that the only error in Mr Simmons's conclusions was that the amount in question was not small.

442 Mr Simmons included in his report a number of recommendations for tightening controls on the handling of SBO. Mr Meldrum called two meetings of officials in the course of April to consider these. Some of the deficiencies in the animal SBO Regulations which we have described above were recognised – it seems for the first time – namely:

- The animal SBO ban did not require the separation of SBO from other material at all stages.
- The 1989 human SBO Regulations did not apply to knacker's yards.

443 Various measures were considered, including the requirement that SBO should be stained with a special dye.

444 On 3 May Mr Eddy, who had taken over as head of the Animal Health (Disease Control) Division in June 1993, chaired a meeting to consider the way ahead. He later wrote that at this meeting:

> We spent a great deal of time clarifying in our own minds how the current arrangements work.

445 This was an exercise that should have been done in 1990 when the animal SBO ban was initially introduced. It was only from 1994 onwards that suitable legislative changes were prepared, including the requirement for a special dye for SBO, a requirement for SBO to be kept separate from non-SBO material, an approved system of movement permits, and a requirement that renderers handling SBO should be licensed.

The penny drops

446 It was at about this time that MAFF officials began to appreciate the true significance of breaches of the animal SBO ban. The numbers of BABs were soaring. In September 1993 the total had exceeded 4,000; by September 1994 it was to reach nearly 13,000. It was apparent that some had been born after the animal SBO ban had come into force. MAFF officials, including Mr Wilesmith, Mr Bradley and Mr Meldrum, were reaching the conclusion that the likely cause was double contamination:

   i.   contamination of MBM used for pig and poultry feed with SBO; and
   ii. contamination of cattle feed with pig and poultry feed.

447 This led Mr Meldrum to initiate a review of arrangements for the disposal of SBO. Subsequent developments were attributable in large measure to the commendable lead of Mr Meldrum.

448 In July 1994, because of pressure of work in Mr Eddy's division, Dr Richard Cawthorne, the head of the Animal Health (Zoonoses) Division, was asked to 'assume overall responsibility for progressing changes to the SBO controls and produce an action plan'. He was assisted by Mr Fleetwood, a Senior Veterinary Officer (SVO) in his division. Mr Fleetwood carried out an informal telephone survey of the quantities of SBO received by the major UK rendering plants. He compared this with the amount that ought to have been generated from the cattle being slaughtered. The weekly total was 400 tonnes short of the 1,200 tonnes which was his average estimate for the time of year. He concluded that the 'SBO controls were not working'.

449 Two further factors added to the gravity of the situation:

- In March 1994 preliminary results of a European study on the effect of the rendering process on inactivating BSE had demonstrated that the three systems that collectively provided most of the UK rendering capacity did not provide effective inactivation.

- In the summer of 1994 initial results of the CVL's attack rate experiment indicated that as little as 1 gram of infective material was capable of transmitting BSE orally to a cow.[60]

The potential consequences of cross-contamination at the renderers and in the feedmill were all too plain.

450 On 10 August the new Minister of Agriculture, Mr William Waldegrave, received a submission proposing radical changes to the animal SBO ban (along with changes to the human SBO ban). In agreeing that they should go to consultation, the Minister expressed concern that 'the controls should be made as simple as possible'. A lengthy consultation period then ensued, which resulted in the introduction of new provisions after the Meat Hygiene Service had replaced the local authorities in the slaughterhouses.

The Meat Hygiene Service takes over and a new SBO stain is introduced

451 On 1 April 1995 the Meat Hygiene Service (MHS) was launched as an Executive Agency of MAFF. It took over from local authorities responsibility for meat inspection and enforcement of the legislation relating to meat hygiene and SBO controls in slaughterhouses and head-boning plants. At the same time, Regulations were introduced which required SBO to be stained with a new distinctive food colour, Patent Blue V, instead of the previous black stain, which was used for other unfit meat.[61] This new stain had been identified as suitable the previous autumn, following instructions given by Mr Meldrum.

More shortcomings revealed

452 Most of those who had worked in slaughterhouses for local authorities as Meat Inspectors, EHOs or OVSs transferred their employment to the MHS. With the

---

[60] The fact that the NPU had already transmitted BSE to a sheep with an oral dose of only ½ gram of infective material appears to have been overlooked

[61] The Bovine Offal (Prohibition) (Amendment) Regulations 1995

MHS in place it was possible for both the MHS and the VFS to carry out rigorous monitoring of the standards of enforcement of the Regulations applied by the staff that the MHS had inherited. It quickly became apparent that there were widespread failures to dye SBO with the new dye, or indeed with the old one. Mr Peter Hewson, a senior official in the Meat Hygiene Veterinary Section, commented in a minute:

> It is clear to us that the local authorities were not implementing the staining requirements of the SBO regulations with the diligence we would have expected.

453 In the first three weeks of June, under the leadership of Mr Fleetwood, the VFS carried out a period of national surveillance, in the course of which every slaughterhouse known to handle bovine material received an unannounced visit on which a thorough inspection was carried out. Mr Fleetwood summarised the result:

> The overall impression of this snapshot view of the industry is that there is widespread and flagrant infringement of the regulations requiring staining of SBO. Insofar as this may reflect the general attitude of the industry to controls on SBO, it is of concern. Although the problems with separation are less extensive, there are grounds for suspecting that the highest risk tissues (brain and spinal cord) have been mixed with other by-products and processed for animal consumption . . . a careless attitude to separation and disposal seems to be prevalent and it is probably leading to accidents during disposal.[62]

454 It is right that we should emphasise here a point made by a number of MAFF witnesses. Responsibility for implementing the human SBO Regulations lay with the operators of the slaughterhouses. Responsibility for enforcing the Regulations rested fairly and squarely on the local authorities, not on MAFF. The legislation did not even provide for MAFF to exercise a monitoring role.

455 We have drawn attention to the fact that the human SBO Regulations did not require SBO to be kept separate from other unfit material. They did, however, require SBO to be stained, whether or not mixed with other unfit material. This requirement was frequently disregarded. Slaughterhouse operators were not fulfilling their statutory obligations and local authorities were not enforcing them.

456 We have suggested that one reason for this was that the Regulations were designed for the protection of human health, and there were no concerns that failure to stain might result in SBO getting into the human food chain. This may explain, but cannot excuse, breaches of statutory duty. There were many other reasons for these:

- Budgeting constraints meant that some local authorities did not employ sufficient staff to carry out slaughterhouse inspection duties satisfactorily. Nor was it easy to recruit staff. This was particularly difficult in the case of the OVSs – who should have been the most important members of the team. Veterinarians tended not to relish slaughterhouse duties and we had evidence that, when it was possible to recruit these, often from overseas, their quality was sometimes poor.

[62] YB95/7.04/3.3, para. 5

- There was often resentment on the part of slaughterhouse operators, Meat Inspectors or both at the imposition at the top of the inspection hierarchy of the OVS, whose need and function was considered to be open to question.

- There was a lack of effective line management. Meat Inspectors were often left to their own devices, without supervision, and tended to become 'almost part of the plant staff', getting involved in trimming and perhaps even dressing, rather than keeping themselves removed and recognising their roles as enforcement officers.

- Local authorities were often reluctant to be over-exacting in respect of slaughterhouses that provided local employment and a local service to farmers, but were operating on the margin of solvency.

- Under the Government's deregulation initiative, there was a culture of 'light touch' regulation. At the same time there was a media campaign which pilloried local authority enforcement officials as 'bureaucrats from hell' or 'little Hitlers'.

457 More than half the plants visited in June 1995 were not meeting the statutory requirements on staining, and 60 out of 435 plants were not separating SBO correctly from other material. Sixteen were not separating it at all. This was the lamentable state of affairs that confronted Mr Douglas Hogg when he replaced Mr Waldegrave as Minister of Agriculture on 5 July 1995.

The new Order

458 One of Mr Hogg's first actions was to approve the terms of the proposed new Order,[63] after having carefully discussed their implications with the Parliamentary Secretary, Mrs Angela Browning, and with his officials.

459 The Order was admirably comprehensive and yet satisfied Mr Waldegrave's request that it should be simple. In a single piece of legislation, enacted under the Animal Health Act 1981, it contained provisions aimed at protecting both human and animal health. Those aimed primarily at protecting animal health included the following:

- a ban on feeding SBO to animals;
- a ban on using SBO in the preparation of animal feed;
- a ban on selling SBO for feeding to animals or for use in the preparation of animal feed;
- a requirement that brain and eyes should not be removed from the head and that the head should be disposed of as SBO;
- a requirement that SBO should not come into contact with any other animal material in the slaughterhouse;
- a requirement that SBO be marked with Patent Blue V;
- a requirement that SBO be removed to an approved collection centre, rendering plant or incinerator;

[63] The Specified Bovine Offal Order 1995

- a requirement that SBO be kept separate from all other animal material in transit, at the collection centre and at the renderer;
- a requirement for weight-recording and record-keeping by all those generating or disposing of SBO;
- a requirement for dedicated SBO facilities at rendering plants.

The last requirement had proved controversial during consultation. It was Mr Meldrum who insisted on its inclusion. Renderers were granted a period of six months to introduce new dedicated lines. The new Order came into force on 15 August 1995.

460 Meanwhile the VFS had carried out two further rounds of intensive unannounced visits, and the MHS management introduced training and awareness-raising for the staff that they had inherited to rectify the shortcomings that had been disclosed.

461 On 29 September Mr Fleetwood was able to report that the amount of SBO being processed had increased by over 100 per cent. A minor part of this increase was attributable to the fact that whole heads were now treated as SBO. The balance is indicative of the extent of the previous evasion of the Regulations. It suggests that the $33^1/_3$ per cent shortfall identified by Mr Fleetwood's telephone survey was probably not far short of the mark.

462 Although there had been a significant improvement by September, the VFS was still finding widespread failure to comply with the Regulations. Up to this point MAFF officials and Ministers had been comforted by the belief that the shortcomings discovered did not endanger human health. However, towards the end of October 1995 Mr Meldrum had the unenviable task of informing the Chief Medical Officer, Dr Kenneth Calman, of four instances where spinal cord had been found in carcasses that had been health stamped by Meat Inspectors. Mr Packer suggested that Mr Hogg should call in the slaughterhouse owners and 'read the riot act'.

463 Mr Hogg did just that. On 8 November he issued 'formal instructions' to Mr Johnston McNeill, Chief Executive of the MHS, calling upon him to 'make every effort to secure 100% compliance' with the Regulations. This was an extreme step for a Minister to take in relation to an Executive Agency.

464 On the following day, Mr Hogg called in slaughterhouse operators and read the riot act. He told them that he would only be satisfied with 100 per cent compliance with the rules. This ambitious goal was not achieved, but the concerted efforts of the slaughterhouse operators, the MHS and the VFS produced impressive results. Whereas in October the VFS visits had disclosed that 31 per cent of slaughterhouses had failed to comply with the Regulations in one respect or another, by November this proportion had dropped to 13 per cent. By the beginning of January, Mr Fleetwood was able to report that:

> Very few problems are now being recorded other than a few lingering defects in staining and record keeping.

Of 344 visits made, only 5 per cent were recorded as unsatisfactory.

465 Knacker's yards and hunt kennels were included in the enforcement campaign. Mrs Browning met with their representatives to emphasise the need for improvement. Once again a remarkable improvement was produced. In one month alone, the proportion of visits to these premises which proved unsatisfactory dropped from 65 to 29 per cent.

466 By the end of 1995 MAFF had published a list of head-boning plants, incinerators and collection centres which had been inspected and which were approved under the 1995 Order to receive SBO. Renderers proceeded to upgrade their plant in order to provide dedicated lines for processing SBO as required by the Order. In some cases, short extensions of the six-month deadline had to be granted. On 13 March 1996 MAFF published a list of renderers approved to handle SBO together with a further list of head-boning plants, incinerators and collection centres.

467 Thus, by the end of the period with which we are concerned, there was at last in place a sound set of Regulations, imposing an effective animal SBO ban which was being properly implemented and monitored. At this point the abrupt change in perception of the risk that BSE posed to humans led to the imposition of a blanket ban on feeding animal protein to animals. The animal SBO ban became history.

Did the provisions of the animal SBO ban matter?

468 To what extent were the shortcomings that we have described attributable to the defects in the provisions of the animal SBO ban that we identified at the outset of this section?

469 Mrs Attridge, who was head of the Animal Health and Veterinary Group which had responsibility for the animal SBO ban, and Mr Meldrum each submitted to us that improvements in the Regulations would have had no significant effect on enforcement in slaughterhouses so long as the District Councils remained responsible for this. They suggested that it was the introduction of the MHS that enabled the tightening of standards in slaughterhouses in and after 1995. This was achieved by consolidating all the Regulations into a single instrument under the Animal Health Act 1981. The MHS enforced the consolidated Regulations in the slaughterhouses and the County Councils enforced them elsewhere. Before the MHS took over, there was no practical way of ensuring that separation of SBO from other material was enforced in the slaughterhouse. Mrs Attridge added that if there had been problems in getting slaughterhouses to apply a single black stain – which there certainly had – a requirement that SBO should be marked with a separate blue stain would have been likely to have compounded those problems.

470 There is force in these points. The regime under which some 300 different councils throughout Great Britain shared responsibility for enforcing Regulations in slaughterhouses had proved to have a severe structural weakness. No changes in Regulations would have overcome that weakness. Furthermore, the District Councils were not concerned with animal health Regulations. The County Councils, which had to enforce these, had no presence in slaughterhouses. Plainly the human SBO ban and the animal SBO ban could not sensibly be consolidated into a single Order so long as this situation prevailed. Nonetheless, we believe that if the animal

SBO ban had been imposed by a detailed code such as that introduced in 1995, the benefits would have been considerable. A statutory obligation to stain SBO with a distinctive stain and keep it separate at all times from all other material would have made it quite clear to slaughterhouse operators what their duty was. Meat Inspectors, EHOs and OVSs employed by the District Councils would in practice have had to have regard to that obligation in the course of enforcing the human SBO ban – indeed the terms of the human SBO Regulations could have been amended to bring them into line. The VFS would have been in no doubt as to the obligations that it was monitoring – and the distinctive stain would have helped it in its task.

471 So far as knacker's yards, hunt kennels, collection centres, transit to renderers, and rendering plants were concerned, there is no doubt that it would have been possible to impose clear and simple statutory obligations to keep SBO separate from other material. The County Councils would have been responsible for enforcing these. We are in no doubt that this would have resulted in significantly more effective enforcement and monitoring of the animal SBO ban.

Why did it take so long?

472 In July 1995 Mr Packer commented in a minute to Ministers:

> The unsatisfactory treatment of specified bovine offal in slaughterhouses reflects an unfortunate state of affairs which has presumably existed for many years. We must expect questions on why we allowed the situation to persist for so long.

473 We asked many witnesses why it was that the VFS did not identify the shortcomings in slaughterhouses earlier than 1994. Most had no answer to make, other than that the shortcomings that were revealed in 1995 were a recent development. This suggestion we reject. We are satisfied that they had persisted throughout.

474 Mr Fleetwood suggested that the problem was that, whether or not visits were made by formal appointment, slaughterhouses would have had advance warning of them. 'Unannounced' visits might have fallen into a pattern so that they were anticipated. Slaughterhouses would have taken steps to ensure that the right bins were in place and liberal quantities of stain being applied when MAFF veterinarians arrived.

475 These suggestions were speculative, but we think that there may be something in them. The VFS had no right of access to slaughterhouses. It would not have been easy simply to turn up to carry out an inspection without liaising with the local authority responsible for enforcement. The truly unannounced and unexpected visit may well have been a rarity.

476 Mr Fleetwood also suggested that animal health officers making the visits may have been fairly recent recruits to the VFS and 'easily browbeaten' by slaughterhouse managers. There may also be some truth in this suggestion. We believe, however, that before 1995 inspections by members of the VFS were much less rigorous than after the MHS had taken over. There were a number of reasons for this.

- Before 1994 the practical importance of the animal SBO ban was not appreciated. It appeared to be a precautionary measure to protect pigs and poultry that was probably unnecessary.

- The growing number of BABs and the result of the attack rate experiment led, in 1994, to the realisation that the animal SBO ban was a crucial element in the eradication of BSE.

- Before 1995 VFS visits were made 'on sufferance'. After 1995 they were made with the support of the MHS.

- Before 1995 the VFS visits were not targeted, for there were no Regulations requiring SBO to be kept separate from other unfit material. After 1995 there were specific statutory requirements to be monitored.

477 We consider that these are all factors which tend to explain why the shortcomings discovered in 1995 were not identified earlier by the VFS. MAFF officials were, however, receiving regular reports from unofficial sources that, contrary to the reports that were being made by the VFS, the animal SBO ban was being evaded. Are they to be criticised for not reacting more rigorously to these reports? Their reaction was steadily to step up the stringency of monitoring by the VFS until, finally, its reports confirmed the unofficial ones. Once again we have concluded that the failure to respond more positively was attributable to the failure to focus at the outset on the possibility that a very small quantity of infectious material might suffice to transmit BSE to cattle. As the years passed without cases of transmission of BSE to pigs and poultry, it must increasingly have seemed that the concerns which had given rise to the animal SBO ban were unfounded.

478 When in 1994 it was appreciated that shortcomings in the enforcement of the animal SBO ban were probably leading to the infection of cattle, Mr Bradley of the CVL concluded: 'We have to quickly and effectively re-assess and, if necessary, improve the policing of the controls both via MAFF and the Local authorities.' We believe that Mr Meldrum and his colleagues reached the same conclusion. Are they to be criticised for not reaching it sooner? Once again we have concluded that the failure to respond more positively was attributable to the original failure to explore the minimum amount that might infect and thus to focus at the outset on the danger of cross-contamination at the time of introduction of the ruminant feed ban. Given that failure, we do not consider that the manner in which MAFF officials performed their role of policing the animal SBO ban fell outside the range of acceptable responses to the facts as they appeared at the time.

Two fundamental issues

479 The story that we have set out raises two fundamental issues:

- should the feeding of all animal protein to animals have been banned from the outset? If not,

- should the requirement that SBO be processed in dedicated rendering facilities have been imposed from the outset?

480 The practice of feeding animal protein to animals was considered, in the context of BSE, by the Southwood Working Party, by SEAC and by the Lamming Committee. None considered that the practice should be stopped, or even that the practice of feeding ruminant protein to pigs and poultry should be stopped. The total ban on feeding animal protein to animals that was imposed pursuant to SEAC's recommendation in March 1996 was a reaction, and a reasonable reaction, to the horror of discovering that BSE was probably transmissible to humans. Its consequence was to turn renderers into a waste disposal industry rather than producers of a valuable animal by-product. We do not consider that it is cause for criticism that MAFF officials, MAFF Ministers and MAFF's expert advisers did not consider that this step was justified prior to 1996.

481 Had the possibility that a very small amount of infective material in feed would suffice to transmit BSE been appreciated, we feel that this should have led to the conclusion that it was unsatisfactory to use the same plant to render sequentially SBO and offal for incorporation in animal feed. We have already criticised the failure to give consideration to the possibility that a small quantity would infect at the time of the introduction of the ruminant feed ban.

482 Given that failure, we would not criticise MAFF officials for not insisting that SBO should be rendered in dedicated facilities. The considerable cost that this would have imposed on renderers could reasonably have been considered disproportionate if its only purpose was to enhance the protection of pigs and poultry against what was no more than a possible risk. Once perceptions had changed in 1994, Mr Meldrum is to be commended for having insisted that renderers should be required to provide dedicated facilities if they were to be permitted to process SBO.

Conclusions

483 We have reached the end of a black chapter in the BSE story. There are lessons to be learned from it, which we consider later. At this point we have a few concluding remarks.

484 Mr Meldrum was correct to stress the structural problems prior to 1995 of enforcing Regulations in slaughterhouses. The MHS was not introduced as a response to the problems of BSE. Its introduction was, nonetheless, of the greatest significance in addressing the dangers that BSE posed to the human and animal food chains.

485 The SVS, of which the VFS was one arm, had no statutory role in relation to the enforcement of the SBO Regulations. The monitoring role that it had undertaken was essential. Statutory recognition of that role, and statutory power of entry in support, would have been desirable.

486 In the event, largely as a result of the direction of Mr Meldrum, the SVS found itself increasingly filling the gaps in the statutory machinery for enforcing the animal SBO ban. One example was the monitoring that the VFS undertook of collection centres. Another was the negotiation with UKRA of the Code of Practice that was introduced in order to reduce cross-contamination at the renderers.

487 Finally, we should recognise the credit due to the continued efforts of the MHS, the SVS and slaughterhouse operators themselves, spurred on by the vigorous intervention of Mr Hogg, in turning round in 1995 what, up to then, had been a most unsatisfactory state of affairs. They were assisted in doing so by the belated introduction of an excellent regulatory scheme.

Cattle-tracking

488 There are two other topics which properly fall within the context of animal health. The first of these is cattle-tracking. Had MAFF had in place a computerised system under which the movements of cattle could be traced back to their place of birth, and their dams identified, this would have been of great benefit in satisfying European requirements that beef exported should have a BSE-free provenance. This we can see with hindsight. When BSE emerged, however, the immediate question was whether such a system needed to be put in place either to meet the demands of controlling BSE, or to meet the demands of disease control that might arise from the emergence of other new diseases.

489 That question was considered in 1990 by the Agriculture Committee of the House of Commons and answered in the affirmative. It was subsequently explored by MAFF officials in the context of a wider consideration of future information technology requirements. Officials concluded that neither the demands of BSE, nor those of disease control in relation to any foreseeable new disease, could justify the expense of introducing a computerised animal-tracking system. In vol. 5: *Animal Health, 1989–96* we have reviewed that conclusion and decided that it is not one we would criticise. We make no comment, for it is not within our terms of reference, on MAFF officials' response to the wider demands and possibilities of information technology.

Breeding

490 The other topic which falls within the context of animal health is that of breeding. In 1990, when it was unclear whether, or to what extent, BSE was a disease which would be maternally transmitted, a practical problem arose of concern to farmers: should they use the offspring of BSE cattle for breeding?

491 The British Veterinary Association and the MAFF veterinarians, headed by Mr Meldrum, were of one mind. Farmers should be advised that it was not desirable to breed from the progeny of BSE victims. Dr Pickles, who led in relation to BSE on behalf of DH, learned of this proposed advice. She considered that it was open to a number of objections, more political than veterinary, which MAFF officials had overlooked. SEAC had just been set up, and Dr Pickles succeeded in persuading Ministers that the new committee should be requested to consider the proposed advice.

492 SEAC did so at its first meeting, and expressed agreement, then and there, with Dr Pickles's reservations. We for our part had reservations about the use that was made of SEAC on this occasion and its outcome, though they did not lead us to criticise anyone involved. A full discussion of this matter is to be found in vol. 11: *Scientists after Southwood*.

493 The result was that MAFF did not advise farmers against breeding from the offspring of cattle which had been affected with BSE. An Advisory Note to farmers, which was issued in 1990, simply recommended that, if in doubt, farmers should consult their vet. Such a recommendation was not very helpful to farmers who received it. Given SEAC's advice that farmers should not be advised not to use the offspring of BSE cattle for breeding, we do not criticise the approach adopted in MAFF's Advisory Note.

6. Protecting human health

Introduction

494 We now turn to that part of the BSE story that has direct relevance to human health. There are many aspects to this part of the story. The main part of this chapter will follow a chronological sequence. However, we propose to introduce at the outset the CJD Surveillance Unit, which was to play a key role in the latter stages of the human health story; and to discuss at the outset as a discrete topic the slaughter and destruction of animals showing signs of BSE and the compensation paid to the owners of those animals.

495 The most obvious pathway by which BSE might be transmitted from cattle to humans was by the food chain. It was that pathway which caused concern to the public. And it was the public's concern about that pathway which was of concern to the Government. The Government was anxious to do all that it believed to be necessary to protect human health. But having taken that action, it was anxious to reassure members of the public that their health was not at risk. MAFF had a dual role. It had to make sure that meat which left a slaughterhouse was safe to eat. That was its prime concern. But it also had to have regard to the interests of the farming industry. There was a continuous concern on the part of MAFF officials and Ministers that the agricultural industry would be damaged by reactions to BSE on the part of the public that were irrational. This concern did not lead them to conceal information from the public. It did, however, lead them to attempt to ensure that information was presented in a manner that would not cause alarm. This sometimes involved delaying disclosure of information. It involved repeated statements that there was no evidence that BSE was transmissible to humans. It involved attempts to present to the public in the most compelling way the message that it was safe to eat beef.

496 This part of our narrative will follow the BSE story of which the public were aware: the events which provoked apprehension on their part and the statements that were made to them about the risk posed by BSE. It will examine the policy decisions that the Government had to take in relation to potential dangers posed by BSE to the human food chain. It will look in particular at public pronouncements and government action in the final months leading up to 20 March 1996.

497 We shall deal later, as separate topics, with aspects of the BSE story of which the general public were unaware:

- Action taken in relation to human and veterinary medicines.
- Action taken in relation to cosmetics.
- International trade.

498 Finally we shall consider the experience of those young victims who were struck down by vCJD and of their families, in order to see what lessons can be learned about dealing with this terrible disease.

CJD surveillance

Surveillance recommended by the Southwood Working Party and the Tyrrell Committee

499 Although the Southwood Working Party thought that it was most unlikely that BSE would have any implications for human health, they considered how BSE might appear, and be recognisable, if it did transmit to humans.

500 The Southwood Working Party noted in their Report that it was a reasonable assumption that, were BSE to be transmitted to humans, the clinical disorder would closely resemble CJD. They suggested that consideration be given to whether specialist branches of the medical profession, such as neurologists, should be made aware of the emergence of BSE so that they could report any atypical cases or changing patterns in the incidence of CJD. They also suggested that epidemiologists should be advised to watch for any such changing patterns.

501 CJD surveillance was also considered by the Tyrrell Committee. The *Tyrrell Report* gave the highest priority to the monitoring of all UK cases of CJD over the following two decades.

The CJD Surveillance Unit established

502 In December 1989 Dr Robert Will, then a consultant neurologist, applied to the Department of Health for a research grant for a project on CJD surveillance. Between 1979 and 1982, Dr Will had worked with Professor Bryan Matthews on various studies relating to the surveillance and analysis of CJD cases in England and Wales. Dr Will's proposal was accepted and the CJD surveillance project began on 1 May 1990 at the Western General Hospital in Edinburgh. It covered the whole of the UK and developed links with the surveillance networks of other countries.

503 The main objectives of the CJD Surveillance Unit (CJDSU) study were to identify any change in the epidemiological characteristics of CJD and to assess the extent to which any such changes were linked to the occurrence of BSE. The CJDSU was expected to document and publish any changes in the clinical or other characteristics of CJD, or in the epidemiology of the disease, and conduct investigations into the cause of these changes. The CJDSU summarised its progress and findings in a series of annual reports. These annual reports were supplemented by Dr Will informing SEAC and DH of developments.

How the surveillance system worked

504 The CJDSU needed to establish a system for the surveillance of CJD that would be able to detect any changes in epidemiology or clinical characteristics, as a result of the emergence of BSE. The main factors investigated included the number of cases of CJD, geographical distribution of cases and occupational incidence.

FINDINGS AND CONCLUSIONS

505 Primarily, this surveillance was achieved by seeking and obtaining direct referral of any suspect cases of CJD from neurologists. These professionals were also asked to report all cases of subacute dementing illnesses or progressive cerebellar dysfunction in specific occupational groups (including farmers and slaughtermen). However, as a precaution, all death certificates mentioning CJD were also obtained and assessed.

506 A standard questionnaire was used to obtain data relevant to diagnosis and ascertainment of possible risk factors. The questionnaire used by the CJDSU was based on the previous one developed by Dr Will for his work with Professor Matthews. It included sections on patients' initial symptoms, past medical history, family history, social history (residential, occupation, diet), exposure to animals, clinical history and results of diagnostic investigation. Minor changes were made to it before it was used in 1991 and subsequent alterations were made throughout the period 1991–95, as knowledge of CJD developed.

507 Unlike BSE, CJD was not made a notifiable disease. The possibility of making CJD a notifiable disease was not supported by either the Chief Medical Officer or Dr Will. Dr Will considered that in order to make CJD a notifiable disease, specific diagnostic criteria would have to be established. Some cases might then be missed as there might be a reluctance to notify cases that did not fulfil the criteria absolutely. Dr Will's view was supported by the European Union Surveillance Group in 1994. Recent data from this Group have lent some further support to Dr Will's view. The introduction of notification in Slovakia resulted in a decrease in the number of referrals.

PHLS excluded from CJD surveillance

508 The Public Health Laboratory Service (PHLS) did not become involved in CJD surveillance until after 20 March 1996. The PHLS is a public body with responsibility for providing a microbiological and epidemiological service to health authorities and local authorities for the diagnosis, control and prevention of infection and communicable diseases. It operates in England and Wales only, but has close working links with the parallel arrangements in Scotland and Northern Ireland.

509 PHLS officials repeatedly raised concerns with DH about the exclusion of their service from CJD surveillance. Since the PHLS's expertise was in communicable diseases, DH officials were concerned that PHLS involvement in the CJD monitoring process might indicate a belief that CJD could be spread from person to person. However, several other reasons were also given to the PHLS by DH for the decision. These included the possibility of unnecessary duplication of work and concern about PHLS priorities.

510 The decision to place the responsibility for surveillance with a small research team of dedicated medical scientists headed by a clinical neurologist with extensive experience in CJD was entirely correct. In 1989 the PHLS did not have expertise in CJD and, most importantly, there was (and still is) no established laboratory test for either CJD screening or for diagnosis in suspect cases. We commend the sterling work of the CJDSU team, who so promptly detected the emergence of vCJD and so efficiently established the clinical and pathological characteristics of the disease.

While we have formed the view that the PHLS could have contributed to various aspects of the task assigned to the CJDSU, assistance from the PHLS would not have enabled identification of vCJD at any earlier date. We do not criticise those who concluded that the task of monitoring CJD should be left to the Surveillance Unit set up for that purpose.

Slaughter and compensation

511 The slaughter and compensation scheme was designed to ensure that animals sick with BSE were destroyed so that there was no way in which they could transmit the disease to humans or to animals. It was a vitally important measure. We have been concerned to investigate allegations that some farmers sent animals showing early signs of BSE to the slaughterhouse in deliberate breach of the Regulations, and that the reason that they did so was because the level of compensation set by MAFF was inadequate.

512 We have seen above the circumstances in which the Government decided to introduce compulsory slaughter of animals showing signs of BSE and the destruction of their carcasses. It received advice that it should do this from the Southwood Working Party on 21 June 1988. Under the Animal Health Act 1981 compensation would have to be paid for compulsory slaughter on grounds of human or animal health. Ministers determined the level of compensation payable but had to have the agreement of the Treasury. Exploratory discussions with the farming industry indicated that payment of 50 per cent of market value might be considered acceptable, provided that 100 per cent was paid in respect of any animal which, after slaughter, was found not to have been suffering from the disease.

513 On 29 June Mr MacGregor wrote to Mr Major, the Chief Secretary to the Treasury, seeking approval for the payment of compensation at 50 per cent of market value. He estimated that on the basis of 60 cases a month this would cost about £200,000 to £250,000 a year. Mr Major agreed to this on 6 July, emphasising that he only did so because of the need to protect human health. Two Orders[64] were drafted by 22 July, and were made on 28 July and brought into force on 8 August, abridging the three weeks that normally elapse before Orders subject to negative resolution procedure come into force. It can be seen that no time was lost in implementing the recommendation of the Southwood Working Party.

514 The formula for determining compensation was complicated. Broadly, but not precisely:

- When the slaughtered animal proved to have BSE, the lesser of:
    i. 50 per cent of the value of that animal (in good health); or
    ii. 62½ per cent of the value of an average animal was payable.
- When the slaughtered animal proved not to have BSE the lesser of:
    i. 100 per cent of the value of that animal; or
    ii. 125 per cent of the value of an average animal was payable.

---

[64] The Bovine Spongiform Encephalopathy (Amendment) Order 1988 and the Bovine Spongiform Encephalopathy Compensation Order 1988

FINDINGS AND CONCLUSIONS

515 When an owner declared to MAFF that an animal was suspected of having BSE, but the animal died or was put down before a MAFF veterinarian confirmed that it appeared to have the disease, no compensation fell to be paid under the Order. On the recommendation of Mr Kevin Taylor and Mr Meldrum, it was agreed that normal compensation be paid on an *ex gratia* basis in those circumstances, provided that the animal was shown to have been suffering from BSE. When the animal did not have BSE, £50 was paid. This arrangement seems to us fair and we commend it.

516 Although industry soundings made by MAFF officials had suggested that the level of compensation would be acceptable, it in fact provoked a sustained barrage of attack:

- 8 July 1988: the National Farmers' Union (NFU) in a press release expressed the view that 100 per cent compensation should be paid for all slaughtered cattle.

- 2 September 1988: Mr Gordon Gresty, the County Trading Standards Officer of North Yorkshire County Council, expressed concern that compensation was only 50 per cent of market value. This might deter farmers from notifying suspect cases.

- 27 September 1988: the Milk and Dairy Produce Committee of the NFU stressed that compensation should be 100 per cent of market value.

- 23 January 1989: the Farmers' Union of Wales expressed 'complete dissatisfaction' with the compensation arrangements, suggesting that the low level of compensation might encourage less scrupulous farmers to dispose of animals showing signs of BSE on the open market.

- 17 February 1989: the first of a series of Parliamentary Questions from the Opposition suggesting that compensation should be raised to 100 per cent.

- 5 May 1989: Mr Peter Walker, Secretary of State for Wales, wrote to Mr MacGregor passing on concerns of his Agriculture Advisory Panel that the level of compensation was leading to evasion of reporting. He suggested reviewing the position.

- 14 June 1989: the National Consumer Council wrote to Mr MacGregor suggesting that, with compensation at 50 per cent, there was 'every incentive for farmers to send a cow for slaughter at the earliest sign of disease . . . the compensation arrangements must be reviewed'.

- 14 June 1989: the NFU wrote asking for a review of the level of compensation which, in their view, should be 100 per cent.

517 To all of these submissions MAFF made the same reply. Compensation at 50 per cent of the market value was fair. That compensation was payable for animals suffering from a terminal illness. The cattle were valued for the purposes of compensation, not as terminally ill, but as if they were unaffected with disease. Furthermore there was no evidence of any farmers attempting to evade the law.

518 This response reflected the advice being given to Mr MacGregor by his officials.

519 In July 1989 ministerial changes brought about a change in attitude in respect of compensation levels. On 6 September 1989 Mr David Curry, one of the new Parliamentary Secretaries, put an aide-mémoire to Mr Gummer, the new Minister, expressing the view that 50 per cent compensation was inadequate, and observing that the possibility of a farmer slipping a diseased animal into the food chain could not be absolutely denied. Officials responded recommending against increasing the level of compensation. Mr Lowson pointed out that only 52 suspect cases had been detected at abattoirs in the first six months of the year, of which by no means all would have resulted from deliberate deception. Mr Curry was not persuaded, but accepted that there was little chance of changing the position in the light of financial constraints.

520 Pressure for an increase in compensation then intensified:

- 4 December 1989: Mr R Cooper, a Director of Sainsburys, wrote saying that his company felt that 'full compensation' should be given for any BSE-infected cattle rather than 50 per cent in order to give the farming community every incentive to isolate diseased cattle.

- 4 January 1990: *The Times* reported that 'farmers are attempting to pass off diseased cattle as healthy because the Ministry of Agriculture will only compensate them for 50 per cent of the value of an infected beast once it is destroyed'.

- The Consumers' Committee of the Meat and Livestock Commission (MLC) expressed the view that compensation should be increased.

- 15 January 1990: a meeting of Dorset farmers expressed concern that failure to pay full compensation was giving the wrong message to consumers and could damage meat consumption.

- 25 January 1990: the President of the NFU wrote to Mr Gummer suggesting that 'raising the compensation to a more realistic level would be the most effective way of reassuring the public that there is no temptation for any farmer deliberately to send to market an animal with incipient BSE'.

521 Up to this point Ministers had continued to advance the same reasons as before for rejecting calls for higher compensation. Mr Gummer now decided that it would be politic to increase compensation. In a meeting with Mrs Thatcher on 30 January 1990, he suggested that compensation for the slaughter of diseased animals should be increased to 100 per cent for two reasons. First, losses were increasing, and some farmers were having a hard time. Second, full compensation would demonstrate that the Government was doing everything possible to keep BSE-infected cattle out of the food chain. The Prime Minister felt that the second was the better case and agreed that Mr Gummer should work up a proposal for increasing the rate of compensation, in consultation with the Treasury, which could then be put to ministerial colleagues.

522 On 7 February 1990, after discussing the matter with his colleagues, Mr Gummer wrote to Mr Norman Lamont, Chief Secretary to the Treasury, proposing an increase in compensation. He stated that he did not believe that farmers were sending BSE suspects to slaughter to any great extent, but that the possibility that they might do so must be growing. The principal case that he made for the increase was that this would allay public concern.

523 A submission to Mr Lamont from a Treasury official in respect of Mr Gummer's proposal observed:

> This is essentially a political matter, and on this basis you may wish to agree. The Prime Minister is thought to be sympathetic to Mr Gummer.

524 On 9 February 1990 Mr Lamont wrote to Mr Gummer reluctantly agreeing to his proposal.

525 On 13 February 1990 Mr Gummer announced the change in policy on compensation to the Annual General Meeting of the NFU. The change that he announced was brought into force the following day.[65] The new level of compensation for confirmed BSE cases was the lesser of 100 per cent of the animal's sound market value, or 100 per cent of the average cattle value.

Was compensation too low?

526 We have carefully considered the level of compensation originally paid to farmers for the slaughter of BSE suspects. It seems to us that the compensation bore a reasonable relationship to the loss caused by the slaughter, and on that basis was fair. We would emphasise that the loss in question was not the loss consequent upon having a cow affected, or suspected of being affected, with BSE. The loss was that experienced as a result of the deprivation of such a cow. To offer 50 per cent of the value of a healthy cow does not seem unreasonable for an animal showing signs of a terminal disease.

527 Nor would we have expected the level of compensation to have resulted in widespread evasion of the duty to notify. We would hope that most farmers would have been sufficiently principled not to seek to put into the food chain an animal that might endanger human life. Furthermore, to send a sick animal off to the market would be a chancy business, for the stress would be likely to make the symptoms more apparent.

528 The evidence that we received suggests that there was not significant evasion of the duty to notify during the period that compensation for infected animals remained at 50 per cent. During December and January MAFF veterinary staff made nearly 300 random visits to over 180 slaughterhouses. Of 1,663 animals sent for slaughter that were inspected, only one suspect case was identified.

529 Leaders within the farming industry, who gave evidence to us, expressed a firm belief that there was no, or negligible, failure to report suspect cases. Farmers gave evidence to the same effect, as did veterinarians.

530 The 1990 Agriculture Committee in its Report commented:

> The introduction of full compensation produced no very dramatic increase in the number of BSE cases being reported but, in view of the general perception that there may be under-reporting of such diseases where farmers are not fully compensated, it might have been prudent, for reasons of public reassurance, to have introduced it earlier.

[65] By the Bovine Spongiform Encephalopathy Compensation Order 1990

531 We agree with the Agriculture Committee that the justification for raising compensation was the desirability of providing reassurance to the public that cattle affected by BSE were not being slaughtered for food, rather than a need to provide a better financial inducement to farmers to obey the law. Mr Gummer's decision was, essentially, a political decision. We have no criticism to make either of that decision or of its timing.

Ante-mortem inspection

532 We have referred to random slaughterhouse inspections in December 1989 and January 1990. These were carried out at the suggestion of Mr Meldrum, who believed it was desirable to check that farmers were not sending off for slaughter cattle that showed signs of BSE. Mr Gummer agreed with Mr Meldrum's proposal. Initially these inspections were carried out by State Veterinary Service (SVS) staff, but from 5 February 1990 this function was transferred to Local Veterinary Inspectors (LVIs). In 1990 LVIs inspected over 31,000 animals at slaughterhouses, among which they identified just 29 suspects, of which only 14 were confirmed. This certainly indicates that after compensation for BSE casualties was raised to 100 per cent, there were at most only a few deliberate attempts to send suspect animals for human consumption. We consider that ante-mortem inspections at domestic slaughterhouses were desirable as a check that the Regulations were being complied with, and we commend Mr Meldrum for promoting them.

Compensation changed again

533 On 1 April 1994 a new formula for calculating compensation was introduced.[66] The change related to the method of calculating the market price element of the formula. This was adjusted downwards to reflect the fact that a large proportion of the cows developing BSE were older animals at the end of their working life. The motive for this change was to save money – it was calculated that it would reduce compensation payable by approximately £5 million in 1994/95. We have no criticism to make of this change or of the reason for it.

Unanticipated burdens

534 When the slaughter and compensation scheme was introduced, it was anticipated that it would apply to about 60 cattle a month. At the height of the BSE epidemic 8,000 suspects were notified in a single month. The task of diagnosing whether or not the suspects were infected with BSE was enormous. It was achieved by performing histopathology on a single section of the bovine brain (the obex section) and sharing the task of analysis between a number of Veterinary Investigation Centres. We commend the Veterinary Investigation Service for the efficiency with which this task was performed.

535 The other unforeseen consequence of the slaughter and compensation policy was the horrific problem of disposing of the carcasses of thousands of slaughtered cattle. This was a major element in the waste disposal problem to which BSE gave rise. We shall revert to the problem of waste disposal later in this volume.

---

[66] By the Bovine Spongiform Encephalopathy Compensation Order 1994

Introduction of the ban on Specified Bovine Offal (SBO) in human food

536 We have seen that the Southwood Working Party drew a sharp distinction between the possible risk to those who ate food derived from a cow with clinical signs of BSE and the risk from eating food derived from a cow incubating the disease, but not yet showing clinical signs (a 'subclinical'). Clinically ill cattle had to be destroyed. The tissues of a subclinical were not regarded by the Working Party as likely to be sufficiently infective to pose a threat – except perhaps to babies.[67]

537 With hindsight, we can see just how dangerous it can be to eat some of the tissues of a subclinical, at least for cattle, where no species barrier is involved. On 8 August 1988 compulsory slaughter and destruction of all cattle showing signs of BSE was introduced. Some 40,000 cattle born since that date have contracted BSE and lived to develop the clinical signs. A multiple of that figure will have been infected but slaughtered before clinical signs developed. The vast majority of those cases are likely to have been infected as a result of eating feed contaminated by very small quantities of infective tissues of subclinicals. These had been through the rendering process. We have seen above how this material got into cattle feed.

538 Since 13 November 1989, the tissues of subclinicals most likely to carry infectivity should not have been fed to humans. On that day a ban on using them for human food was introduced ('the human SBO ban'). The introduction of that ban at a time when most considered it highly unlikely that BSE could be transmitted to humans was one of the most far-sighted measures introduced in response to BSE – or it would have been had it been introduced as a result of foresight. As we shall see, however, the process that led to its introduction was haphazard rather than the result of rigorous risk evaluation. Mr MacGregor, who was responsible for the measure that Mr Meldrum described to us as 'inspirational', was at pains to emphasise to us that scientific considerations were not the primary factor which motivated him. Did it matter that the process was haphazard? We think that it did. First, it meant that the process was protracted. Second, it contributed to a failure to emphasise the importance of the measure, which detracted from the rigour of its implementation. In this chapter we shall describe how the policy decision to introduce the human SBO ban came to be taken, the reasons that were given for that decision and the manner in which it was translated into statutory Regulations.

Government response to the *Southwood Report*

539 Good government does not blindly follow the advice of scientific experts. Before doing so, it must evaluate the advice to make sure that it appears sound. In the case of the *Southwood Report* this was not easy. The Working Party had not expressed their reasons for concluding:

- that all clinically sick animals should be destroyed;
- that the risk that BSE posed to humans was remote;
- that manufacturers of baby food should exclude certain bovine offal; and

---

[67] See paragraph 264 above for the baby food recommendation

- that no measures were justified to prevent others from eating offal from subclinicals.

540 Nor had the Working Party made it plain that they were attempting to apply the ALARP principle.

541 Dr Hilary Pickles had the lead for DH in relation to BSE. She had been DH secretary to the Southwood Working Party and had drafted some of the most important parts of their Report. She wrote to Sir Donald Acheson on 6 February 1989 saying that the Report should be with him in a day or two. She commented:

> In my view DH can be very pleased with the way the report has turned out. Sir Richard and his team are to be congratulated.

542 Dr Pickles did, however, inform Sir Donald of one concern that was not reflected in the Report. She was worried about the safety of bovine-based vaccines. Sir Donald minuted Dr E L Harris, the Deputy CMO, to ask him to look into this. Sir Donald told us that he also asked Dr Harris to conduct a complete review of the *Southwood Report*. Dr Harris has died, so we could not ask him about this, but our analysis of the evidence set out in vol. 6: *Human Health, 1989–96* has satisfied us that Sir Donald's recollection is at fault here. He should have ensured that the Report was reviewed by his Department, but he did not do so. No doubt he placed confidence in the views of Dr Pickles. She was someone who inspired confidence. But because of her involvement she was not in a position to review the Report.

543 Sir Donald forwarded a copy of the *Southwood Report* to the Secretary of State for Health, Mr Kenneth Clarke, on 9 February. He commented:

> I regard it as a thorough study of the subject with sound and balanced conclusions.

He also expressed the view that, with one possible exception:[68]

> Every reasonable step has been taken to minimise any theoretical risk of transmission by destruction of affected cattle.

Sir Donald said nothing about the baby food recommendation.

544 When Mr Lawrence, MAFF's secretary to the Working Party, presented the Report to MAFF Ministers, he identified in a covering note a number of areas of interest to MAFF. One of these was the baby food recommendation. He sent a copy of his note, together with the Report, to 'interested Divisions within the Department'. Mr MacGregor raised the question of baby food at a meeting with Sir Richard Southwood a few days later. Sir Richard commented that the point in the Report in relation to baby food was not a specific recommendation, but a counsel of 'extreme prudence'.

545 The baby food recommendation was, however, causing concern to MAFF officials, in particular to Dr Mark Woolfe of the Food Science Division, who considered that identification of babies as a high-risk category did not appear to have been 'well thought out', and to Mrs Attridge, the head of Emergencies, Food

---

[68] This was a reference to Dr Pickles's concern about vaccines

Quality and Pest Control Group. Mrs Attridge was concerned because her responsibilities included the composition of food, and cow's liver and kidney were a valuable source of nutrition for babies. She was concerned that the baby food recommendation was based not on consideration of all the relevant science, but on 'poorly substantiated speculation'. Although Mrs Attridge's concern was that the baby food recommendation might result, without good reason, in the removal from babies of valuable nutrition, she commented in minutes to Mr Cruickshank, the Under Secretary in charge of the Animal Health Group, that MAFF would be asked why action should be taken on baby food but not on other food.

546 At a Cabinet meeting on 23 February to discuss the response to the *Southwood Report*, there was lively debate about the baby food recommendation. Mr Clarke, supported by Mr MacGregor, urged that the Report should be published and the baby food recommendation accepted. Other Ministers were concerned that publication of the recommendation would lead to a baby food scare. The decision was taken that the Report should be published after Mr MacGregor and Mr Clarke had prepared, with the help of the CMO, a clear and accurate statement of the Government's response to the baby food recommendation.

547 After the Cabinet meeting Sir Richard Southwood was contacted by Sir Donald Acheson. Sir Richard said that the baby food recommendation should only be treated as applying to brain, spinal cord, spleen, intestine and thymus, and not to heart, liver and kidney. This took the heat out of the situation. None of the former types of offal was included in manufactured baby food. The recommendation would not be likely to give rise to a boycott of baby food.

548 On 27 February 1989 the *Southwood Report* was published. In a written announcement, Mr MacGregor explained that none of the types of offal, which were the subject of the baby food recommendation, were used in the manufacture of baby food, but that as a precautionary measure he intended to make it illegal for anyone to sell baby food containing such products in the future.

549 No one in either DH or MAFF gave thought to the question that Mrs Attridge had warned would be raised. If these types of offal could not safely be fed to babies, why was it safe to feed them to children and adults? This important question was one that any thorough departmental review of the *Southwood Report* should have addressed. Another, linked, question that needed to be addressed was why the Working Party were so concerned about animals showing clinical signs of BSE, but not concerned, at least so far as safety of food was concerned, with the subclinicals.

550 We have already rejected Sir Donald Acheson's evidence that a full review of the Report was carried out by Dr Harris. Mr Clarke told us that in his Department there had been a very great deal of copious review, correspondence and discussion about the Report, which would have included the questions raised above, although he could not now remember the details of these. He also referred to an 'amazing quantity of exchanges' going on between his Department and Mr MacGregor's. We did not accept this evidence. As Secretary of State for Health, Mr Clarke needed to be in a position to answer the question 'If offal is not safe for babies, why is it safe for adults?' He should have ensured that his Department reviewed the Report and provided an answer – if there was one. He did not.

551 At Prime Minister's Questions on 28 February, Mr John Evans, from the Opposition benches, asked Mrs Thatcher:

> If, as appears likely to the Secretary of State for Health, BSE is a threat to humanity, why not ban the use of this offal for all human consumption? If according to the Minister of Agriculture, it is not a danger, why was it banned for babies?

She replied:

> We set up a committee of experts under Professor Southwood. We published the report in full. We referred it to the Chief Medical Officer of Health and we accepted the recommendations of both, precisely. There is no point whatsoever in setting up a committee of experts, in having a Chief Medical Officer of Health, in receiving their advice and then not accepting it. We would rather accept their advice than that of the hon. Gentleman.

Her Secretary of State for Health would not have been in a position to give a more informative reply.

552 What of MAFF? Dr Woolfe and Mrs Attridge had directed attention to the questions raised by the baby food recommendation, and are to be commended for this. But after the Cabinet meeting the questions were not pursued. We have concluded that there were a number of officials who should have made sure that the outstanding questions were answered. First of all, we think that Mrs Attridge herself, being concerned for composition of food, should have pursued the question of 'why should we take action on baby food and not on hamburgers', which was one that she had raised earlier. We consider that Mr Cruickshank should have taken steps to find out why the Southwood Working Party had drawn a distinction between babies and others, and between clinical and subclinical animals. We think that Mr Meldrum should have pursued these questions. The former distinction involved consideration of analogies with matters within the expertise of the veterinarians, such as the apparent susceptibility of calves to BSE. The latter was quite plainly a matter of veterinarian expertise.

553 Mr Andrews, the Permanent Secretary, had received a copy of one of the minutes in which Mrs Attridge raised the question of why action should be taken on baby food and not other food. He should have raised with Mr MacGregor the need to have an answer to this question. Mr MacGregor himself had been alerted to Mrs Attridge's concerns and should have seen that the question of 'why babies and not adults' was pursued.

554 In short, there was at MAFF, as at DH, a team failure to subject the *Southwood Report* to a proper review in order to evaluate whether the unexplained differences in approach to the food risks posed by BSE had explanations that appeared to be sound.

The decision to introduce the human SBO ban

555 In the months that followed the publication of the *Southwood Report*, a number of influences combined to drive MAFF towards the decision to introduce a ban on using for human food those types of offal that were most likely to carry BSE infectivity.

556 In the first place there was the public reaction to the Report. This started with a broadcast on the day the Report was published from Dr Helen Grant, a consultant neuropathologist at Charing Cross Hospital in London, who commented on the risk posed by cattle brains that were going into the human food chain. In an article in *The Guardian* on 2 March 1989, she suggested that the Government was concentrating on baby food 'to divert the public from thinking about other foods and thus to imply that they are safe, which they are not'.

557 In May three articles appeared in *The Times,* suggesting that sausages and meat pies were a risk to health and that the Government should ban the use in food of potentially infected organs. On 24 May the Woman's Farming Union issued a press release calling for a ban on the inclusion of brain and spinal cord in products for human consumption. This theme was taken up the next day by delegates when Mr MacGregor attended the Conservative Women's Conference. On the same day the Bacon and Meat Manufacturers' Association advised its members to exclude bovine pancreas, brain, intestine, spinal cord and spleen from their products. The Meat and Livestock Commission (MLC), which was being advised by Dr Kimberlin (whom we have already met as a witness to the Southwood Working Party[69] and a member of the Tyrrell Committee and SEAC),[70] wrote to Mr MacGregor urging him to introduce a general ban on the use of bovine offal for human consumption for the sake of public perception.

558 The Parliamentary Secretary at MAFF, Mr Donald Thompson, had started his working life in his father's butchery business. He told us that he had all along been worried about the brains of subclinical animals entering the human food chain. In March he made the suggestion that cull cows might be excluded from the human food chain. This received short shrift from MAFF officials, but Mr Thompson returned to the charge, seeking advice on removing brains and certain other types of offal of cull cows from the human food chain, a measure that he subsequently supported. We commend him for this.

559 From the middle of 1988 the pet food industry had begun to address the possible infectivity of bovine raw materials incorporated in pet food. In July 1988 Pedigree Master Foods commissioned Dr Kimberlin to advise on whether their raw materials might carry the BSE agent. What he had to tell them they considered to have wider significance and they offered to share the information with MAFF. On 16 May 1989 Pedigree Pet Foods invited Mr Meldrum and other MAFF officials to meet Dr Kimberlin. Dr Kimberlin gave Mr Meldrum details of the advice that he had given to Pedigree, including the categorisation of offal into four categories of risk. The highest was brain and spinal cord and the next consisted of ileum, lymph nodes, proximal colon, spleen and tonsil.[71] Mr Meldrum told us that it was clear to him that Dr Kimberlin thought it a good idea to keep the more infective offal out of

---

[69] See para. 255 above
[70] See paras 286–95 above
[71] This was based on studies on the infectivity of the tissues from cases of natural scrapie carried out by Dr William Hadlow

the human food chain. He left the meeting converted to this viewpoint. Dr Kimberlin's analysis had added a huge amount to his knowledge. We wish to commend Pedigree for their initiative in seeing that this information was provided to MAFF.

560 Meanwhile MAFF officials had been preparing draft Regulations and a consultative paper in respect of the proposed ban on offal in baby food. Mr Andrews warned Mr MacGregor that this would lead to pressure to extend the ban to all human food. Mr MacGregor was already under pressure in Parliament from Mr Ron Davies, the Opposition spokesman on Agriculture, to do just this. Mr MacGregor then met with Mr Meldrum. Mr Meldrum told him of what he had learned from Dr Kimberlin. This did not persuade Mr MacGregor that the Southwood Working Party's assessment of risk was unsound. He told us that what it did was to provide him with 'a scientific underpinning for the selection of tissues if Ministers were to adopt a policy to further reduce the remote risk of transmission of BSE to humans'. He told us, 'I had some concern about this. Most of the scientists were telling me that this concern was unjustified, but there was just beginning to emerge some body of scientific opinion that there may be something in it, so it had the merit of dealing with that risk, if there was a risk.'

561 Within days Mr MacGregor had decided to go ahead with a ban. He told us that his reasons for this decision were:

- He wished to reassure the public.
- It was easier to introduce a general ban than a baby food ban.
- It would deal with any clinical animals that might slip through the net.
- It would deal with any risk from tissues from subclinical animals.

562 There was one practical difficulty. It was desirable to get Sir Richard Southwood's approval to this course. This called for diplomacy as MAFF proposed to go beyond the measures that his Working Party had advised.

563 On 6 June Mr MacGregor had a meeting with his officials, to which Dr Jeremy Metters of DH was invited, in order to prepare for a meeting with Sir Richard Southwood the following day. Sir Donald Acheson had got wind of what was afoot and was unhappy about it, fearing that it might raise concerns about the safety of vaccines. He briefed Dr Metters to resist the move, at least for the time being. Dr Metters was Senior Principal Medical Officer in DH who had recently become involved in BSE matters. In August he became Deputy CMO. Dr Metters raised the concern about the vaccines at the meeting, but reported that this 'cut little ice' with MAFF officials. Mr MacGregor did not refer at the meeting to Dr Kimberlin's analysis of the infectivity of tissues in subclinical animals. He left those present with the impression that his motive for the ban was simply a wish to allay the public concern which had developed.

564 On the next day the meeting reconvened with Sir Richard Southwood. Dr Pickles was also present. When told of the proposed ban, Sir Richard made the point that the scientific evidence had not changed, but accepted the 'political necessity for action'. Mrs Attridge then made a suggestion about presentation. As she reported later:

i. Professor Southwood maintained his position that there was no scientific evidence to support the belief that offal presented a human health hazard (DOH Dr Metters did not dissent).

ii. The Minister maintained his view that presentationally something had to be done to allay public concern.

iii. The CVO pointed out that the easiest way to ensure any ban was operated was to remove offal (brains, spinal cord, spleen, tonsils, thymus) that were to be covered in the baby food regulations at the slaughterhouse.

iv. I suggested that the way to proceed was to say that the Minister considered the easier and more enforceable way to implement the Southwood recommendation on baby foods was to remove the offal at slaughterhouses and there it would be dyed and used for fertiliser and that the Minister would thereby not be appearing to contradict the scientific evidence in the Southwood report by taking more comprehensive action than recommended and there would be no need to proceed with consultations under the Food Act.

565 To those unaware of the potential infectivity of subclinical animals, Mrs Attridge's suggestion on presentation must have seemed attractive. If there was no scientific justification for the ban, it would do no harm to suggest that its introduction was no more than an administratively convenient way of introducing the ban on baby food. The vice of this presentation was, however, that it suggested that the ban was unnecessary. It would not encourage those who had to implement the ban to take it seriously. Unfortunately, Mr MacGregor agreed to Mrs Attridge's suggestion as to how the ban should be presented.

566 The presentation of the ban suggested by Mrs Attridge was widely disseminated. When Mr Lowson was preparing a briefing for incoming Ministers in July after a reshuffle, he included it as the reason for the decision to introduce the SBO ban. We were concerned about this, for he did not mention what he thought to be the true reason, namely to allay public anxiety as to the risk from subclinical animals. But given the pressure of time within which such briefings have to be prepared, and their ephemeral nature, we think it would be wrong to criticise Mr Lowson's draftsmanship. Mr Gummer, Mr Maclean and Mr Curry all told us that Ministers do not place great weight on such briefings, but Mr Gummer subsequently passed on the presentation. At a meeting with UKASTA in October 1989, and again before the Agriculture Committee in 1990, he emphasised that the ban went beyond what the Southwood Working Party had advised was necessary, but was introduced as a practical way of giving effect to their baby food recommendation. Mr Lawrence included the presentation as the reason for the ban in the submission to Mr Gummer that he prepared in November 1989 seeking approval of the terms of the draft Regulations. This submission was widely circulated within MAFF, DH and the Territorial Departments.

567 In his press release announcing the ban, Mr MacGregor referred to the Government's undertaking to implement the Southwood baby food recommendation. He then added:

> In working out the details, I have concluded that a better way of dealing with
> this would be to ensure that the relevant types of bovine offals should be
> rejected at the slaughterhouses for all cattle so that they cannot be used for
> human consumption in any way . . . This approach also deals with a separate
> problem, namely ensuring that if there is any risk that there are cattle
> incubating the disease but not showing clinical symptoms which are not
> being slaughtered and destroyed, their offals do not enter the food chain
> either.

568 This at least referred to the subclinical animals, but in terms that suggested that there was no more than a risk that some of these might go for slaughter. In fact this was inevitably happening on a substantial scale.

569 How far the presentation, which played down the importance of the human SBO ban, influenced people's attitudes we shall never know. We had evidence from many sources, however, of a perception that the ban was not really necessary as a public health measure. We do not criticise Mrs Attridge for her suggestion, made in ignorance of the science that underpinned the ban, nor those who repeated what Mr MacGregor had agreed should be the public presentation of the reason for the ban. Mr MacGregor is to be commended for introducing a ban which was to prove such a vital element in guarding against the risk that BSE posed to humans. However, he should not have agreed to a presentation which played down the importance of the ban as a protection for human health.

570 One person who thought that the human SBO ban was an unnecessary precaution was Dr Pickles. She remained of the view that the Southwood Working Party had recommended all that science justified. She suggested that MAFF should be left to introduce the ban on its own. Sir Donald Acheson had by now decided, however, that DH should support the ban. This attitude was shared by Mr Clarke, although his understanding was that MAFF was motivated by a desire to restore consumer confidence rather than by any scientific consideration. Mrs Thatcher approved the ban. She informed us that she did not believe that she would have accepted the need for the ban *solely* for public reassurance.

Preparation of the Regulations

571 The ban was announced on 13 June 1989. Five months were to pass before it was brought into force.[72] The Agriculture Committee criticised this delay. We have considered why it occurred and concluded that it would not be fair to criticise either MAFF or DH for not moving faster. The ban was introduced under the Food Act 1984 and made use of procedures and mechanisms for dealing with unfit meat that were already in place under the Meat (Sterilisation and Staining) Regulations 1982 (MSSR). This made good sense, but it carried with it a statutory obligation to consult. Regulations requiring the removal of tissues from apparently healthy animals on the ground that a small minority would be incubating a disease that carried a remote possibility of transmission to humans were novel. They were quite complex. They carried serious economic consequences for some. We think that consultation was desirable. What took longer than anticipated was the task of identifying which offal should be subject to the ban. This was not due to any lack of diligence, but to the complexity of some of the technical issues that arose. It

[72] We have described the Regulations at paras 386–7 above

would have been better to have introduced a ban on those tissues which were known to be high risk and added to them later by amendment, but that is to use hindsight.

572 From the outset it was the intention that the ban should apply to brain, spinal cord, tonsils, spleen, thymus and intestines, which were recognised as high-risk tissues. The principal issues as to the ambit of the ban were whether:

- it should include tripe and rennet;
- it should include mesenteric fat;
- it should include intestines which had been processed to make casings for sausages and other meat products;
- it should apply to tissues of calves under the age of six months.

573 Resolving those issues required research, consultation with the industries involved and discussion between MAFF and DH. All of this took time.

574 Mr Bradley, who had been placed in charge of BSE research work at the CVL, carried out the research. So far as the first three issues were concerned, his task was to ascertain the extent to which lymphoid tissue would remain after the industrial processes that were involved. He set about this task with characteristic diligence.

575 Discussion between MAFF and DH involved Mr Meldrum on the one hand and Dr Pickles and Dr Metters on the other. Mr Meldrum's approach was one of reluctance, without good reason, to countenance extending the ban to the detriment of established sectors of the food industry. This was a proper approach provided that he did not permit concern for the food industry to prejudice the safeguarding of public health. We were conscious of accusations that MAFF had done precisely that, so we scrutinised this part of the story with particular care. We concluded that Mr Meldrum adopted a conscientious and objective approach to his task.

576 Neither Dr Pickles nor Dr Metters believed that there was any justification for the human SBO ban. They saw it as an exercise carried out by MAFF in order to improve public confidence in the safety of beef. We were concerned to see whether this perception led to any lack of rigour on their part in considering what should and what should not be included in the ban. We concluded that it did not. Dr Pickles told us that if Ministers, for all sorts of good reasons, wished to do something that was not strictly necessary, she would support them. Her aim was to ensure that all the bits of offal that might be of concern were removed from the food chain.

577 Mr Gummer was appointed Minister of Agriculture in July, in the course of the preparation of the SBO Regulations. He gave Mr Maclean, one of the new Parliamentary Secretaries, special responsibility for food safety. We are satisfied that Mr Gummer and Mr Maclean gave careful consideration to the terms of the human SBO Regulations. They did not rubber-stamp their officials' proposals, but sought and considered the reasons behind the inclusion or exclusion of various types of offal from the ban.

578 Notwithstanding the diligence that was applied to most aspects of the preparation of the SBO ban, it was inevitable that borderline decisions would be influenced by the general belief that the ban was being imposed as a measure of extreme prudence which went beyond the recommendations of the expert scientists. While those involved made no conscious application of the ALARP principle, the exercise that they were engaged in entailed weighing perceptions of risk on the one hand against the economic consequences of banning particular tissues on the other.

579 We turn to record briefly the decisions that were reached as to the ambit of the ban.

Brain, spinal cord, thymus, spleen and tonsils

580 These 'high risk' tissues were intended to be covered by the ban from the outset. Dr Kimberlin, whose advice was sought by MAFF on the ambit of the ban, advised that the proposed ban on these tissues was well founded.

Tripe and rennet

581 Rennet was extracted from the abomasum, the fourth stomach of the cow, and was used for making cheese. One form of tripe was also made from the abomasum. Concern about these products arose from the fact that the abomasum contained significantly more lymphoid tissue than the other stomachs. In relation to lymphoid tissue Mr Bradley proposed a pragmatic test. Lymphoid tissue would be banned only when macroscopically visible, that is, to the naked eye. On this approach, the abomasum and its products did not fall within the ban. This approach was approved by Dr Kimberlin and accepted by Dr Pickles and Dr Metters. Very careful consideration was given to this issue, which involved a question of nice judgement as to where the borderline should be drawn. The decision not to include tripe and rennet in the ban was endorsed by Mr Maclean and Mr Gummer.

Mesenteric fat

582 This was fat that was originally attached to the intestine and which contained lymphoid tissue. It was excluded from the ban on the basis that, in the course of processing, the protein containing the BSE agent would fractionate with the solids rather than with the fat. Similar reasoning had led to the conclusion that tallow need not be subject to the SBO ban. Ministers initially queried the exclusion of mesenteric fat from the ban, but on being given this explanation were satisfied with it.

Casings

583 MAFF officials initially believed that the cleaning of intestines, which were used as sausage casings, would remove all but an insignificant quantity of lymphoid tissue and proposed that casings should be excepted from the ban. Dr Pickles challenged this assumption, whereupon the CVL confirmed that the processing of sausage casings removed lymphoid tissue. This conflicted with information that DH had obtained from a medicinal company in relation to the manufacture of sutures from intestines. Mr Bradley carried out further research, which revealed that lymphoid tissue remained in casings after processing. Mr Meldrum reported this, but suggested that casings could be excluded from the ban because they were only used on black and white puddings, were cooked and were usually discarded at the table. Sir Richard Southwood and Dr Tyrrell were then consulted, and both indicated acceptance of Mr Meldrum's reasoning. Dr Metters expressed continuing reservations, but added that DH was content for MAFF to proceed as it thought fit. Mr Meldrum then had second thoughts. He advised Ministers *not* to exempt casings and wrote to Dr Metters explaining: 'I believe it is most important that we have a fully agreed position on this most important area.' Ministers accepted Mr Meldrum's advice.

Calves under 6 months of age

584 There were a number of reasons why Mr Meldrum was anxious to exclude offal from calves aged less than six months from the human SBO ban:

- It was not slaughterhouse practice to split the carcass of calves, so a requirement to remove spinal cord would raise practical problems.

- Feed compounders were threatening to boycott SBO-derived MBM. An exemption from the ban in respect of calves might discourage them from this step.

- A ban on SBO from calves would add to the waste disposal problem.

- A ban on SBO from calves might provoke export restrictions. The UK had a large trade in the export of veal calves.

585 There were two arguments that could be advanced to justify an exception from the ban in respect of the offal of calves:

- Calves would have been born after the ruminant feed ban came into operation and therefore should not have been infected from feed. The weakness of this argument was that it was possible – and Mr Meldrum thought it was likely – that calves would be infected with BSE as a result of maternal transmission.

- Analogy with scrapie research suggested that infectivity would not reach the brain or spinal cord of cattle in the first six months of life. This was a cogent argument for exempting brain and spinal cord of calves from the ban. Dr Kimberlin was, however, concerned that the lymphoreticular system (LRS), and in particular the spleen and thymus, might be infective at any age.

586 Dr Metters indicated that DH could not agree to an exemption in relation to calves in the absence of scientific advice justifying this.

587 Mr Meldrum made enquiries of the trade and was informed that spleen and thymus did not enter the human food chain. He passed this information to Dr Kimberlin and to Dr Metters, adding that very few calves were slaughtered in the United Kingdom each year. Dr Kimberlin then reconsidered the issue and indicated that he would be content with an exemption in respect of calves. Sir Richard Southwood and Dr Tyrrell were both consulted, and accepted that Mr Meldrum had demonstrated valid reasons for an exemption in respect of calves. Finally Dr Metters indicated DH agreement to this, adding that the position would have to be reviewed if maternal transmission were established. Ministers accepted advice that offal from calves should be excluded from the ban.

588 The facts that we have outlined above caused us concern. While only 25,000 calves were slaughtered each year in the UK, 250,000 were exported to the Continent to be slaughtered for veal. Furthermore, thymus, or 'ris de veau', was a prized delicacy on the Continent. In such circumstances, to have sought to disguise the risk posed by thymus by exempting calves from the ban, with the motive of protecting our export market, would have been scandalous. We explored our concerns with the witnesses. Dr Kimberlin assured us that his advice had not been influenced by export considerations. He also said that he was not overly concerned about the thymus because scrapie research indicated that thymus was lower risk than other LRS tissues.

589 Sir Derek Andrews summarised the factors which had satisfied him that the exemption in respect of calves was justified:

- The SBO ban was a measure of extreme prudence.
- The risk of transmission to humans was considered remote.
- Calves had not been fed MBM.
- Scrapie research indicated that calves under 6 months would not contain the agent.

590 The evidence we received satisfied us that those involved in the decision to exempt offal from calves from the ban were not improperly motivated by a concern to preserve exports and that the exception could be justified on an objective appraisal of such facts as were known at the time.

Mechanically recovered meat (MRM)

591 We now come to a topic which we have identified as a serious flaw in MAFF's precautions to prevent SBO from entering the human food chain. Once all meat had been removed from the carcass, it was often the practice to subject the bare bones to the process of mechanical recovery of meat. High pressure was applied to the bones to separate from them anything that was still adhering. The resultant slurry was used in a range of meat products for human consumption, including lower grade sausages, burgers and pies. The major source of bovine MRM was the spinal column.

592 Spinal cord, together with the brain, was identified as the tissue which contained the highest titre of BSE infection. It had long been the usual practice in slaughterhouses for bovine spinal cord to be removed and sent for rendering as part of the meat-dressing process. That is not to say that it used to be cleanly removed. We received evidence that before the human SBO ban it was common for sizeable sections of spinal cord to be left in the spinal column. In that event it would be sucked out as a constituent of MRM.

593 Once spinal cord was prescribed as an SBO, standards of removal of the spinal cord in slaughterhouses improved. In 1995, however, it was discovered that slaughterhouses were, on occasion, leaving small portions of spinal cord attached to or trapped within the spinal column. We are satisfied that that was a state of affairs which had persisted ever since the human SBO ban was introduced. Portions of spinal cord will have gone into MRM.

FINDINGS AND CONCLUSIONS

594 When the human SBO Regulations were being formulated, peripheral nervous tissue was believed not to be high risk. Since 1996 experiments have shown one respect in which that belief was fallacious. The autonomic nervous system is linked to the central nervous system at junction boxes, consisting of clusters of nerve cells, alongside each vertebral body. These are known as dorsal root ganglia.[73] This tissue has now been shown to develop high infectivity between 32 and 40 months after a cow is infected with BSE. Dorsal root ganglia will also have been sucked out by the MRM process.

595 On more than one occasion, consideration was given to the question of whether it was satisfactory to continue the practice of extracting MRM from the spinal column of cattle. Not until late 1995 was it decided that this practice should be banned. The first occasion on which the question arose was when the SBO Regulations were being prepared.

596 In June 1989 a minute circulated within MAFF recording that a slaughterhouse had given up producing MRM from cattle bones because it could not guarantee that all central nervous system (CNS) tissue would be removed from the backbone. This did not stimulate any detailed consideration. Views were expressed that the quantity of CNS material involved was unlikely to be significant. Mr Bradley responded with a warning that the vertebral column might be contaminated with spinal cord and commented, 'Clearly spinal cord must be removed before processing to produce MRM should this be allowed to continue.'

597 The consultation process in relation to the human SBO ban provided further warning of the danger that spinal column would be contaminated with residues of spinal cord. Some of those consulted responded that total removal of the spinal cord was impractical. One pointed out that 'the residual bone treated hydraulically to produce re-claimed meat [would] include spinal cord pieces'.

598 A meeting was held in MAFF on 27 September 1989 to consider the responses to the consultation letter. It was chaired by Mr Cruickshank and attended by, among others, Mr Kevin Taylor, Mr David Taylor, Mr Lowson, Mr Lawrence, Mr Maslin, Mr Wilesmith, Mr Duncan Fry and representatives from the Territorial Departments. Dr Pickles had been given short notice of the meeting and was unable to be present. There was no representation from DH. No witness had any recollection of what transpired in relation to MRM at this meeting. MAFF's note of the meeting recorded:

> The proposed ban on specified offals was in itself a measure of extreme prudence, going beyond what Southwood recommended. Though some tissue would be contained in MRM it would be minimal and not present a significant risk. No action should be taken on MRM.

599 Mr Ron Martin, Deputy CVO at the Department of Agriculture for Northern Ireland (DANI), also made a note of the meeting, which recorded the discussion of MRM as follows:

> The possible danger raised by several of those consulted was recognised and during discussion there was an expression of the illogicality of what was being done and, in particular, how easy it would be to have to concede the

---

[73] See the illustration in Chapter 5 of vol. 16: *Reference Material*

> possible dangers of material other than those listed in the proposed ban. It was agreed not to raise it.

600 The issue of MRM was a complex one. In the following year, as we shall see, MAFF prepared a paper on it for consideration by SEAC. The amount of work that went into that paper is illustrative of what was required if the matter was to be properly considered in 1989. To make a reasoned decision about MRM, it was necessary to assess:

- the amount of spinal cord that might be left attached to the spinal column and recovered as MRM; and
- the minimum quantity of spinal cord that might be capable of carrying an infective dose for humans.

601 Those present at the meeting were not in a position to provide definitive answers to those questions, but they were in a position to identify that such questions needed to be addressed. They did not identify them. Part of the problem appears to have been that no one took on personal responsibility for addressing the question of whether MRM posed a risk to human health. Responsibility for producing the human SBO Regulations had been shared. MAFF's Meat Hygiene Division had agreed to be responsible for the mechanics of drawing up the Regulations, but considered that the Animal Health Division had retained responsibility for policy. It was the Animal Health Division that had charge of the consultation exercise.

602 Mr Cruickshank said that he relied on the veterinary judgement that MRM was acceptable. Mr Kevin Taylor said that he had no responsibility for matters relating to human as opposed to animal health. Mr Lowson said that his divisional responsibilities were limited to animal health. He also said that the Meat Hygiene Division had taken the lead in preparing the Regulations. Mr Lowson said he had to rely on Mr David Taylor, the SVO dealing with meat hygiene issues. Mr Keith Baker, for whom Mr David Taylor was deputising, told us that it was not for his section to advise on the implications of infective dose for the safety of MRM.

603 We found this evidence confusing and unsatisfactory, bearing in mind that all present on 27 September were participating in an exercise that had only one object – the protection of human health.

604 The decision on MRM depended critically on a combination of knowledge of the processes of carcass-splitting and removal of the spinal cord; knowledge of the processes of extracting MRM; knowledge of standards of operation, inspection and monitoring of abattoirs; and an understanding of what was known, and what was not known, about infective dose in relation to TSEs.

605 No one before or after the meeting of 27 September set about collecting this information and presenting it in a form that would enable an informed policy decision to be taken. There appears to have been a general assumption that, if any spinal cord were to get into MRM it would do so in quantities too small to represent a threat. Some failed to appreciate the extent to which spinal cord might get into MRM. Some seem likely to have made unwarranted assumptions about the minimum effective dose.

FINDINGS AND CONCLUSIONS

606 Consideration of the proposed SBO Regulations was a team exercise and the failure to give rigorous analysis to MRM was a team failure. We believe that this failure is explained in large part, and mitigated, by the general belief that the SBO ban was a measure of extreme caution that went beyond the recommendations of the scientists. In the circumstances it is easy to understand the reaction that if there was a failure on occasion to remove a little bit of spinal cord, it was unlikely to matter. This does not, however, excuse the failure to carry out the rigorous risk evaluation that was required in order to reach a sound decision on policy.

607 The problem posed by MRM should not have been dismissed at the meeting on 27 September 1989. It should at least have been identified as calling for further consideration. However, no witnesses could remember any relevant detail as to the information or views contributed on this subject at that meeting. It would not, in these circumstances, be fair to criticise any individual for the conclusion that was reached. Nor would it be fair to criticise those who placed reliance on that conclusion. We are simply not in a position where it would be fair to allocate blame to any individual for the failure to give rigorous analysis to MRM in 1989.

608 Dr Metters and Dr Pickles of DH received copies of MAFF's note of the meeting on 27 September. They had no knowledge of the nature of MRM. They read the statement that the amount of nervous tissue that it would contain would be minimal, and were content with that.

609 Those who relied upon the outcome of the meeting included Mr Lawrence. He advised Mr Gummer that any nervous tissue in MRM would be minimal and that the ban should not extend to MRM. Ministers questioned this advice. Mr Maclean asked how they could be sure that all abattoirs removed the spinal cord cleanly before MRM production took place. Mr Meldrum reassured Ministers that the risk from MRM was no greater than that in other cases where an exclusion from the ban had been agreed. Mr Meldrum told us that he was not concerned about spinal cord. He believed that any fragments would be removed at the dressing stage. He had concerns about peripheral nervous tissue, but Dr Kimberlin had provided reassurance about this. Mr Meldrum also relied on the conclusions reached at the meeting on 27 September.

610 On this occasion the chance to identify the danger posed by MRM was lost. What would have transpired had that danger been identified? We do not think it likely that it would have led officials to advise, or Ministers to decide, that the practice of extracting MRM from the spinal column of cattle should be banned. Mr Cruickshank told us that officials were conscious that the ban went beyond what the scientists had advised was necessary for the protection of public health, and were apprehensive that action that appeared disproportionate would provoke a judicial review. Had the danger of MRM been recognised, we think that this would have led MAFF to emphasise to slaughterhouse operators and local authorities that it was essential to remove spinal cord in its entirety and to monitor the extent that this was achieved once the ban was in force.

611 In the event, when the ban was introduced, no guidance was given to slaughterhouse operators or to the local authorities who had to enforce it. Nor were any instructions given to the veterinarians in the VFS, whose job it was to monitor the enforcement of the Regulations, that it was important to check that all spinal cord was being removed from carcasses.

BSE and human health in 1990

612 1990 was an eventful year in the BSE story. It saw a number of practical problems raised in relation to the implementation of the human SBO ban and the manner in which government addressed these. It saw restrictions placed on the export of beef by the EU, and their implications for the United Kingdom. It saw the natural transmission of BSE to cats, the alarm that this caused, and the response of government to that alarm. It saw the extension of scientific knowledge about BSE, with experimental transmission to mice, to cattle and subsequently to a pig. These latter events led to the introduction of the animal SBO ban, which we have described in the previous chapter. In this chapter we shall be looking at events that had relevance to the implications of BSE for human health.

613 In 1990 Mr Gummer completed his first year as Minister of Agriculture, Fisheries and Food. He had brought with him a new broom. He sought to draw a clear distinction, within the Ministry, between looking after the interests of the industry and looking after the interests of the consumer. The former he entrusted to Mr Curry; the latter to Mr Maclean. As Minister for Food Safety, Mr Maclean presided over the newly formed Food Safety Directorate. He also chaired a new Consumer Panel. Mr Gummer made it plain that his Ministry would be following a policy of openness of information about food safety. He also announced that the results of all research into BSE would be made public.

614 The same year saw the setting up of SEAC. Mr Gummer was a firm believer in taking the advice of experts and then following that advice. As soon as SEAC was set up he began to seek its advice on a wide variety of topics.

Implementation, enforcement and monitoring of the human SBO ban

615 In the previous chapter we looked at what happened to SBO once it had been removed from the carcass. This assumed importance in relation to animal health. There was never any apprehension that, once removed, SBO would find its way into the human food chain. So far as human health was concerned, the important thing was that the SBO should be cleanly removed from the carcass without contaminating the meat.

616 We have already commented on the poor standards of hygiene prevalent in UK slaughterhouses and the fact that the manner and rigour of enforcement of the Regulations varied from one local authority to the next.[74] Happily these standards were not generally reflected in the diligence with which the Meat Inspectors set about their task of ensuring that SBO, and in particular spinal cord, was removed from the carcass. This was not, however, an easy task. The operation involved sawing the carcass in half down the backbone with a power saw, thus exposing the spinal cord, and then removing the cord. It was inevitable that in the process the spinal cord would sometimes get damaged and that portions of it would remain trapped or hidden within the vertebrae. It would have needed the most meticulous skill and care on the part of the Meat Inspectors to make sure that no carcass that received the health stamp contained any remnants of spinal cord. Skill and care to that degree was not shown during the period with which we are concerned.

[74] See Chapter 3, para. 389 above

Meat Inspectors were often rushed, and holding up a production line for inspection was not popular. No one emphasised that removing all the spinal cord could be a matter of life and death, and it was not so regarded. As a result the occasional portion of spinal cord would pass through, undetected, with the health-stamped carcass, and be destined in many cases to be extracted as MRM.

617 We have described earlier the monitoring role of the VFS in respect of compliance with a large number of Regulations applicable in a slaughterhouse. The removal of spinal cord from the carcass was only one of many of the statutory requirements that they had to monitor. They were not instructed to give this particular attention. On the contrary, insofar as they received instructions, these focused on the disposal of the SBO after removal from the carcass, and we had evidence that this aspect of the SBO Regulations was the one with which they were more concerned.

618 In these circumstances we can understand why it is that, prior to 1995, there is only one recorded occasion on which a member of the VFS identified health-stamped meat that contained spinal cord. Only during the national surveillance in 1995, when unannounced inspections were carried out and when VOs were instructed to pay particular attention to the removal of spinal cord, did the fact that there were shortcomings in this respect come to light.

Bovine brains

619 One slaughterhouse problem that did quickly become apparent after the SBO ban was introduced related to bovine brains. Before the SBO ban the head meat would normally be removed at a slaughterhouse or head-boning plant, after which the head would be sent off with the brain inside to be rendered. Under the SBO Regulations a head with brains inside had to be treated as SBO. A practice started almost immediately of splitting the skull and removing the brain, so that the head could then be despatched, free of regulation, as BSE-free material. This practice created an obvious contamination hazard in the slaughterhouse.

620 No sooner had the ban come into force than Environmental Health Officers (EHOs) began to raise with MAFF concerns about the risk of contamination as a result of head-splitting and brain removal. There were a number of different techniques for splitting the skull and one method of removing brain that avoided this was by blasting the brain out of the base of the skull with a high pressure jet of water or air. The Institution of Environmental Health Officers expressed concern that all methods involved the risk of contaminating the head meat and urged that the practice of removing the brain be forbidden. A Liberal Democrat MP, Mr Matthew Taylor, took up this cause. MAFF officials took the view that any contamination was likely to be too small to worry about. Mr Hutchins, an SVO in the Meat Hygiene Veterinary Section, carried out a survey. He advised that there was no reason to prohibit the open-skull method of brain removal, although he had reservations about the high-pressure method. Mr Gummer was not persuaded, and promised Mr Taylor that he would ask an outside expert to consider the matter.

621 The chosen expert, Mr A M Johnston, expressed reservations about all the methods of brain removal and advised that, whenever possible, head meat should be removed before any cut was made in the skull. Both Mr Maclean and Mr Gummer expressed continued concerns about brain removal. Officials reassured them that draft guidelines on the techniques of brain removal were being prepared which would reduce contamination to a minimum. The problem was minuscule. The financial consequences of restrictions would be considerable. None should be imposed. Ministers were minded to accept this advice, but there followed a further spate of protests about the practice from many quarters. On 21 May 1989, pressed about the practice in a parliamentary debate, Mr Gummer stated that it would be referred for consideration by SEAC.

622 SEAC considered draft guidelines prepared by the Meat Hygiene Division at their meeting on 13 June and gave them short shrift. They advised that it was not consistent with common sense to permit the removal of the brain before the head meat was harvested. Mr Gummer directed that guidelines be issued reflecting this advice. Mr Meldrum sent them out the following day. They directed that bovine head meat had to be recovered from the intact skull before the brain was removed.

623 On 10 July 1990 the Agriculture Committee published its report on BSE. One recommendation was that MAFF's guidelines on head-splitting should be enshrined in legislation at an early opportunity. Ministers accepted this recommendation. On 12 March 1992 Regulations were introduced which:

i. prohibited the removal of head meat after the skull had been opened or brain removed; and

ii. prohibited the removal of brain in a slaughterhouse or boning plant except in a special area at no time used for food for human consumption.

624 No reasoned application of the ALARP principle was carried out by MAFF. MAFF officials assumed that contamination would be too minuscule to matter. Ministers were justified in their reservations about this, and did well to call for independent advice. SEAC was not an appropriate body to consider technical questions of head-splitting techniques. It was, however, well qualified to express a view as to whether risks of contamination from such practices were acceptable. SEAC did not attempt any quantification of the amount of contamination liable to result from brain removal. Nor did it weigh in the balance the financial consequences of the various options. The Committee applied a robust common sense in assuming that contamination was liable to be significant and advising accordingly. The outcome was satisfactory. The same cannot be said of SEAC's next venture into the world of the slaughterhouse.

Slaughterhouse practices and mechanically recovered meat

625 We have referred to concerns expressed about the removal of spinal cord and MRM in the course of consultation about the proposed SBO Regulations. These continued after the Regulations were brought into force. Mr Corbally of the Institution of Environmental Health Officers expressed the concern of its members about this. On 18 April 1990 he wrote to Mr Keith Baker:[75]

> Do you consider that the continued use of mechanically recovered meat from bovines is acceptable? . . . MRM could contain significant quantities of spinal cord nervous tissue.

[75] Assistant CVO, Meat Hygiene

626 On 21 May Dr John Godfrey of the Consumers in the European Community Group, at a meeting with Mr Meldrum and Mr Maclean, questioned whether dorsal root ganglia might not be as infectious as spinal cord. Two weeks later, Mr Meldrum wrote to Mrs Attridge expressing concern that MRM might be significantly contaminated. He told us that it was peripheral nervous tissue that had given rise to his concern.

627 Calls for the banning of the practice of recovering MRM from the spinal column of cattle came from:

- the Consumers' Association; and
- the MLC Consumer Committee.

628 Concerns about the practice were expressed to the Agriculture Committee from a number of quarters. Of particular note was a submission from Dr Gerald Forbes, Director of the Environmental Health (Scotland) Unit, who wrote of MRM:

> Can any guarantee be given that parts of the central nervous system of cattle do not enter this product? I would suggest that this is not possible and whether or not the practice of producing mechanically recovered meat can be considered safe is very much open to doubt.

629 As we have seen, Mr Gummer decided in May that slaughterhouse practices should be referred to SEAC. MAFF set about preparing a paper that would provide SEAC with the information that it would need to consider these. The drafting of this paper was a major undertaking involving input from the Meat Hygiene Division, the Food Standards Division, the Food Science Division, Mr Meldrum and officials in the Animal Health Division. The final draft was not produced until October. The paper gave SEAC the following information about slaughterhouse practices:

- The spinal cord will inevitably receive some damage during carcass-splitting.
- Inevitably some nervous tissue can remain and some contamination of the vertebrae with CNS tissue can occur as a result of:
    a. small pieces of spinal cord inadvertently remaining in the spinal column
    b. contamination from carcass-splitting
    c. the failure to remove nerves from between the vertebrae.

630 Those responsible for preparing the paper had reached the conclusion that some action was called for. Originally they had been prepared to place before SEAC a series of alternative options:

   a. issue guidance to the trade on minimising contamination;
   b. request local authorities to ensure spinal cord had been removed;
   c. ban the extraction of MRM from the bovine vertebrae;
   d. ban manufacture of MRM from bovine carcasses.

Of these, option (c) was to be advanced as the preferred option, coupled with a recommendation that certain specified research be carried out to ascertain the extent of the contamination of MRM that was occurring.

631 In the event it was decided not to refer to these options, but simply to ask SEAC to advise:

> . . . whether any action or guidance is required in relation to slaughterhouse practices, and whether any new R&D is needed.

632 What then occurred was this. SEAC members decided that they would visit a slaughterhouse and see for themselves the procedures involved. Most of them did so and were given a 'Rolls-Royce' demonstration of carcass-splitting and removal of spinal cord. Those who saw this concluded that spinal cord could be extracted from the carcass without difficulty. At SEAC's next meeting, slaughterhouse practices was one item of an over-charged agenda. SEAC dealt with that item by advising, in the case of some members on the basis of what they had seen, that so long as the rules were properly observed and proper supervision was maintained, there was no need to recommend further measures on grounds of food safety. MAFF officials and Ministers treated this as reassurance that all was well, and no further consideration was given to MRM for some years to come.

633 It does not seem that there was any discussion at the meeting about MRM. Dr Tyrrell suggested to us:

> I suspect that what happened was that we reckoned there was not really a problem with MRM if the vertebral column was being cleanly cut and dissected.

634 The events that we have summarised demonstrate a serious breakdown of communication. MAFF officials knew, as their paper expressly stated, that a degree of contamination of the spinal column with spinal cord was inevitable. Some members of SEAC, Dr Tyrrell among them, proceeded on the basis that clean removal of spinal cord was easy and thus something that could be achieved in practice. It was on the basis of that assumption that they advised that there was no need for any action. MAFF officials, however, understood that SEAC was indicating that the degree of contamination described in the paper as 'inevitable' was no cause for concern.

635 We do not consider that this sorry story is a matter for individual criticism. There are, however, lessons to be learned from it. What went wrong?

- SEAC had too much on its plate. The agenda did not allow sufficient time for a detailed discussion of MAFF's paper on slaughterhouse practices.
- The advice sought from SEAC was not targeted. SEAC's expertise lay not in slaughterhouse practices but in the potential consequences of consumption of spinal cord. As we shall see, the Committee had been considering infectious dose for the purpose of advising the CMO. It based its advice not on this consideration, but on its conclusion about slaughterhouse practices. SEAC should have been asked expressly whether the contamination described in MAFF's paper was cause for concern.

- SEAC was not informed of the options which MAFF officials had identified. We consider that it would have been helpful if SEAC had been told about these.

- SEAC was unaware of the concerns that had been expressed about the removal of spinal cord and the safety of MRM.

636 Had SEAC been aware of all these matters, we think it likely that it would have endorsed the suggestion that further research be carried out in order to quantify the amount of spinal cord material getting into MRM. This might have led to SEAC endorsing the further option of recommending a ban on the extraction of MRM from the bovine vertebrae. There can be no certainty that it would have done so.

637 Had MAFF officials been left to advise Ministers unaided by SEAC, we think it likely that they would have recommended option c) of those they had identified, as set out in paragraph 630 above. If not, they would surely have recommended options a) and b). It was unfortunate – and possibly tragic – that the intervention of SEAC should, as a result of a breakdown of communications, have left MAFF officials and Ministers falsely reassured about the safety of MRM.

Europe and lymphoid tissue

638 The slaughter and compensation policy and the human SBO ban protected consumers of beef products in both the United Kingdom and countries to which these were exported. The European Commission decided, however, to take additional measures to protect Continental purchasers of British beef. These included a requirement[76] made in June 1990 that the UK should certify all boneless beef for export to other Member States as being 'fresh meat from which during the cutting process obvious nervous and lymphatic tissue has been removed'.

639 MAFF carried out a survey to discover the extent to which the cutting procedures employed in UK plants satisfied this requirement. It was discovered that the procedures varied widely from those plants which removed virtually all lymph nodes to those which removed very few. Alarmingly, 'healthy' lymph nodes which had been removed were used in meat products for human consumption or rendered for either human food or animal feed.

640 Consideration was given to legislating to add lymph nodes to the list of SBO. There were, however, intractable problems with such a course. Not all lymph nodes could be prescribed, for they were to be found throughout the carcass. It would not be practicable to have Regulations which prescribed 'obvious lymphatic tissue', for this would lack certainty. Furthermore, lymph nodes were often not removed until meat was being dressed in the butcher's shop, and it would be difficult to devise Regulations that would cover that situation.

641 In the event it was decided to issue guidelines, designed both to enable the UK to comply with the EC Decision and to set a common standard for beef, whether it was to be consumed in the United Kingdom or exported.

---

[76] Introduced by European Commission Decision 90/261/EC

642 On 16 June guidelines were issued which provided that:

> All lymphatic and nervous tissue that is exposed during normal cutting operations must be trimmed off, so that such material is not visible on the cut surfaces of the meat.
>
> Lymphatic and nervous tissue that is removed must not be used in meat preparations or products that are intended for human consumption.[77]

643 We consider that the response to the Commission Decision was reasonable. It had, however, one consequence which we do not believe was appreciated. Because lymphoid tissue was not brought within the definition of SBO, it continued to be available for rendering for animal feed after the animal SBO ban was introduced.

Alarms and reassurances

644 We now turn to a quite different topic, one of great interest to our Inquiry – the communication of risk to the public. By 1990 BSE had been transmitted to a number of different species – for the most part experimentally. Transmission naturally, through feed, had occurred in a number of exotic species in zoos. The range of species in which transmission had occurred was wider than that observed with scrapie. These transmissions were, to put it neutrally, consistent with the possibility that BSE was transmissible to humans. Few put it neutrally, however. The media, focusing on the comments of some independent scientists, were quick to draw the conclusion that instances of cross-species transmission demonstrated that humans were at risk. Government officials were at pains to emphasise that experimental conditions were not reproduced in nature and that no implications as to human risks could be drawn from transmission to animals. Reassurances were given about the safety of beef. The Meat and Livestock Commission (MLC) regarded its principal role as the support of the meat and livestock industry. The MLC was particularly assiduous in seeking to counter the suggestion that it might be dangerous to eat beef. Regrettably this enthusiasm led on occasion to statements which were not scientifically correct.

645 In January, *The Independent* quoted scientists at the NPU acknowledging a 'remote possibility' that BSE might move from cows to people, and the comment from one of them that nothing would induce him to eat sweetbreads, spleen or brain. 'A human would have to eat an impossible amount of pure cow brain at the height of infection' to reach an equivalent dose to that needed to infect a cow, riposted Mr Colin Maclean, Technical Director of the MLC. He should have resisted this absurd exaggeration.

646 By this time Professor R M Barlow at the Royal Veterinary College had succeeded in effecting oral transmission of BSE to mice, and preliminary results of experiments at the CVL had demonstrated that inoculation of cattle with BSE-infected material had transmitted the disease. MAFF delayed making public the results of the mouse experiment until 1 February 1990 for presentational reasons. They considered it essential for the results of both sets of experiments to be announced at the same time. MAFF's press release received consideration by Mr Andrews and by Mr Gummer. It included this comment:

[77] YB90/6.14/3.3

> The BSE results therefore provide further evidence that BSE behaves like scrapie, a disease which has been in the sheep population for over two centuries without any evidence whatsoever of being a risk to human health.

Thus the first oral transmission of BSE to another species was presented as reassuring. Not everyone found it so. An official who visited the NPU in January reported:

> The researchers I spoke to are obviously very troubled about the ability of this disease to jump species. If it can be passed from cattle to mice, then what about humans?[78]

The press contrasted MAFF's statement with views expressed by Dr Helen Grant, Consultant Neuropathologist:

> My gut feeling is that some genetically susceptible people may have become infected with material by eating meat products.

647 From March 1990 the media began to give prominence to the views of Professor Richard Lacey, a Professor of Clinical Microbiology at Leeds University. *Today* reported him as predicting:

> In the years to come our hospitals will be filled with thousands of people going slowly and painfully mad before dying.

648 In April Humberside County Council banned beef from school meals. Other local authorities were to follow their example. Then came the cat.

The cat

649 On 6 May 1990 officials at MAFF and DH reported to their Ministers that Bristol University had diagnosed a 'scrapie-like' spongiform encephalopathy in a domestic cat. Here was a bombshell. The public was likely to conclude that the cat had caught BSE from eating contaminated beef. And if this could happen to a cat, why should not human beings suffer the same fate? Yet it was far too soon to jump to any such conclusion. It was possible that there had always been the occasional case of feline spongiform encephalopathy (FSE) which had gone unrecognised. Nonetheless, if a cat had caught BSE from food, it was cause for concern. CJD had been transmitted experimentally to a cat by inoculation, but attempts to transmit scrapie had not succeeded. Here was an indication that BSE might be more virulent than scrapie.

650 On 10 May Mr Gummer and Mr David Maclean, the Parliamentary Secretary, met with officials to discuss how to make public the news of the cat. A note of the meeting prepared by Mr Gummer's Principal Private Secretary recorded that Mr Meldrum 'confirmed the Minister's assumption that there was no likely connection between this case and BSE'. We have already noted (paragraph 363) that there was no basis for this degree of reassurance and Mr Meldrum should have been more cautious.

[78] YB90/1.9/3.1

651 Mr Meldrum found himself under pressure from the media to comment on the implications of the cat. He emphasised that this was the first known case of FSE and that there was no known connection with other animal encephalopathies, but that investigations into the case were continuing. The risk to humans was no greater than before the diagnosis; the cat was no cause for concern.

652 We think that Mr Meldrum played down the potential significance of the cat more than an objective appraisal would have justified. But he no doubt had in mind the part played by the media in previous 'food scares', such as salmonella in eggs and listeria, and was seeking to counter extreme statements about the implication of the cat which went much further than justified on what was then known. In the circumstances we do not think it would be fair to criticise him for his defensive public stance.

653 Intense media coverage followed. *The Sun* published an article stating that BSE could be the biggest threat to human health since the Black Death plague. British beef was reported to have been banned in Russia and in schools up and down the country. Professor Lacey called for the slaughter of every herd with a case of BSE.

654 Again the MLC leapt into the breach with too much vigour. Mr Colin Maclean was responsible for the text of a video to be distributed to local authorities which on one reading erroneously suggested that it would be necessary to eat an impossible amount of brain and spinal cord in order to be at risk. In a press release he stated that 'even if no further action had been taken following the outbreak of the disease there was considered to be no risk to consumers from eating beef'. We do not believe that Mr Maclean intended to mislead, but both these statements were capable of doing so. We think that he should have been more careful.

655 Of more importance were the official statements. MAFF issued two press releases on 15 May, for the terms of which Mr Gummer was himself responsible. These were directed to the safety of beef. Mr Gummer made unequivocal statements that it was safe to eat beef, but he made it plain that he did so on the basis that the slaughter and compensation policy and the SBO ban provided protection for the consumer against any remote risk which might otherwise exist. This qualification was vital and, in the light of it, we would not criticise these press releases.

656 The following day, BBC Newsnight featured television footage of Mr Gummer attempting to feed his four-year-old daughter Cordelia a beefburger. We understand that Mr Gummer had been challenged by a newspaper to demonstrate his confidence in beef in this way. Mr Gummer was faced with choosing between two unattractive alternatives. It may seem with hindsight that, caught in a 'no win' situation, he chose the wrong option, but it is not a matter for which he ought to be criticised.

657 Sir Donald Acheson was pressed by MAFF to add his reassurance that it was safe to eat beef. His press officer told him that, having regard to the media pressure, it was essential that he should make a statement. He managed to discuss the terms of his statement with three members of SEAC – Dr Tyrrell, Dr Will and Dr Kimberlin. He then issued the following press release on 16 May:

> I have taken advice from the leading scientific and medical experts in this field. I have checked with them again today. They have consistently advised me in the past that there is no scientific justification for not eating British beef and this continues to be their advice. I therefore have no hesitation in saying that beef can be eaten safely by everyone, both adults and children, including patients in hospital.

Later, in a television interview, he stated that 'there is no risk associated with eating British beef'.

658 Sir Donald told us that when he learned of the cat he 'remained deeply concerned about the possible implications of a further 'transpecies "jump" of BSE'. He told us that his statement about the safety of beef was made, as were Mr Gummer's, 'on the confident assumption that the SBO ban was already fully implemented'.

659 In contrast to the press statements made by Mr Gummer, Sir Donald's statement did not explain that his confidence in the safety of beef was premised on the removal of all SBO. It gave no indication of any concern about the cat. It was, we feel, a statement that was likely to convey the message not merely that 'beef is safe', but that 'BSE is no risk to human health'.

660 We do not consider that, as Chief Medical Officer, Sir Donald should have restricted his public statement in the way that he did. The development of a spongiform encephalopathy in a cat had raised a concern that BSE might be transmissible in a way that scrapie was not. Sir Donald was in no position to allay that concern. He avoided addressing it by limiting his statement to the safety of beef. He did not explain that he considered beef safe only because the parts of the cow that might be infective were being removed from the food chain. His statement was likely to give false reassurance about the possibility that BSE might be transmissible to humans and we think that he should have appreciated this. The possibility that BSE might have been transmitted to a cat was cause for concern and needed to be investigated by the scientists. He should have explained that he believed that beef was safe to eat because of the precautionary steps that had been taken to guard against the possibility that BSE might be transmissible in food.

661 Sir Donald's unqualified statement that it was safe to eat beef was to set a pattern. Public concerns about the dangers arising from BSE were met by statements limited to giving assurance that it was safe to eat beef. Members of the public tended to equate those statements with assurances that BSE posed no risk to humans. It was natural that they should do so. It is no wonder that when, on 20 March 1996, the Government announced that there was probably a link between BSE and vCJD, many felt that they had been deceived.

The Agriculture Committee

662 On 16 May 1990 the public concern generated by the cat led the Agriculture Committee of the House of Commons to institute an inquiry into BSE. Over a period of just over a month an impressive body of evidence, both oral and written, was received. The Committee reported on 18 July. The Committee observed that while scientists believed that there were too many unknowns to say anything about the disease with absolute certainty, no evidence had been forthcoming that it did pose a risk to human health. It concluded:

> The Government has already acted to cut off the presumed source of the disease in cattle and has banned the sale of all specified cattle offals for human consumption. We believe these measures should reassure people that eating beef is safe.
>
> If the ban on the sale of specified cattle offals for human consumption is properly policed in slaughterhouses, full public confidence can be maintained.

SEAC considers the safety of beef

663 At the request of Sir Donald Acheson, SEAC held an emergency meeting on 17 May 1990 to consider the implications of the cat. Sir Donald had hoped that SEAC would produce a letter endorsing the statement that he had made about the safety of beef. At their meeting the Committee members found themselves unable to agree on the terms of this. Not until 24 July were they able to give final agreement to the terms of a letter to the CMO and an accompanying annex dealing with the safety of beef.

664 There were unsatisfactory features both about the manner in which these documents were prepared and about the terms in which they set out SEAC's advice. The letter set out briefly the reasons for SEAC's conclusion that:

> In our judgement any risk as a result of eating beef or beef products is minute. Thus we believe that there is no scientific justification for not eating British beef and that it can be eaten by everyone.

The annex spelt out in greater detail the reasons for that conclusion.

665 The origin of the annex was a paper that Dr Pickles had prepared to brief the CMO before his appearance before the Agriculture Committee. She explained to him:

> The arguments are those that have or should have been discussed by the Tyrrell Committee [ie, SEAC].

666 It was subsequently adopted by SEAC as the basis for their advice to the CMO. The draft annex was, however, circulated widely by Dr Pickles and Mr Lowson within DH and MAFF, so that officials could suggest amendments to the draft. Mr Thomas Murray[79] of DH expressed concern that 'the Annex will give us considerable presentational problems and do little/nothing to reassure the public about the safety of British beef'. In MAFF it was forwarded to Mr Gummer and Mr Maclean for approval, but only after a process which had led Mr Lowson to note that 'the most inflammatory pieces of drafting in earlier versions have now been edited out'.

---

[79] Head of Section, Environmental Health and Food Safety Division

667 We were unhappy about this editorial process. It seemed to us that there might well be a conflict between officials' desire that the annex should not contain inflammatory matter and the desirability that the annex should fairly and objectively summarise SEAC's views on risk.

668 Dr Tyrrell accepted that, had there been time, it would have been preferable for the Committee to have formulated its own view, but defended what had occurred because SEAC was under time constraints. We do not believe that the editorial process resulted in any distortion of SEAC's views, but remain of the opinion that it would have been preferable if the Committee had been left to do its own editing of the draft annex.

669 We turn to the substance of SEAC's advice. The passages that gave us concern were those that dealt with dose. The question of the amount of infective material that might suffice to transmit the disease was of practical importance when considering the precautions that needed to be taken against transmission, whether to other animals or to humans. SEAC commented more than once that 'very large doses' were needed for oral transmission. The Committee members explained to us that they were speaking of the titre of infectivity, not the quantity of physical material that held the dose. Once this was explained, we could follow SEAC's reasoning. Nonetheless, we felt that the language that they had used tended to suggest that they were speaking of the amount of infective material. Here is an example:

> . . . the incubation period in mice was longer after large oral doses of BSE-infected cattle brain than after much smaller parenteral injections – in these, as in other animal experiments, large doses appear to be needed for successful disease transmission.

670 SEAC submitted to us that the letter and its annex were prepared for the CMO and would have been likely to circulate among readers who were familiar with the concept of dose. We accept that point and have concluded that it would not be right to criticise SEAC for the language used. We believe, however, that the annex was circulated within MAFF and fear that it may have given rise to misunderstanding. The evidence shows that in 1990, and indeed for some years thereafter, there was a perception on the part of many within government that a substantial quantity of infective material would be required orally to transmit BSE to a cow and that the same would be true of transmission from cow to human, if indeed such transmission was possible. It is at least possible that SEAC's annex contributed to this belief.

A look ahead

671 In the period up to 1990, MAFF had taken the lead in addressing the possibility that BSE posed a risk to the safety of human food. Although Dr Metters and Dr Pickles had played a diligent role, albeit a secondary one, in considering which tissues should be included in the human SBO ban, they had done so in the belief that the ban was not scientifically justified.

672 The attitude of Dr Metters at this time was demonstrated by a response that he sent in October 1990 in answer to a suggestion by Mr Murray that DH should ensure that a continuous flow of appropriate BSE information should be sent to Directors of Public Health, Consultants in Communicable Disease Control and Environmental Health Officers. Dr Metters wrote that he was concerned that such activity might raise the implication that:

> . . . somehow the disease poses a risk to human health. Every effort has thus far been made to underline the Government's position, based on advice from the Southwood and Tyrrell Committees that the disease is not a risk to humans. That principle lies behind this Department's low-key approach to publicity.

Dr Metters should not have given this response, which seems to us to convey quite the wrong message.

673 In the years ahead DH continued to play a subordinate role in addressing the food risks relating to BSE – so much so that, in the final days before 20 March 1996, it did not occur to Mr Hogg and Mrs Browning that Health Ministers should even be consulted about appropriate measures to enhance the protection of human health.

674 The first case of FSE was not merely of concern to the general public. It was of concern to SEAC. The Committee was unable to draw conclusions without knowing whether the cat had contracted the disease from BSE. It advised that there was an urgent need for research. In due course, as the number of cases of FSE grew, it became accepted that they had probably caught the disease from eating bovine offal infected with BSE. Mr Meldrum commented in evidence to us that no specific observations or recommendations were ever made on the effect of FSE on the risk to humans. In this he is correct. We had evidence from a number of scientists that transmission of BSE to cats was an event which altered their belief that BSE posed no greater risk to humans than scrapie. The public were never told that scientists' appraisal of that risk had changed. On each occasion that public concerns were raised about BSE, they were met with the same refrains – 'There is no evidence that BSE is transmissible to humans'; 'It is safe to eat beef'. Risk communication in relation to BSE was flawed.

The false peace – 1 January 1991 to 31 March 1995

675 In this section we take the story on to 1 April 1995, when the national Meat Hygiene Service (MHS) took over the enforcement of slaughterhouse Regulations from the local authorities. This was a watershed event in the BSE story. It led to discovery of the scale of the inadequacies of the implementation and enforcement of the animal SBO ban. This we have described in Chapter 5. It led to the discovery of shortcomings in the clean removal from the carcass of all spinal cord. This we shall consider in the next section. This section covers a period of relative inactivity in the BSE story.[80]

676 We shall begin with a short description of the hygiene standards in slaughterhouses that led to the setting up of the MHS. We shall also describe shortcomings in the regulatory structure which the MHS inherited. These are of

---

[80] Changes in the MAFF and DH teams during this period included the following: Mrs Gillian Shephard succeeded Mr Gummer as Minister of Agriculture, Fisheries and Food on 27 March 1993 and she, in her turn, was succeded by Mr William Waldegrave on 20 July 1994. Mr Richard Packer succeeded Sir Derek Andrews as Permanent Secretary at MAFF on 17 February 1993. In DH Mr Waldegrave was succeeded as Secretary of State in 1992 by Mrs Virginia Bottomley, and Dr Kenneth Calman took over from Sir Donald Acheson as CMO in September 1991

relevance in helping to understand why there were failures in implementing and enforcing the obligations to remove spinal cord. They also explain the much more serious inadequacies in the handling of SBO once it had been removed, which we have looked at earlier in this volume. We shall, in addition, describe briefly the political process which led to the setting up of the MHS.

677 Next we shall look at the evidence relating to monitoring of the human SBO Regulations up to April 1995, and at some further consideration that was given to MRM. We shall note an important amendment to those Regulations.

678 During this period knowledge about BSE advanced as results began to be received from the research projects that had been undertaken. We shall consider the extent to which this knowledge was communicated to the public. Events which caused concern to the public, and to government, were the incidence of two cases of CJD in dairy farmers and the first case of a teenager to suffer from this disease. We shall look at the media reaction to these events and the official response.

Slaughterhouse standards

679 In an era of deregulation, a convincing case had to be made out for the introduction of the centralised MHS. Standards of hygiene in British slaughterhouses provided that case. Mr Gummer gave this vignette to the House of Commons Agriculture Committee in October 1992:

> 'Slaughter hall floor heavily soiled with blood, gut contents and other debris – no attempt to clean up between carcasses. Car cleaning brush heavily contaminated with blood and fat being used to wash carcases. Knives and utensils not being sterilised. Offal rack and carcase rails encrusted with dirt. Missing window panes in roof – birds, flies and vermin entering'. Another slaughterhouse report: 'Filthy equipment and surfaces – congealed and dry blood on offal racks. Effluent discharging across floor under dressed carcasses – risk of contamination. Slaughterman at cattle sticking point not sterilising knife. No sterilisers to wash basins in pig slaughter hall. No fly screening on open windows'.

680 The previous year Mr Gummer had reported to the Prime Minister that 60 per cent of red meat slaughterhouses did not meet European standards. Many plants recorded as satisfactory were only just acceptable. On the introduction of the Single European Market on 1 January 1993, 544 British slaughterhouses sought a temporary derogation from compliance with European hygiene requirements. When EU Veterinary Inspectors carried out surveillance of these establishments in 1994, they found that 68.5 per cent were of concern or of grave concern.

681 MAFF officials initially had little knowledge of how local authorities set about complying with their obligations to enforce Regulations in slaughterhouses. In 1992 Mr Lawrence was appointed to lead an MHS Project Team to investigate this. He discovered an unsatisfactory state of affairs. There were instances of animosity between plant management and Inspectors, and between Official Veterinary Surgeons who oversaw enforcement, usually under contract, and the Inspectors and EHOs on the staff of the Environmental Health Departments of the local authorities. In many cases there was an unclear management chain and lack of teamwork.

682 In January 1992, Mrs Jane Brown, Head of Meat Hygiene Division, forwarded a paper to the Cabinet Office as a basis for discussion by officials of the proposal to create a national Meat Hygiene Service. This recorded:

> The State Veterinary Service, who monitor standards, have no real control over LAs. The Official Veterinary Surgeon . . . has little real management control over the meat inspectors in the plant . . . standards of enforcement are uneven across the country.

683 A review in 1992–93 of hygiene standards in a sample of slaughterhouses confirmed this picture and commented: 'In many cases, the Local Authority appeared disinterested.' Many witnesses gave evidence to us to similar effect.

684 We asked MAFF officials whether evidence of poor hygiene standards in slaughterhouses did not raise concerns about the standard of enforcement of the duty to remove spinal cord from the carcass. Each replied that it did not. Some commented that they had imagined that this was a simple operation. Others said that removal of unfit meat from the carcass was so important that they believed Meat Inspectors gave priority to strict enforcement of that obligation.

685 We were at first inclined to believe that poor standards of general hygiene would inevitably go hand in hand with poor standards of compliance with the SBO Regulations. So far as concerned the formalities of disposal of SBO once it had been removed from the carcass, we were proved right. Standards of removal of spinal cord do not, however, appear to have reflected the poor standards prevailing elsewhere in the slaughterhouse. After the MHS took over, inspections disclosed that failure to remove all spinal cord before meat was health-stamped had probably been occurring on average in four cases out of a thousand. Although this level of failure was not satisfactory, it suggests that in general the operation of removing the spinal cord was carried out efficiently and effectively. The occasional failure to remove all the spinal cord had been described in MAFF's paper to SEAC in 1990 as inevitable. Under the structure in place before the MHS took over we believe that it was. After the MHS was in place, by adding resources and monitoring a campaign aimed at ensuring 100 per cent removal of spinal cord, MAFF and the MHS appear to have come close to achieving this goal.

History of the setting up of the Meat Hygiene Service

686 In July 1991 Mr Gummer wrote to Mr Waldegrave, who was at that time Secretary of State for Health, to propose the setting up of what was to become the MHS. Mr Waldegrave replied that he was 'content' with the proposal. In November the proposal was placed before the Prime Minister, who wished to know the reaction of the Treasury. Mr Mellor, Chief Secretary to the Treasury, at first had reservations, but those were dispelled and Mr Major announced on 9 March 1992 that a new Meat Hygiene Service was to be set up.

687 The decision proved controversial. When the Conservative Party was returned to office after the General Election with a greatly reduced majority, there was backbench opposition from its own MPs to the need for additional hygiene measures. Many, including the meat industry, major retailers and some journalists, considered that MAFF was going too far in pandering to what they saw as European over-regulation.

688 When Mrs Shephard succeeded Mr Gummer, she took a fresh look at the proposal for the MHS. Although she had initial misgivings, she was persuaded by her officials that it was an essential measure. She ran into opposition, however, from Mr John Redwood, who had been appointed Secretary of State for Wales. In October 1993 Mr Michael Portillo, who had been appointed Chief Secretary to the Treasury, also suggested that she should look again at the proposal. Mrs Shephard stood firm, supported by Mr Ian Lang, Secretary of State for Scotland. The following month Mr Redwood and Mr Portillo indicated their acceptance of the project.

689 In 1994 the work of establishing the MHS proceeded. Mr Johnston McNeill was appointed Chief Executive. The new Agency was to inherit the staff in the case of 176 of the local authorities; their existing terms and conditions differed and had to be renegotiated in each instance. In July 1994 Mr Waldegrave succeeded Mrs Shephard as Minister of Agriculture. Once again he satisfied himself of the merits of the scheme. The MHS replaced the local authorities on 1 April 1995.

690 The establishment of the MHS was not a measure taken in response to the emergence of BSE. Accordingly it has not fallen within our terms of reference to consider why so long elapsed between the decision to introduce the Service and the implementation of that decision. The establishment of the MHS had a beneficial impact on the implementation of both the human and the animal SBO ban. It is unfortunate that this was so long delayed.

Monitoring compliance with the SBO Regulations

691 In Chapter 5 we saw how monitoring of the SBO Regulations in slaughterhouses was intensified between 1991 and 1995. This was, however, in response to concerns about the animal SBO ban. The instructions received by the Veterinary Field Service (VFS) required it to concentrate on the handling of SBO after removal from the carcass. The focus of attention was the gut room, not the 'clean' side of the slaughterhouse. The only specific question on the SVS pro forma covering slaughterhouse visits that related to human health asked whether removal of bovine brains involved contamination risk. There was no mention of spinal cord.

692 Records of slaughterhouse visits have been lost for large parts of the period between 1991 and 1995. In 1990 there had been one report of a failure to remove spinal cord from the carcass. That is the only such report of which we are aware. Apart from a few early reports about brain removal, there was nothing to suggest that slaughterhouse operations involved any risk to human health.

693 We have already discussed why it was that the VFS did not discover the deficiencies in compliance with the Regulations in the gut room until after the MHS had taken over. The same reasons apply in relation to the removal of spinal cord. We believe that the principal reason was the difference in rigour of the inspections before and after the MHS took over.

694 Mr Christopher Clarke, who had served as a Meat Hygiene Inspector, told us that it was typical for MAFF Veterinary Officers on their periodic inspections to arrive mid-morning and depart a few hours later, after discussion with the management of the plant and the principal Environmental Health Officer. Such a visit was unlikely to detect the occasional failure to remove a segment of spinal cord, particularly if the focus of the visit was what was taking place in the gut room.

695 It may well be that there was, on occasion, a lack of diligence on the part of the Veterinary Officer making the monitoring visit. It was regrettable that the need to give specific instructions to monitor the removal of spinal cord was not identified when the Regulations were being introduced and particularly unfortunate that, when SEAC was asked to look at slaughterhouse practices, its response was understood to signify that these were not cause for concern. We have no criticism to make of Mr Hutchins, Mr Simmons or their superiors in relation to this aspect of the monitoring duties of the SVS.

MRM on the agenda again

696 On 8 April 1994 Mr Meldrum called a meeting of MAFF officials to review arrangements for disposal of SBO. Although the primary concern seems to have been enforcement of the animal SBO ban, Mr Meldrum suggested that 'one way to increase security would be to prohibit the use of spinal column for MRM'. Impetus was given to this suggestion when, in July, the European Commission's Scientific Veterinary Committee recommended that vertebrae from cattle killed in the UK should no longer be used for the production of MRM. This recommendation was not pursued, but MAFF prepared a paper on MRM for SEAC to consider at its meeting on 30 August 1994. The Committee was asked to advise on the use of spinal column for the production of MRM. Not for the first time SEAC had a heavy agenda, and this item was deferred, to be restored in June the following year.

The distal ileum of calves

697 One experiment carried out by the CVL[81] involved feeding calves with BSE-infected brain and then slaughtering an animal every four months (after the first two months had passed) and testing 44 tissues for infectivity by injecting them into the brains of susceptible mice. In June 1994 a positive result was obtained from the distal ileum (small intestine) of a calf slaughtered only six months into the experiment. This was an event of some significance. Hitherto only brain and spinal cord of BSE victims had been found to be infective. Furthermore, tissues from calves of less than six months of age had been excluded from the SBO ban. MAFF Ministers and officials were informed of the result and Mrs Bottomley, the Secretary of State for Health, was informed the same day.

698 It was agreed between the two Departments that SEAC's advice should be obtained before this experimental result was made public. An 'exceptional meeting' was called on 25 June 1994. SEAC expressed the view that any risk to humans from food derived from calves was minuscule, but added that it was not possible to give a definitive answer:

---

[81] The pathogenesis experiment

> There is a theoretical risk and Government could respond by a limited SBO ban for calves to exclude the intestines.

699 Over the weekend Mr Meldrum and MAFF officials held lengthy meetings with Dr Calman, the CMO. Dr Calman said that he would be advising Ministers that the distal ileum and thymus of calves should be proscribed as SBO. Those present agreed with his conclusion. Officials met with MAFF Ministers the next day. The point was made that the proposed ban would have a serious effect on the export of calves and have a knock-on effect on the price of beef. Mrs Shephard responded that where public health was concerned, trade was the least important consideration. She later met with Dr Calman to discuss the terms of the ban.

700 MAFF at once sent letters to operators of all slaughterhouses, telling them of the proposed extension of the SBO ban and asking them to give effect to it on a voluntary basis, pending amendment of the Regulations.

701 How the news of the experiment result and the action to be taken should be made public was the subject of discussion in the Cabinet. A draft press release prepared by the CMO was considered. It included a statement that the risk to human health was considered to be 'minuscule'. In discussion it was suggested that this should be deleted, so that the statement would indicate that there was no risk at all. Mr Major, in summing up, said that Mrs Shephard should proceed with the announcement as planned.

702 A lengthy press release was issued on 30 June, accurately describing the course of events, and setting out SEAC's advice in full.

703 This decision was a model of how government ought to handle such an issue.

- SEAC's advice was sought as to the implications of the finding in the pathogenesis experiment.
- SEAC limited its advice to the effect this result had on the question of risk of transmission to humans and did not recommend the appropriate policy decision.
- MAFF and DH worked closely together in considering the appropriate response.
- The issue was discussed with Mrs Shephard at a meeting at which the CMO expressed his advice in favour of an extension to the SBO ban.
- The effect that such an extension would have on trade was considered.
- The Minister and the Parliamentary Secretary were in agreement that 'protecting the public health was the first of MAFF's aims'. The CMO's advice would be followed, notwithstanding the potential for serious impact on trade.
- The practical implications were considered.
- The results of the experiment and the Government's response were announced without delay.

PROTECTING HUMAN HEALTH

- There was swift consultation and prompt action. Slaughterhouses, local authorities and bodies consulted were individually informed of the extension of the Regulations.

Advances in knowledge of BSE

704 Between 1991 and 1995 a lot more was learned about BSE. Advances in knowledge up to about September 1994 were summarised in a Report produced by SEAC in September 1994 and published in February the following year.[82] The following we find particularly significant:

- By September 1994, 57 cats had been confirmed as having contracted FSE, presumptively from feed containing the BSE agent.

- The following animals had contracted spongiform encephalopathies (SEs), in most cases presumptively from feed containing the BSE agent:

    – Nyala

    – Gemsbok

    – Arabian oryx

    – Greater kudu

    – Eland

    – Moufflon

    – Puma

    – Cheetah

    – Scimitar-horned oryx.

705 Strain-typing showed that, in contrast to scrapie, which had a number of different strains, cases of BSE from different parts of the United Kingdom and in different years were indistinguishable from each other but distinct from all previously studied laboratory strains of scrapie.

706 In addition to the natural transmissions set out above, on 14 February 1992 BSE was found to have been successfully transmitted to a marmoset by cerebral inoculation. This was the first transmission to a primate. A meeting of SEAC was immediately called to consider the implications of this. SEAC concluded that as marmosets had in the past been infected with SEs, including scrapie, using similar methods, the results were not surprising and had *no implications for the safeguards already in place for human and animal health*.

707 We have emphasised those last words, for they were significant. SEAC's 'public advices' on risk tended to focus on the question of whether the

[82] *Transmissible Spongiform Encephalopathies: A Summary of Present Knowledge and Research*

precautionary safeguards in place were adequate to protect the public. They did not comment on the effect that events had on the assessment of the risk that BSE might be transmissible to humans. Thus the impression was given that that risk never changed. There is no better illustration of this than the following passage of oral evidence given to us by Mr Gummer:

> ... during the period of time in which I was Minister and my junior Ministers were with me, that science was tested all the time, but it did not change. The advice was and continued to be that the risk to human beings was remote ...[83]

708 To the casual reader of SEAC's 1994 Report, nothing had changed. Thus, under the heading of risk assessment, SEAC wrote:

> Our conclusion therefore is that, as the Southwood Working Party determined, taking all the available evidence together, the risk to man from BSE is remote.

709 The careful reader, however, might have noted this passage which followed:

> In conclusion, therefore, our scientific assessment is that the risk to man and other species from BSE is remote because the control measures now in place are adequate to eliminate or reduce any risk to a negligible level. We do however point out that any species exposed already and before any bans were effective could be incubating disease, and therefore continuous monitoring is very important until any possible incubation period has been exceeded.

710 SEAC only evaluated the risk as still remote because precautionary measures, and in particular the human SBO ban, had been put in place. The Southwood Working Party, however, had not taken that view – at least in relation to human food, where they considered the risk remote even without an SBO ban.

711 The advances in knowledge by September 1994 significantly altered the scientific evaluation of the risk that BSE might be transmissible to humans. Professor John Collinge[84] told us:

> Certainly the appearance in domestic and captive wild cats was a very important development. It demonstrated that you could no longer really plausibly argue that BSE was just scrapie in cows with all the same properties. This agent, wherever it had originated from, had quite different biological properties to scrapie as manifested by the extended host range of affected species, including things like nyala and kudu as well as the cats that had not been affected by scrapie before, so far as we were aware.

712 Dr Tyrrell confirmed that the transmission of BSE to cats and wild cats had shifted his perception of the risk of transmissibility 'a bit'. Dr Kimberlin said that his reaction to the cat was:

---

[83] T94 pp. 75–6
[84] Professor of Molecular Neurogenetics at St Mary's Hospital, London; a member of SEAC since December 1995

> Thank God we have got the SBO ban because if it should so happen that the species barrier between cattle and humans is no higher than between cattle and cats . . . then we would have a problem.

713 We do not criticise SEAC for what was a detailed and careful analysis of the existing data. Nonetheless we think it a pity that its Report did not spell out more clearly and simply the fact that perception of risk had changed since Southwood. Had the Committee done so, its Report might have attracted some attention and resulted in the public being better informed about risk. As it was, the Report appears to have attracted no press coverage.

Knowledge about dose

714 One important experimental result did not receive comment in SEAC's 1994 Report. The NPU had succeeded in transmitting BSE to sheep using an oral dose of no more than ½ gram of BSE-infected brain. What is more, the sheep infected were of a breed not susceptible to scrapie. The interim result of this experiment was known in November 1990 and published in the *Veterinary Record* in October 1993. The significance of this experiment seems to have been totally overlooked by MAFF officials, and indeed by SEAC. We have not been able to discover why this was.

715 The CVL had, in January 1992, initiated an 'attack rate' experiment under which they had fed different quantities of BSE brain to cattle. The smallest quantity was 1 gram and, in September 1994, MAFF officials learned that this had transmitted the disease. There was general surprise and concern that such a small quantity had proved infective. This result demonstrated the importance of avoiding:

- contamination of MBM designed for animal feed with SBO in the course of rendering; and
- contamination of cattle feed with pig and poultry feed containing that contaminated MBM in the feedmills.

716 Had the significance of the NPU experiment been drawn to the attention of MAFF officials in November 1990, the extent of the danger of cross-contamination might have been appreciated four years earlier.

Two dairy farmers die from CJD

717 In May 1990, in accordance with a recommendation of the Southwood Working Party, the CJD Surveillance Unit (CJDSU) had been set up under Dr Robert Will.[85] Its main objective was to identify any change in the epidemiological characteristics of CJD cases and to assess the extent to which they were linked to the occurrence of BSE. The CJDSU summarised its progress and findings in a series of annual reports, and Dr Will submitted articles about these to *The Lancet*. Dr Will was a member of SEAC, and findings of the CJDSU were reported to SEAC when they met.

---

[85] Consultant Neurologist at the Department of Clinical Neurosciences, Western General Hospital, Edinburgh. Volume 8 gives a fuller description of the establishment and work of the CJDSU

FINDINGS AND CONCLUSIONS

718 There was a more immediate link with DH through Dr Ailsa Wight who, in September 1991, took over from Dr Pickles the responsibility for provision within DH of medical advice in relation to BSE and CJD and was DH's observer on SEAC. Thus DH and, through DH, MAFF usually received confidential information about victims of CJD well before news of them became public. There was ample time to decide upon the appropriate official response to such news.

719 On 6 March 1993 *The Lancet* published an article by Dr Will on the first recorded case of CJD in a dairy farmer. He had died the previous October. He had had BSE in his herd. The article concluded that the case was most likely to have been a chance finding and that 'a causal link with BSE is at most conjectural'. The media naturally developed the conjecture that there might be a link between this case and BSE. Professor Lacey did not think that there was. Interviewed on the radio, he gave his opinion that the case had occurred too soon to have been contracted from BSE.

720 The media interest led Mr Gummer to discuss a press release with Dr Calman, who agreed that it was necessary to reassure the public. On 11 March the CMO issued a public statement. This repeated the assurance about the safety of beef given by his predecessor, Sir Donald Acheson, in 1990 that we have criticised above.[86]

721 We found it open to precisely the same criticism. Dr Calman was seeking to address fears that a farmer had somehow caught BSE from his cattle. Responding to such fears by emphasising that it was safe to eat beef naturally carried the inference that transmission of the disease from cow to human was impossible. That Dr Calman's statement was in fact misinterpreted in this way is demonstrated by *The Mirror*'s report that:

> Chief Medical Officer Dr Kenneth Calman had insisted that BSE could not cause a related brain disease in humans.

722 Dr Calman should have been careful not to make a statement in terms that suggested such a belief, for he considered that there was a real potential for BSE to move from cows to humans.

723 On 23 March Mr Lowson commented in a minute for Mr Gummer's attention:

> It was not easy to get the CMO to make a statement in response to recent press speculation about a possible link between BSE and human disease.

724 The reason why MAFF wished the CMO to make a statement was, no doubt, because of the damage that public concern about BSE might cause to the beef industry. The evidence suggests that Dr Calman had reservations about complying with MAFF's request for assistance. Having decided to comply with that request and make a public statement, he should have taken great care to ensure that his statement fairly reflected his appraisal of the risk posed by BSE.

725 On 12 August 1993 *The Daily Mail* recorded the death from CJD, earlier in the month, of a second dairy farmer, who had had BSE in his herd. The CJDSU had been monitoring this case, and had concluded that there was nothing to suggest that it was other than a case of sporadic CJD. A DH spokesman was quoted by

[86] See para. 657 and following

The Daily Mail as saying that two cases might occur in dairy farmers by chance and that it was not possible to reach any conclusions about a link between BSE and CJD.

Vicky Rimmer

726 Vicky Rimmer fell ill early in the summer of 1993 at the age of 15. She had a neurodegenerative disease which the medical specialists were unable to identify. In mid-September she went blind and fell into a coma. She remained in a coma until she died on 21 November 1997, over four years later. The CJDSU now attributes her death to CJD, but her illness did not have the characteristics of the cases now classified as vCJD. In January 1994 the CJDSU was unsure whether her illness was CJD.

727 It was in January 1994 that the press first started to write about Vicky Rimmer, quoting her grandmother's belief that Vicky had been infected as a result of eating beef infected with 'mad cow disease'. Dr Stephen Dealler and Professor Lacey were reported to have concluded that this was the first case of BSE infecting a member of the human race through food.

728 In response to intense media coverage, Dr Calman released a statement on 26 January. This stated that:

- no one knew what illness the patient was suffering from; and
- on the basis of the work done so far, there was no evidence whatever that BSE caused CJD and, similarly, not the slightest evidence that eating beef or hamburgers caused CJD.

729 We consider that it was reasonable for Dr Calman to make a public statement to counter media reports which suggested that the link between Vicky Rimmer's disease and eating beefburgers was established. The terms in which he did so were somewhat more emphatic than was desirable, but not to the extent that it would be right to criticise him for his choice of language.

730 Dr Dealler's and Professor Lacey's conclusion that Vicky Rimmer had caught BSE through food was speculative. In the next chapter we shall see the first of the cases that have been identified by the CJDSU as cases of vCJD linked to BSE.

Chinks in the armour – April–December 1995

731 In this section we shall consider, from the viewpoint of public health, the revelations that followed the takeover by the MHS of enforcement of Regulations in slaughterhouses. We shall consider how government responded to what was discovered. We shall look at growing concerns caused by further cases of CJD in farmers and in young people and we shall look at official statements and media comment in relation to the risk posed by BSE to humans. We shall cover the period up to the end of the year.[87]

---

[87] Ministerial changes in MAFF and DH during this period included the following: Mr Douglas Hogg succeeded Mr William Waldegrave as Minister of Agriculture, Fisheries and Food on 5 July 1995. Mr Stephen Dorrell succeeded Mrs Bottomley as Secretary of State for Health in July 1995

FINDINGS AND CONCLUSIONS

732 The MHS took over on 1 April 1995 with Mr Johnston McNeill as Chief Executive and Mr Philip Corrigan as Head of Operations. Mr Corrigan was succeeded in August 1995 by Mr Peter Soul. The MHS commissioned a survey of standards at slaughterhouses from Eville & Jones, a firm of private veterinarians which provided Official Veterinary Surgeon (OVS) inspection services. Deficiencies summarised in its report which existed at the time of takeover on 1 April 1995, included widespread lack of awareness of SBO legislative requirements and instances of incomplete removal of spinal cord. The report noted significant improvements over the five months between April and August 1995. When Dr Cawthorne at MAFF learned of this report, he asked himself why these deficiencies had not been drawn to the attention of the SVS or the Meat Hygiene Division. We think that the explanation must have been poor OVS/local authority/VFS liaison.

733 The MHS also organised an internal survey of slaughterhouse standards by its own Hygiene Advice Teams. These teams encountered occasional failures fully to remove tonsils, thymus and spinal cord, but felt able to report that SBO removal in the slaughterhall was carried out in accordance with the legislation.

734 VFS staff were instructed to visit slaughterhouses once every two months and carry out a thorough inspection in company with MHS staff. They were instructed to examine, in particular, methods used to separate SBO from material intended for human consumption as well as staining and disposal of SBO. As we have seen, when looking at animal health, inadequacies in the handling of SBO led to the institution of a period of national surveillance.

735 In May 1995 Mr Meldrum gave instructions that Meat Hygiene Inspectors (MHIs) should be told to take particular note of the operation of removing the spinal cord from the vertebrae. This led to an Information Note being circulated to all MHIs and OVSs instructing them to ensure the complete removal of spinal cord from the vertebral column. In July the question was raised as to whether a Meat Hygiene Inspector could refuse to apply the health stamp on the ground that not all spinal cord had been removed. MAFF lawyers replied in the affirmative. We think it significant that this should be in doubt over five years after the SBO Regulations were introduced.

736 The July report on the results of the first round of national surveillance found widespread deficiencies in the handling of SBO, but made no mention of deficiencies in removing SBO from the carcass. In a submission to Mr Hogg, Mr Packer noted that the implications of the failures in the controls were for animal health, not for human health. Mr Meldrum confirmed that there was no public health problem because there was no question of SBO entering the human food chain.

737 By the time of the second round of national surveillance, the importance of ensuring the complete removal of spinal cord had been specifically drawn to the attention of the VFS in accordance with Mr Meldrum's instructions. On the second round of inspection, three instances were discovered of failure to remove SBO from the carcass. When this was reported to Mrs Browning, the Parliamentary Secretary, and to Mr Hogg, both were perturbed. Mr Richard Carden[88] suggested that enforcement should be tightened up and prosecutions launched where companies repeatedly infringed the Regulations. Mr Hogg agreed that this should be done.

[88] The Grade 2 head of MAFF's Food Safety Directorate

738 The surveillance results were reported to DH. Mr Meldrum assured Dr Metters that specific and detailed instructions had since been issued by the MHS to their staff on the checks necessary to ensure compliance with the legislation. Dr Calman received copies of this correspondence and resolved to look carefully at the next round of surveillance in order to see whether or not the deficiencies that had been discovered were isolated incidents.

739 On 23 October Mr Meldrum wrote to Dr Calman informing him that SVS staff had found a further four cases of health-stamped carcasses with portions of spinal cord attached. He described these results as 'disappointing', but added:

> It is inevitable that instances of the type referred to will continue to be reported albeit at low frequencies since no system operated by humans can deliver at 100 per cent efficiency all the time.

740 Two days later Dr Calman met Mr Packer to 'express disquiet about the position on BSE'. Dr Calman said that he 'could not be so unequivocal as he had been in the past' about the safety of beef. In a confidential file note he recorded:

> The issue remains, however, that the uncertainty has increased, rather than decreased. Urgent action is required to reassure the public that all steps are, and have been, taken to minimise any possible risk.[89]

741 When Mr Hogg learned of Dr Calman's concerns, he called a council of war of his junior Ministers and senior officials. We have already recorded, when looking at animal health, Mr Packer's advice that Mr Hogg should read the riot act to the MHS and the slaughterhouse industry. In the formal instructions that Mr Hogg proceeded to issue to Mr McNeill, he instructed him that his staff:

> . . . must ensure that all SBO is removed from a carcass before they give it a health stamp. Failure to do so should be viewed extremely seriously.

742 This led the MHS management to introduce what one union officer described to us as a 'disciplinary purge'. Immediate and emphatic instructions were issued to the workforce that failure to ensure that all spinal cord was removed would be treated as a serious disciplinary offence. Mr Hogg for his part met with representatives of slaughterhouse operators and told them robustly that he would only be satisfied with 100 per cent compliance with the rules and that those who did not provide this would be prosecuted.

743 On 1 November Mr Don Curry, the chairman of the MLC, wrote a strong letter to Mr Hogg expressing concern at breaches in the integrity of the SBO system, in particular those leading to the four cases in which spinal cord had been found in carcasses that had been passed as fit by meat inspectors for consumption. He wrote:

> We detect an attitude in the industry which says, 'you have told us this disease was not a threat to humans so why do we need all these controls?'. The danger that such an attitude engenders to our market, both at home and overseas, is very worrying indeed.

[89] YB95/10.25/16.1–16.2

744 This was one of a number of occasions in and after 1994 that the MLC commendably urged the importance of compliance with the SBO Regulations both on MAFF and on the industry. We would remark, however, that the attitude of which Mr Curry complained may well have been encouraged by some of the exaggerated reassurances that had been given earlier by the MLC.

745 On 7 November Dr Calman and Dr Metters met Mr Hogg, Mrs Browning and Mr Packer. Dr Calman did not mince his words. He said he found the attitude of the farming industry and the slaughterhouses astonishing. While there was no evidence that meat was not safe, it could not be said with confidence that no contaminated offal had entered the food chain. If pressed on the safety of food containing MRM, he would be in a difficult position.

746 On 20 November 1995 MRM was discussed at a meeting between Dr Calman, Mr Meldrum and other officials from both MAFF and DH. Dr Calman suggested that it was impossible to be 100 per cent certain that spinal cord was not being included in MRM derived from spinal column. Mr Meldrum confirmed that this was the position. It was agreed that SEAC should once again be invited to consider MRM.

747 On this occasion it was DH that had played the lead role in pursuing an issue arising from BSE in respect of the safety of food. Dr (now Sir Kenneth) Calman is to be commended for the vigour of his reaction on learning that segments of spinal cord were escaping the attention of slaughterhouse operatives and meat inspectors. By pursuing this matter with Mr Hogg, and subsequently with Mr Meldrum and other MAFF officials, he was instrumental in ensuring that the question of MRM was brought back before SEAC.

Action at last on MRM

748 We saw that a paper on MRM was placed before SEAC in August 1994 and deferred. A revised paper was prepared for its meeting on 21 June 1995. This annexed MAFF's paper on slaughterhouse practices that had been before SEAC in 1990 and the EU Scientific Veterinary Committee's recommendation that spinal column of cattle slaughtered in the UK should not be used for MRM.

749 The paper informed SEAC that the transfer of responsibility of meat inspection to the MHS:

> ...should ensure that no carcass is permitted to leave the slaughterhouse for human consumption unless the spinal cord has been completely removed.

750 The paper recommended that:

> In the light of the changes which are to be made to the controls on SBO and the methods of enforcing these controls... SEAC is recommended to advise that the use of spinal columns from cattle born and slaughtered in the UK for the mechanical recovery of meat may continue.

751 SEAC duly concluded that:

> ... provided in the slaughtering process the removal of spinal cord was done properly, the MRM process was safe and there was no reason for the Committee to change its advice.[90]

752 Just as in 1990, SEAC's advice was premised on the total removal of spinal cord.

753 When SEAC met on 28 November, it had a new chairman. Professor (later Sir) John Pattison, who had been a member of the Committee since January 1995, had replaced Dr Tyrrell. SEAC was informed that there had been 14 instances, involving at least 25 carcasses, in which SBO had been left attached to carcasses after dressing. The Committee was told of the steps that had been taken to tighten up enforcement of the Regulations. After protracted debate, SEAC decided that until it was clear that removal of spinal cord was being undertaken properly in all cases it would be prudent, as a precaution, to suspend the use of vertebrae from cattle aged over six months in the production of MRM.

754 SEAC's advice was accepted by both Mr Hogg and Mr Dorrell. Despite considerable resistance from the industry, the Order[91] banning the use of bovine vertebral column for the recovery of meat by mechanical means was made on 14 December 1995 and came into force the following day. For practical reasons, no exception was made in respect of calves aged less than six months.

755 The minutes of SEAC's meeting suggest that the decision was a close-run thing, with arguments from Dr Will and Professor Pattison winning the day. Would the decision have been the same, if the Committee had not known about the result of the attack rate experiment and had been unaware of concerns raised by incidents of CJD in farmers and young people? Would SEAC in 1990 have taken the same decision, if aware then of the extent of the failures to remove spinal cord identified in 1995? We do not believe that a confident answer can be given to either question.

756 As to preventing fragments of spinal cord getting into human food, SEAC's decision was to a large extent a case of shutting the stable door. Measures were in hand to ensure effective implementation of the duty to remove all spinal cord from the carcass. The more significant benefit of the new Order was that it kept dorsal root ganglia out of human food. The benefit was not appreciated at the time. The pathogenesis experiment had not yet shown these to be infective – the positive result was to come later.

757 Has MRM infected humans with BSE in the years up to 1995, and if so on what scale? It is too early to attempt to answer this question. What is, we think, now clear is that this was the route by which infectious material was most likely to end up in human food during that period.

Cause for concern

758 In the second half of 1995, the public learned of the death from CJD of a third, and then a fourth, dairy farmer. The third had died in December 1994. There had

[90] YB95/6.21/2.6
[91] The Specified Bovine Offal (Amendment) Order 1995

been two cases of BSE on the farm where he worked. SEAC held a special meeting to consider this case on 13 January 1995. They concluded that the occurrence of three cases of CJD in dairy farmers with BSE in their herds was worrying, but that more information was needed before any conclusions could be drawn. The death of this farmer was reported in the national press on 29 September. On that day the CMO learned of a suspected fourth case.

759 Again SEAC met in special session. The fourth farmer was still alive, but suspected of having CJD. His herd had had a single case of BSE in 1991.

760 At this special meeting, SEAC considered that although four cases were likely to be more than might be expected as a chance phenomenon for the known population frequency of the disease, analysis of CJD in Europe showed that the incidence of the disease in farmers was similar in countries with no or very few cases of BSE. An important factor was that the clinical and pathological features of these cases were no different from those found in classical sporadic CJD. SEAC released a statement of its conclusions.

761 These findings remain unexplained. Among occupational groups exposed to BSE, farmers remain the exception in having such an excess over the incidence of CJD for the population as a whole. Recent transmission studies in mice indicate that the causal agent in these cases has various characteristics, including incubation period and neuropathology, which are distinct from both vCJD and BSE.

762 Thus they appear to have been typical cases of sporadic CJD, although it is not easy to accept that these four cases were simply a statistical anomaly.

763 The farmers were not the only cases of CJD that were causing anxiety. Two more adolescents had been diagnosed as having contracted the disease. SEAC released a statement saying that it was not possible to draw any conclusions from these cases, which needed to be studied in great detail. SEAC added that cases of CJD had been found in the same age-group in other countries. This was true, but such cases were extremely rare. Sporadic CJD almost always attacks the elderly.

764 Further reports of suspected cases of CJD in young people were received by the CJDSU. By the year-end, ten cases of patients aged under 50 had been referred to them. Three of those had been confirmed by neuropathology.

765 The scientists of the CJDSU were not alone in becoming concerned about cases of CJD in young people. Professor Collinge, who was conducting BSE experiments with transgenic mice, recognised these cases as extraordinary and feared that they could represent the transmission of BSE to humans. At a meeting with Dr Calman at the end of October he told him of his fears. In December 1995 Professor Collinge accepted an invitation to become a member of SEAC.

Public debate

766 Other scientists expressed their concerns more publicly. Dr Stephen Dealler and Dr Will Patterson had been carrying out calculations of the number of cattle subclinically infected with BSE that must have been slaughtered and eaten. Their conclusion that these totalled 1.5 million received wide publicity in the press.

'Most beef eaten already exposed to mad cow agent' was the headline in the *Daily Telegraph*.

767 On 1 December Sir Bernard Tomlinson, Emeritus Professor of Pathology at Newcastle University, said in a radio interview that he would not eat a beefburger and that all offal should be kept from public consumption. His views received wide press coverage. In *The Times*, he was quoted as saying:

> I have become more cautious because of recent CJD cases in dairy farmers and teenagers. These seem to be more than coincidences. My feeling is that it is possible that BSE is transmitting to humans.

768 In a television interview on 3 December, Mr Dorrell explained that the Government had removed from the food chain all organs which could possibly carry the risk of transmission of BSE – even if it were transmissible. 'So there is, you are saying, no conceivable risk from what is now in the food chain; that's the position?' asked the interviewer, Jonathan Dimbleby. 'That is the position', confirmed Mr Dorrell. Mr Dorrell told us that he regretted that answer because it went further than the words of his Chief Medical Officer. We think that it was regrettable that he gave a public assurance in terms more extreme than he could justify. He told us that it led to his being quoted in the press the next day as saying that there was no conceivable risk from eating beef.

769 The words of the CMO, to which Mr Dorrell referred, had been included in a press release in October to mark the release of the CJDSU's fourth annual report. Dr Calman stated:

> I continue to be satisfied that there is currently no scientific evidence of a link between meat eating and development of CJD and that beef and other meats are safe to eat. However, in view of the long incubation period of CJD, it is important that the Unit continues its careful surveillance of CJD for some years to come.[92]

770 We do not think that Dr Calman should have gone out of his way on this occasion to volunteer the unqualified statement that he was satisfied that beef and other meats were safe to eat. We believe that at this time Dr Calman had concerns about slaughterhouse practices, which he expressed to Mr Packer later in the month. He also had concerns about the dairy farmers that had contracted BSE. If he was going to make a statement about the safety of beef, he should have made it plain that this depended on an improved standard of compliance with the SBO Regulations by those who worked in slaughterhouses.

771 Neither Dr Calman's assurance about beef in October, nor Mr Dorrell's assertion that there was no conceivable BSE risk from food, did much to quell the alarm raised by Sir Bernard Tomlinson. The Local Authorities Catering Association received hundreds of calls from worried parents and head teachers about school meals, and advised school cooks to substitute turkey, chicken and pork for beef. On 8 December *The Independent* reported that 1,150 schools had taken beef off the menu or were offering alternatives.

[92] YB95/10.05/3.2

772 On learning that schools and caterers were beginning to remove beef from the menu, Dr Robert Kendell, the Chief Medical Officer for Scotland, decided to make a public statement. He did this on 7 December in these terms:

> The Government's independent scientific advisers are saying consistently that there is no evidence at all that eating beef or other foods derived from beef is dangerous. My general advice to people is therefore to carry on eating what you want to eat as you were before.
>
> We have no evidence of any connection between BSE and CJD. However, both conditions are being monitored and studied by scientists, in this country and abroad, as there is much about both that is still unknown.[93]

773 We have the same concerns about this statement that we had about Dr Calman's. Dr Kendell told us that, from early 1995 onwards, he was becoming increasingly concerned that BSE might have implications for human health. He told us that some of his concerns were allayed by Mr Hogg's firm stance on the SBO Regulations and the ban on the use of bovine vertebral column for the recovery of MRM. We think that Dr Kendell should have made it plain in his statement that the safety of eating beef was dependent on strict compliance with the precautionary measures introduced by the Government.

774 BSE was discussed in the Cabinet on 7 December. Mr Hogg explained about the problems discovered in slaughterhouses and the action that he had decided to take in relation to MRM. In summing up the discussion which followed, the Prime Minister said:

> ... that there was a disturbing degree of public anxiety over BSE once more and that the Government must be ready with an immediate and coherent response. The key element in that response should continue to be the assurance from the Government's chief professional advisers that there was no evidence that the disease could be transmitted to humans.[94]

A campaign of reassurance

775 MAFF Ministers and officials met the same afternoon to discuss the way ahead. They decided to use SEAC to try to get the message across that beef was safe. Professor Pattison would be invited to draft a letter to the press. Mr Hogg instructed Mr Eddy to draft a questionnaire for SEAC with the intention that the answers that they gave should be made public.

776 On 8 December *The Independent* published a lengthy article by Dr Will. The tone of this was generally reassuring, although it contained a caveat that the possibility of a link between BSE and CJD could not be excluded for many years because of the long incubation period. It ended:

> I do not believe it is reasonable to conclude that there is significant risk from eating beef. I have therefore not altered my consumption of beef or beef products, and neither have any of my colleagues at the CJD Surveillance Unit.

[93] See also vol. 9: *Wales, Scotland and Northern Ireland*
[94] YB95/12.07/14.5

777 On the same day Professor Pattison and Dr Will, acting on behalf of SEAC, sent a long letter about the safety of beef to *The Times*. *The Times* was only prepared to publish this in an edited form, an offer which was declined. The letter was adapted and turned into a letter to Mr Dorrell and Mr Hogg, and presented to the press at a press conference on 14 December, attended by Mr Hogg, Mrs Browning, Dr Calman, Professor Pattison, and Mr McNeill (of the MHS). The letter, after describing the precautionary measures that the Government had taken, and the strengthening of those measures, stated that:

> On the basis of the measures taken SEAC has a high degree of confidence that the beef reaching the shops is safe to eat.

778 This was a message that those who gave the press conference did their best to reinforce.

779 It is apparent to us that members of SEAC were pressed by government to intervene in the public debate about the safety of beef. We believe that this is something that was likely also to be apparent to members of the public. SEAC's proper role was to provide expert advice to the Government – advice which it was normally desirable to make public. If it appeared to the public that members of SEAC were being used to provide publicity to bolster the beef market, SEAC's credibility was likely to be damaged. We consider there was a danger of that on this occasion. When we look back on events in December 1995, we think that it would have been preferable if SEAC had not become involved in the public debate in this manner.

780 But for the intervention of Mr (now Sir Richard) Packer, Professor Pattison would have become even more embroiled in the 'beef is safe' campaign. After the press conference on 14 December, the MLC filmed an interview with Professor Pattison with the intention of using this as part of its advertisements for beef that were to be televised. When Mr Packer learned of this, he was concerned that it might 'be interpreted as associating Professor Pattison unduly with the beef lobby, or in other words, could be used to justify claims that he lacked independence'. Mr Packer intervened and Mr Colin Maclean of the MLC reluctantly agreed that the recorded interview with Professor Pattison should not be used for advertising purposes.

781 We consider that Mr Packer's concerns were well founded. We commend him for his prompt intervention. This was an incident in a vigorous advertising campaign which the MLC ran in 1995. In the course of that campaign there were occasions when hyperbole displaced accuracy. Our criticisms of these can be found in Chapter 6 of vol. 6: *Human Health, 1989–1996*. Although he was not always personally involved in the choice of wording in the MLC's promotional material, Mr Maclean has accepted that as Director-General he was responsible for it.

The final months

782 We come to the last section of this part of our narrative – the final months leading up to the Government's announcement that young victims of a new variant of CJD had probably caught BSE. In the final days leading up to 20 March 1996, there was frantic activity. In January and February the contemporary documents give no hint that anyone in MAFF or DH appreciated the storm that was gathering. Do they paint an accurate picture? Were MAFF and DH taken by surprise when scientists at the CJDSU identified a new variant of CJD and SEAC concluded that it was probably linked to BSE? Had they given any thought to how they might respond in that eventuality? Should they have done? Was the action taken in the final days an adequate response to the situation? If there was any delay in waking up to the fact that a crisis might be approaching, did it affect the outcome? These are some of the questions that we shall be considering in this section.

783 Before turning to these important matters, we propose to follow a sub-plot of less significance. In the last section, we looked at action being taken by the Government and by the MLC in an attempt to allay concerns about whether it was safe to eat beef. Further steps to achieve this object continued to be taken in 1996. We consider these both with a view to examining whether they were appropriate in the circumstances and for the light they throw on the extent to which those involved appreciated the storm that was about to break.

Mr Hogg's questions

784 In the previous section (paragraph 775) we saw that Mr Hogg decided that SEAC should be asked a number of questions. This was not because he wished to know the answers. It was in the hope that the answers would be suitable to publish in order to give reassurance to those who were worried about the safety of eating beef.

785 This was a venture of which the MLC approved. It also hoped to make use of SEAC's answers in its campaign to restore consumer confidence in beef. Dr Kimberlin, who was a member of SEAC, was also retained as a paid consultant to the MLC. Mr Colin Maclean sent Dr Kimberlin a list of model answers to SEAC's questions. He explained:

> We agree that we need succinct answers to these questions and my colleagues in our PR company . . . have drafted the sort of answers they would like to see (although they cannot put words into SEAC's mouth!). However, this should give you some feel for what we would initially like before you face the questions in SEAC. Anything you can do to help get crisp answers would be a big help.

786 The model answers, as one might expect, all provided the maximum reassurance as to the safety of beef.

787 We do not think that Mr Maclean should have asked Dr Kimberlin to provide this assistance. It put him in a position where his interest in helping the MLC might reasonably have been perceived to conflict with his duties as a member of SEAC. Dr Kimberlin did not perceive that the request created a potential conflict of interest. He told us that when addressing the questions as a member of SEAC he was wearing his SEAC hat, not his MLC hat. He did not inform SEAC of the MLC's request when discussing the answers to the questions.

788 SEAC considered the questions when they met on 5 January 1996 and again on 1 February. The Committee members did not agree on all the answers and the exercise was never completed; it was overtaken by events in March. Dr Kimberlin suggested answers of the kind that the MLC wanted. One was virtually verbatim in the form of the suggested model answers. All were reassuring about the safety of beef. We do not suggest that these represented other than Dr Kimberlin's own opinions. Thus there was in fact no conflict between his duty to advise objectively as a member of SEAC and the interests of his client, the MLC. There was, nonetheless, the appearance of a conflict. Dr Kimberlin should have told the members of SEAC of the request that the MLC had made, so that no one would have been able to suggest at the time or subsequently that he had a hidden agenda.

789 Suggested answers to the questions from other members of SEAC were not succinct or unequivocal. They would have been quite unsuitable for use in support of a 'beef is safe' publicity campaign. We think that these members were not prepared to lend themselves to the exercise that Mr Hogg had planned. With hindsight we can see that it was not a desirable exercise. In the first place, it diverted SEAC from more important work which they might otherwise have been asked to do. In the second place, we consider that the appropriate role for SEAC was to provide advice to the Government, not to provide publicity material to bolster the beef market. In the third place, if SEAC had provided the sound bites which had been wanted, the public would have perceived them for what they were – publicity material – and SEAC's credibility would have been damaged.

790 Mr Hogg and his officials gave further consideration to how to support the beef market at a meeting in the middle of January. Mr Hogg concluded that MAFF's principal role was to put factual information into the public domain and that MAFF should not be involved with the MLC campaign. We think that this was a wise decision. MAFF set about preparing their own information pack and revising two booklets about BSE.

791 By the end of February MAFF had prepared a leaflet entitled 'British Beef and BSE: The Facts', which was intended for a wide distribution. On the front page it stated:

> Two facts should be made absolutely clear at the outset:
>
> Fact 1 There is currently no scientific evidence to indicate a link between BSE and CJD.
>
> Fact 2 The independent expert committee set up to advise the Government on all aspects of BSE is satisfied that British beef is safe to eat.

792 Dr Wight, who was leading for DH on the medical aspects of BSE and CJD and attended SEAC's meetings as an observer, met with MAFF officials on 28 February. The next day she minuted Dr Metters, suggesting that there was 'some merit in the leaflet being issued jointly by both Departments'. Dr Metters did not agree. He replied:

> . . . some statements are too definite and in time may be seen to be wrong. We should not follow MAFF's hyperbole of reassurance. We must leave DH

> Ministers and CMO in particular, an escape route if any of these categorical statements turns out to be WRONG.

793 Mr Richard Carden, Head of the Food Safety Directorate, told us that MAFF's publicity material went to Ministers 'on the precise day when the first report suggesting there was a new variant of CJD came through' and that it was overtaken by events.

794 The fact that at the end of February Dr Wight was in favour of DH collaborating with MAFF in putting out this reassuring material suggests that she had no inkling of the storm that was about to break. The same can be said of the MAFF officials who placed the material before their Ministers at the moment that the thunder began to rumble.

SEAC's meetings on 5 January and 1 February 1996

795 SEAC met on 5 January 1996.[95] Dr Will updated members on the current state of CJD surveillance. He drew attention to the number of cases of CJD diagnosed in young people. Between 1970 and 1989 no one under 30 had contracted CJD in the UK.[96] Since 1990 there had been four definite cases and one possible. Two of the cases had unusual pathology and Dr Will thought that there was a very high chance that they were genetic.

796 The minutes of the meeting, as finally agreed, recorded that:

> Dr Will was not unduly concerned at the overall number of CJD suspect cases that had occurred in the under 30 age bracket. What he did find worrying was that all the cases had occurred over a very short period. Professor Collinge was extremely worried at the occurrence of this number of young cases in such a short period, which could suggest a link to BSE. He requested that a formal statistical analysis be carried out to assess this further. The Committee concluded that the situation demanded the continuation of intensive monitoring of CJD.[97]

797 Following the meeting, Mr Eddy, SEAC secretariat, sent a full note of the entire proceedings to Mr Meldrum, who had asked for this. Mr Meldrum told us that he was already concerned about the cases of CJD in young people and that Mr Eddy's minute indicated that there was no additional reason for concern.

798 Dr Wight told us that it was her practice to circulate a minute of SEAC meetings only if they had raised something that was relevant to public health, or required action that the Department needed to take forward, which senior officers needed to be aware of. On this occasion she sent a minute to Dr (now Sir) Kenneth Calman, which recorded the cases of CJD in young persons and added:

> Although this is a significant increase over the incidence in the UK in this age group during the preceding surveillance period, it is not without precedent worldwide.

---

[95] SEAC's membership had been strengthened by the addition of Professor John Collinge, Dr Michael Painter, Professor Peter Smith and Professor Jeffrey Almond
[96] Save for some young people infected as a result of being injected with contaminated growth hormone
[97] YB96/1.05/1.8 para. 25

799 This was an inadequate report of this important item of SEAC's business. Dr Wight's minute neither suggested that the figures were cause for concern nor disclosed that the head of the CJDSU considered them cause for concern. The cluster of young cases observed within such a short period were without precedent in the world, let alone in a single country, and there had been no such sporadic cases in the UK in the previous surveillance period. Dr Wight's statement that 'it is not without precedent worldwide' was misleading and encouraged false reassurance. When giving evidence, she commented that her statement had been 'not quite correct'. She had meant to say that cases in young persons were not without precedent worldwide. She added, 'I probably dashed this off too quickly.'

800 Insofar as Dr Wight made no mention of the concerns expressed by Dr Will and Professor Collinge, we do not believe that she appreciated the significance of what was said. We consider that she should have communicated Dr Will's concerns to the CMO. In the event, her minute went on to deal with recommendations in relation to research, and she appears to have thought that this was the most important item that arose at the meeting, so far as DH was concerned. Dr Wight's minute was copied to Dr Metters and Dr Eileen Rubery, her immediate superior, among others. It did not alert anyone to the fact that the young victims of CJD were cause for concern. We are not surprised.

801 SEAC met again on 1 February. Dr Will's concerns about the young cases of CJD had increased because they appeared to share both a novel pattern of clinical symptoms and a novel pathology, although it was still too soon to reach a concluded judgement about these. Dr Will informed SEAC of these developments. The minutes record that he:

> reiterated that the crucial issue is not simply the young age or pathology of recent cases but the short time scale in which 5 cases in individuals under 30 years of age had occurred.

802 Dr Will told the Committee that his colleague, Dr James Ironside's,[98] view was that it was premature to decide that these cases were linked with BSE.

803 Professor Smith confirmed Professor Collinge's suggestion that these cases were very significant in statistical terms. Professor Pattison's concern about the cases in young people was minuted. Professor Collinge told us that he reiterated his concerns that this was likely to represent BSE transmission to humans. Dr Will told SEAC that he intended to publish two scientific papers, one being about the young cases of CJD.

804 Mr Eddy circulated a minute about the meeting to Mr Hogg, Mrs Browning, Mr Packer, Mr Carden and Mr Meldrum. We think that he should have included a clear warning of the concerns that had been expressed about the young cases and the possibility that they might prove to be linked to BSE. He did not do so. He referred to the papers to be published by Dr Will as likely to give rise to problems which were essentially presentational.

805 Dr Wight minuted Sir Kenneth Calman about the meeting, with copies to Dr Metters and Dr Rubery, among others. Once again her minute was inadequate in that it failed adequately to express the concerns of members of SEAC about the

[98] Neuropathologist at the CJDSU

young cases. In describing the conclusions that might be drawn from these, she used language which suggested that there was, in reality, no likelihood of a link between BSE and a new variant of CJD. We are inclined to think that this was, in fact, Dr Wight's own understanding. That would explain her support for issuing reassuring publicity, which persisted until the end of the month. Although a careful reading of her minute of the February meeting should have alerted the reader to the fact that young victims were a cause for concern, Dr Wight should have put this beyond doubt by referring to the concerns expressed by Professor Pattison and Professor Collinge.

806 We observed at the start of this section that the contemporary documents gave no indication that either MAFF or DH was aware in February of the storm that was gathering. The evidence given to us by some of the witnesses painted a different picture. We propose to defer our analysis of this evidence until we have taken on the story that is supported by contemporary documents to its close.

The storm clouds gather

807 On 1 March Mr Eddy passed on to Mr Meldrum some disconcerting news that he had just received from Dr Wight. It was looking 'rather firmer' that the cases of CJD in young people represented the emergence of a new sub-population of the disease. Dr Wight had suggested a meeting between MAFF and DH officials and press officers to discuss how it should be handled. He had agreed with Dr Wight that it would be absolutely essential in handling the news to have some form of statement from SEAC as to the implications. He would keep Mr Meldrum posted on developments.

Rumbles of thunder

808 SEAC met on Friday 8 March. Dr Ironside gave a presentation showing that a subset of young people with CJD had been identified with a tendency to a long duration of illness and a unique neuropathology. The pathology differed from the rare young cases of CJD that had occurred abroad.

809 Later that day, Mr Mike Skinner[99] minuted Sir Kenneth Calman and Mr John Horam.[100] He informed them that SEAC had concluded that exposure to BSE in the 1980s was a likely explanation for the novel cases of CJD.

810 Sir Kenneth Calman received Mr Skinner's minute on Monday 11 March. After discussing the position with his colleagues he decided to call a meeting with MAFF. This took place on 13 March.

811 On 11 March some members of SEAC made a visit to a slaughterhouse. They saw SBO being properly removed, identified and treated, and decided that there was no need to recommend any additional precautionary measures at that stage.

812 On 12 March Mr Eddy minuted Mr Packer to tell him of SEAC's conclusions about the novel cases of CJD. Mr Packer told us that from that date the pace of

---

[99] Mr Skinner had succeeded Mr Charles Lister as DH secretary to SEAC in January
[100] Mr Horam became a Parliamentary Under-Secretary for Health on 29 November 1995 and was given responsibility for BSE and CJD from 31 January 1996

events became frenetic as it became more apparent every day that they would shortly be at the centre of a major national crisis.

813 Mr Hogg told us that he learned of the approaching crisis when Mr Packer came to his room one evening and told him that SEAC was coming to the view that BSE was transmissible to humans. There was no record of this visit, but Mr Hogg believed that it must have been sometime after SEAC's meeting on 8 March.

814 On 13 March Sir Kenneth Calman and other DH officials met Mr Packer, Mr Meldrum and MAFF officials. Professor Pattison was also present. Mr Packer advised Professor Pattison that SEAC should consider what action it thought appropriate. If the Committee made a recommendation, the Government would be likely to follow it. He added that it did not follow from the worst case scenario that the current rules needed to be changed. In a minute to Mr Hogg the same day, Mr Packer said that some elements in SEAC were apparently thinking of recommending a ban on the consumption of beef from animals over two years old. He questioned whether the cost of such a measure would be proportionate to any reduction in risk. He added:

> Nevertheless, on the pessimistic scenario worries about the economic consequences of SEAC recommendations would be academic. If SEAC and the CMO issue statements acknowledging the possibility of BSE/CJD transmission I am sure that the public and market reaction would be such that the political and economic effects would be a disaster of unparalleled magnitude so far as UK food scares are concerned. The consumption of beef would be likely to fall immediately to a small proportion of its former level.

815 In discussion on that day and the next, Mr Packer and Mr Hogg agreed that it was necessary to get clear advice from SEAC as to the facts and the steps which the Government should take. They also agreed that they should 'avoid seeking to influence in any way' the conclusions to which SEAC would come. On 14 March Mr Hogg wrote to Professor Pattison asking him to submit SEAC's advice as soon as he was in a position confidently to do so.

The storm breaks

816 SEAC held an emergency meeting on Saturday 16 March. Dr Will gave details of nine confirmed and three suspect cases of CJD in young patients. Three independent neuropathologists had confirmed that these cases formed 'a distinct entity unlike any previously seen CJD'. There was intense discussion of what, if any, additional precautionary measures should be put in place. So far as human health was concerned, options discussed included:

- a ban on cattle aged more than 30 months entering the human food chain; and
- a requirement that meat from animals over 30 months old should be completely deboned and their obvious nervous and lymphatic tissue removed.

817 The discussion was inconclusive, apart from agreement that SEAC should 'recommend that all steps should be taken to ensure that the current SBO ban be enforced completely rigorously'. Finally SEAC agreed on a statement to Ministers.

This noted that it had proved impossible adequately to explain the cases of CJD in young people, and continued:

> This is cause for great concern. On current data the most likely explanation at present is that these cases are linked to exposure to BSE before the introduction of the SBO ban in 1989.

818 Mr Carden told the Inquiry that SEAC's desire to give further thought to the need for new measures caused acute difficulty over the following three days. At meetings during this period, Mr Hogg, Mr Packer and officials explored with Professor Pattison what SEAC's likely recommendations might be, but it became clear that SEAC could not reach a final view until it had fully assessed all the options.

819 On Monday 18 March Mr Hogg discussed with his officials a plan of action that he had decided on. He suggested that there should be a ban on the sale of beef from animals over 30 months old (what became known as 'the Over Thirty Month scheme'), and a judicial inquiry into the Government's handling of BSE. Both Mr Packer and Mr Meldrum questioned whether the 30 months scheme was proportionate and cautioned against taking action ahead of advice from SEAC. Mr Hogg said that he was not prepared to rely on the SBO ban as the sole line of defence when the controls were not being implemented perfectly. He wanted 'belt and braces'. Furthermore he was minded to recall all beef products from the food chain. In the early afternoon Mr Hogg had a meeting with Professor Pattison, who said that SEAC would not be in a position to advise until after its next meeting, which was scheduled for 23/24 March. He expressed a personal view that Mr Hogg's proposal of a 30 months scheme was 'justifiable, logical and not irrational'.

820 Before his meeting with Professor Pattison, Mr Hogg had signed a letter to the Prime Minister, to be sent jointly by himself and Mr Dorrell. This explained what had occurred to date and said that a detailed analysis of what would need to be done would depend in part on SEAC's recommendations and the policy conclusions that would flow from them. Before Mr Major had seen this, Mr Hogg told Mr Michael Heseltine, the Deputy Prime Minister, about the information it contained. Mr Heseltine was plainly horrified. He asked about the implications of slaughtering the entire national herd, and interrupted a meeting that Mr Major was holding to draw his attention to the joint letter.

821 Later in the day Mr Hogg sent a second letter to the Prime Minister. This set out his proposal for the 30 month scheme. It raised the possibility of withdrawing all beef products from the food chain and proposed a judicial inquiry into the Government's reaction to BSE.

822 In the early evening Mr Hogg and Mr Dorrell met, accompanied by their officials. Mr Hogg told Mr Dorrell of his proposal for a ban on beef from animals over 30 months old and for a judicial inquiry. The implications for DH of SEAC's findings were discussed. These included investigations into the safety of products other than food which had bovine content, such as vaccines.

823 Late in the evening Mr Hogg, Mr Dorrell and other members of the Cabinet met the Prime Minister. It was decided to call a ministerial meeting the following day and invite the CMO, the CVO and Professor Pattison to give their advice.

824 At the ministerial meeting on the morning of Tuesday 19 March, Mr Hogg told us that his recommendations were comprehensively rejected by his colleagues and that he accepted the decision of the meeting, although he believed it to be mistaken. This rejection is not clearly apparent from the contemporary record of the meeting. What is clear is that Professor Pattison would not be drawn into giving specific advice in advance of SEAC's meeting, scheduled for the weekend. After lengthy discussion it was decided that further information from SEAC was necessary in order to enable the Government to make a statement that included something of substance. 'An early meeting of SEAC would therefore be encouraged.'

825 Encouragement resulted, by 4.00 in the afternoon, in the assembling of Professors Pattison, Almond, Smith and Collinge and Dr Will in London, and the establishment of telephone linkage with Mr Bradley and Dr Kimberlin in Paris, where they had been attending a meeting of the Office International des Epizooties. Different options were discussed at length. By late in the evening no conclusion had been reached, but the meeting had received a message that the Government needed advice by 1030 the next morning. The meeting adjourned until 0800 the next day.

826 On 20 March it became clear that the news about BSE had leaked. 'Official: Mad cow can kill you', announced the headline of *The Mirror*. Other newspapers also carried the story that the Government was to announce the possibility that BSE could be transmitted to humans.

827 SEAC reconvened at 0800. By 0930 the Committee had agreed a statement. After saying that 10 cases of CJD in young people had been identified, this continued:

> On current data and in the absence of any credible alternative the most likely explanation at present is that these cases are linked to exposure to BSE before the introduction of the SBO ban in 1989.
>
> CJD remains a rare disease and it is too early to predict how many further cases, if any, there will be of this new form.

The Committee went on to make the following recommendations:

> a. that carcasses from cattle aged over 30 months must be deboned in licensed plants supervised by the Meat Hygiene Service and the trimmings must be classified as SBO.
>
> b. prohibition on the use of mammalian meat and bonemeal in feed for all farm animals.
>
> c. that HSE and ACDP, in consultation with SEAC, should urgently review their advice in the light of these findings.
>
> d. that the Committee urgently consider what further research is necessary.

> The Committee does not consider that these findings lead it to revise its advice on the safety of milk.
>
> If the recommendations set out above are carried out the Committee concluded that the risk from eating beef is now likely to be extremely small.

828 The Cabinet met at 1045 to consider SEAC's statement and a statement that Sir Kenneth Calman had prepared. It was decided that SEAC's recommendations would be accepted in full. It was also agreed that both Mr Dorrell and Mr Hogg should make statements to the House of Commons.

829 That afternoon Mr Dorrell made the first statement to the House. He described the CJD Surveillance Unit's findings of a new variant of CJD in young people and SEAC's conclusion that the most likely explanation was that those cases were linked to exposure to BSE before the introduction of the SBO ban in 1989. He explained the recommendations that SEAC had made and said that the Government had accepted them in full and would implement them as soon as possible. He then turned to a question that Sir Kenneth Calman had raised that morning – the question of whether children were more at risk than adults of contracting CJD. He stated:

> There is at present no evidence for age sensitivity and the scientific evidence for the risks of developing CJD in those who eat meat in childhood has not changed as a result of the new findings. However, parents will be concerned about the implications for their children, and I have asked the advisory committee to provide specific advice on that issue following its next meeting.

830 Mr Hogg followed with his statement. He confirmed that the Government had accepted SEAC's recommendation that carcasses from cattle over 30 months must be deboned in specially licensed plants supervised by the MHS, and that any trimmings would be kept out of both the human and the animal food chains.
In addition, Mr Hogg explained that he had instructed that existing controls in slaughterhouses and other meat plants and in feedmills should be more rigorously enforced. He emphasised that if the public accepted 'the best opinion that we have' that beef and beef products could be eaten with confidence, then he believed there would be no damage to the British beef market.

Postscript

831 This brings the period with which this Inquiry is concerned to an end. We should, however, record that on 3 April 1996 Mr Hogg announced to Parliament that the 30 month scheme that he had favoured would be put in place rather than the deboning scheme that SEAC had recommended. The principal reason for this change of policy was that the deboning scheme did not suffice to allay the anxieties of the consumer. Furthermore, within 24 hours of the Government's announcement accepting SEAC's advice, supermarkets made it clear that they would not be willing to sell meat from animals aged more than 30 months. A further, though subsidiary, problem was that the capacity of deboning plants was not enough to provide for the deboning under official supervision of all beef. It may be that a further motivation for the change was that it might help to persuade the EU to reverse the ban, which it had just imposed, on all British beef.

832 We have asked ourselves whether these problems that confronted the Government in its choice of policy option could not and should not have been foreseen. This leads us to the question of the extent to which there was any contingency planning in the months leading up to 20 March.

Contingency planning

833 At the meeting of the MAFF Consumer Panel, set up by Mr Gummer, of 24 January 1996, MAFF tabled a paper which included details of the recent young victims of CJD. Dr Godfrey, a member of the Panel, wrote a response, dealing with what he accepted was the unlikely possibility that they might prove to have been infected by BSE. He commented:

> If the tiny cluster is due to people having been infected, further cases are likely, perhaps many of them. It seems best for government to plan now for this highly improbable possibility. This should include: (a) taking statistical advice on what will be taken as significant evidence, leading to action; (b) what advice should be given to consumers. It should be the aim to get advice across to us before the predictable reactions to what would be major tragedy, but also a major news story; (c) what action should be taken, in this hypothetical situation, to make the beef that could be eaten by consumers in the future safe again. This would obviously cost a lot, and be technically difficult, but possible.

His observations made sound sense.

834 In his first witness statement to us, Mr Carden gave this account of the reaction within MAFF to Mr Eddy's minute of 6 February:

> Those of us who received Mr Eddy's 6 February report were aware that we could be on the edge of a very far-reaching change in the picture we had of BSE. My recollection is that from then on until SEAC reached a concluded view on 20 March 1996, we felt in a state of high alert. We – I am referring to myself and the circle of people within Government to whom the news at that state was deliberately confined – paid extremely close attention to each new indication from the leading experts. But for more than a month the tentative indications from SEAC's 1 February meeting were all we had to go on. The hints of bad news remained tentative, and we lived in suspense.

835 In a subsequent statement, he added:

> Dr Will's findings were the first firm indications that the balance of probability might be shifting in favour of BSE actually being transmissible to man (contrary to what had generally been believed in MAFF up till then), and that one suspected means of transmissibility – ingestion of beef – had suddenly gained ground over the others that had been attracting more attention in autumn 1995 . . .
>
> I and my colleagues in MAFF devoted much time and energy in the first months of 1996 to watching every new indication of what was going on; we moved into a state of high alert as events unfolded, and discussed and

> evaluated each new development intensively; with MAFF and DH in very close touch both at official and ministerial level at all key stages.

836 This is precisely what we would have expected to have happened on receipt of Mr Eddy's minute. We have criticised Mr Eddy for not drawing attention in it to the concerns expressed by members of SEAC about the implications of the young victims of CJD. Despite this, we consider that the contents of his minute should have put those who read it on alert in the manner described by Mr Carden. It did not. Mr Carden's recollection of the reaction to Mr Eddy's minute is mistaken. Whatever impression Mr Eddy's minute made on those who read it, it did not lead any of them to take any action.

837 Despite the shortcomings in Mr Eddy's minute, on reading that minute Mr Hogg and Mrs Browning should have sought to discuss its implications with Mr Packer, Mr Carden and Mr Meldrum. Similarly, on reading that minute, those officials, after discussion among themselves, ought to have raised its implications with Mrs Browning and Mr Hogg. Each of these five individuals should have considered the action that might be required should the scientists advise that BSE had probably been transmitted to humans, and they should have recognised the need for MAFF and DH to address the implications in conjunction, for example by seeking the views of Sir Kenneth Calman and by discussion between Mr Hogg and Mr Dorrell. In the event Mr Eddy's minute seems to have been treated by all simply as information on matters that called neither for action nor for discussion.

838 Mr Hogg told us, on the basis not of recollection, but of reconstruction, that he believed that he must have developed his 30 month scheme over a period of months, and discussed it with Mr Packer and other officials. Mr Packer gave this evidence some faint support when speaking of dim recollections of discussions with Ministers and others on a 'what if' basis. We are satisfied that there were no such discussions about Mr Hogg's 30 months scheme. Mr Hogg did not decide on this until shortly before he presented it to his officials on 18 March. There was no discussion between Mr Hogg and his officials prior to 8 March as to the options that would need to be considered should it prove that BSE had been transmitted to humans.

839 The position was precisely the same in DH. Sir Kenneth Calman made it plain that he was not himself involved in any contingency planning or discussions before March 1996. He added:

> After the meeting in February, clearly both the Department of Health and MAFF, particularly through Dr Rubery's Division, were and should have been looking at these issues; indeed, as MAFF were; and clearly Ministers would be informed, as they always are when things are changing.

840 Dr Rubery, Dr Wight's superior, told us that she was worried about the cases of CJD in young people. She spoke of having frequent meetings with Dr Roger Skinner, a Principal Medical Officer at DH, which reflected her and her Department's growing concern about them. She said that this concern was also reflected in 'many informal discussions with Dr Wight, Dr Skinner, Dr Metters, the CMO and the Permanent Secretary', although she could not recall any further details of these informal meetings. We are satisfied that Dr Rubery's recollection that such meetings took place in February is mistaken. DH was not on a state of alert about the implications of these cases prior to March. Mr Dorrell was not even notified of the findings reported to SEAC at its February meeting. Dr Metters gave us some additional written evidence after he had appeared in Phase 2 of the Inquiry, in which he spoke of discussing prevention, care and treatment options with the Permanent Secretary and with Sir Kenneth Calman in mid-February. We do not believe that these discussions can have taken place before March.

841 Mr Carden stated that MAFF and DH were in very close touch at both official and ministerial level at all key stages. We have found that there were no interdepartmental discussions about the possible implications of the findings of the CJDSU in either January or February. Indeed, the Departments do not seem to have started to work together to address these until the meeting called by Sir Kenneth Calman on 13 March. Even then Mr Hogg proceeded to decide on the response that he considered appropriate without reference to Mr Dorrell or Sir Kenneth Calman. When we asked him whether he should not have discussed the 30 month scheme with Mr Dorrell, he replied:

> No, forgive me, the 30 month rule was down to me; that was my policy; it was something for which MAFF was answerable.

842 We have already expressed the view that MAFF officials and Ministers should have consulted Sir Kenneth Calman when they learned about the content of the SEAC meeting in February. Equally we consider that when Sir Kenneth and Dr Metters received Dr Wight's minute of that meeting, albeit that it was couched in sedative terms, they should have initiated discussions with MAFF officials to discuss the implications of the new evidence, and Sir Kenneth should have alerted Mr Dorrell.

843 What was the reason for the inertia on the part of both Departments prior to March? Mr Carden gave this answer when asked why there had not been contact between MAFF and DH after SEAC's meeting of 1 February:

> I think that both Departments will have been looking to SEAC to bring forward a firmer scientific view.

844 It was not merely SEAC's scientific view that the two Departments were awaiting. By 1996 the practice had become firmly established of looking to SEAC to advise on policy decisions – to an extent that came close to delegating them to SEAC. Witnesses told us that as the Government would not be prepared to take a decision without the advice of SEAC, contingency planning was a waste of time until SEAC's advice had been received.

845 Waiting for SEAC was not a satisfactory alternative to examining policy options. The choice between those options did not turn simply on matters falling within SEAC's areas of expertise. Wider political considerations needed to be taken into account, and these could well have been identified and discussed, on a contingency basis, in February. Nor was there any reason why SEAC should not have been asked to consider the various options that might be adopted to reduce risk of transmission further, and comment on their efficacy.

What would contingency planning have achieved?

846 The major policy decision taken on 20 March proved almost immediately not to be viable. The deboning option was not acceptable to the market, nor was it practicable. This option was recommended by SEAC under enormous pressure and instantly adopted by the Government, with no time to consider its implications. Mr Hogg took the view that it was not safe to rely on the proper performance of slaughterhouse operations to guarantee the safety of food. He wanted belt and braces. The supermarkets took the same view. Had MAFF, with the assistance of SEAC, begun to consider the options in February on a contingency basis, it is at least possible that they would have anticipated the problems which resulted in the choice of the deboning option being reversed almost as soon as it was made.

847 When Mr Dorrell made his statement to Parliament, he was unable to answer an obvious question. Were children more susceptible than adults to BSE? All that he could say was that he had asked SEAC to advise on this. In the event SEAC advised that there was no reason to believe that children were particularly susceptible. Contingency planning should have led to the anticipation of that question. SEAC could have been requested to answer it. Had its advice been obtained before 20 March, parents could have been reassured rather than alarmed.

848 There is a more fundamental question. One body of opinion considers that the over 30 months scheme was an over-reaction and that the risk that BSE was shown to pose to humans would have been adequately addressed by SEAC's deboning recommendation. We have asked ourselves whether the announcement of 20 March would have come as less of a shock:

- if the communication of risk to the public had not suffered from the defects that we have described;
- if successive CMOs and SEAC had stated plainly that they had growing concerns that BSE might be transmissible and that some humans might have been infected before the various precautions were introduced; and
- if those officials who commented on risk had frankly stated that the cases of CJD in farmers and in young persons were cause for concern, rather than emphasising that it was safe to eat beef.

Would the public have accepted that SEAC's deboning recommendation was an adequate response, so that beef from cattle aged over 30 months, removed from the bone, could have continued to be sold and eaten?

849 We have no doubt that had the approach to risk communication been that suggested above, the announcement of 20 March would have been less of a shock, and the public would not have felt that they had been deceived about the risk posed by BSE. But we do not believe the outcome would have been different. In March 1996 it was not clear how and to what extent the ruminant feed ban and the animal SBO ban had cut the rate of infection in cattle. No one knew, or could reliably calculate, how many cattle subclinically infected with BSE were entering the food chain. The improvement in slaughterhouse standards of removal of SBO was not yet clear. We believe that the public would inevitably have shared Mr Hogg's reaction that belt and braces were needed. Even today, over four years on, when these matters can much more readily be evaluated, the Over Thirty Months Scheme remains in place.

7. Medicines and cosmetics

Medicines

850 We turn now to the major topic of the safety of medicines and medical devices that use bovine tissues. Unlike food products, these did not attract a great deal of public attention and debate in connection with BSE. No doubt this was because their provenance was far less apparent.

851 As indicated in Chapter 2, bovine material was used in a variety of ways in the manufacture of medicines and medical devices. Some, like insulin, hormone treatments and sutures, contained bovine material as an ingredient. Others, in particular vaccines, were rather different. Although these did not directly use bovine ingredients, bovine material was widely used to grow cells and viruses. This material did not form part of the final product, but it was not known if its use at the earlier stages of preparation could transmit infection.

852 Officials speedily realised that medicines might offer a pathway for infection either between animals, or from animals to humans. Scrapie had in the past been inadvertently transmitted between sheep through a vaccine containing contaminated brain material. Pooled pituitary glands used to derive human growth hormone had also transmitted CJD between humans. Risk from 'biologicals'[101] immediately occurred to the Chief Medical Officer (CMO) when he was told about BSE in March 1988.

853 We devote a large part of vol. 7: *Medicines and Cosmetics* to examining in detail the way matters were handled by the medicines licensing divisions in DH and MAFF.

854 There has recently been lively public interest in action on vaccines and the fate of existing stocks when their formulation was being changed so as to substitute non-UK for UK-sourced material. This interest seems to have been stimulated by the documents and statements collected and published by our Inquiry. From the documents made available to us, it was not possible to determine precise dates on which stocks of vaccines sourced from UK bovine material were used up. Although there is no evidence at this stage that medicinal products were implicated in transmitting the disease, the possibility cannot be ruled out. Accurate tracing of available products would then be helpful. We found frustrating the gaps in records and recollections about this.

855 We recognise that the relevant documents were bulky, highly technical and confidential. Witnesses spoke of files piled room high on individual products. The paper trail would have been difficult to follow at the best of times. However, matters were made worse by defects in the record-keeping systems used at the time that the implications of BSE were being considered. Questionnaires had to be sent out to all licence holders in 1989 seeking fresh information about the use of animal materials. The Medicines Control Agency (MCA) appears to have taken some years to put

---

[101] Biological material used in the production of human and veterinary medicines, and in medical devices

matters right and to have had difficulties keeping material up to date. In 1994 it was discovered that, although the information obtained via the questionnaire had been recorded on the database, it had not been updated with information from new licence applications received after that time.

856 We were able to piece together the main bones of the story from contemporary papers and minutes, together with evidence from witnesses. What follows looks at the most significant aspects of what happened. It begins with a brief outline of the medicines licensing system, which is very different from that covering food safety. Fuller details can be found in Volume 7.

857 We have recently seen papers from DH concerning a review by the Committee on Safety of Medicines (CSM) of BSE-related issues associated with the use of seedlots[102] in the manufacture of vaccines. It will be apparent that a number of assumptions made by the CSM are open to question for reasons we have set out in our Report (see vol. 8: *Variant CJD*, Chapter 5). We hope that government will look at the topic again in the light of what we have said.

The medicines licensing system

858 Under the Medicines Act 1968, medicinal products could not be sold in the UK without a 'product licence' from the 'licensing authority'. The Secretary of State for Health carried out this role for the UK as a whole in respect of human medicines and the Minister of Agriculture, Fisheries and Food carried out the equivalent role in respect of veterinary medicines. In order to be granted a licence, a product had to satisfy criteria of safety, quality and efficacy. The licensing authority also had power to revoke, vary and suspend product licences.

859 Licensing decisions on human products were handled on Ministers' behalf by officials in the Medicines Division (MD) of DH, and from 1989 by the Medicines Control Agency (MCA). Those on veterinary medicines were handled in MAFF's Animal Medicines Division (part of the Animal Health Group) advised by the Medicines Unit and the Biological Products and Standards Department of the Central Veterinary Laboratory (CVL), amalgamated in 1989 as the Veterinary Medicines Directorate (VMD). These officials were a mixture of administrators, doctors, pharmacists and toxicologists. Ministers were consulted over controversial decisions.

860 Individual licensing decisions could be appealed against and legal challenges mounted. The burden of proof lay with the licensing authority to justify its decisions. Decision-making thus had to be based on proper evidence and be demonstrably untainted by departmental and political interests. Officials and Ministers relied heavily on advice from several committees of outside experts set up under section 4 of the Medicines Act and known as 'section 4 committees'. Many of the members were of great eminence in their field and their advice was almost invariably followed. This was certainly the case in dealing with BSE.

861 The main section 4 committees that advised on human medicinal products at risk from BSE were the Committee on Safety of Medicines (CSM), chaired by

---

[102] Master stocks from which each batch of vaccines is derived

Professor (later Sir) William Asscher; the Committee on Dental and Surgical Materials (CDSM), chaired by Professor (later Sir) Colin Berry; and the Committee on Review of Medicines (CRM), chaired by Professor David Lawson. Two subcommittees of the CSM played a key role: the Biologicals Sub-Committee (BSC) and the specially constituted BSE Working Group (BSEWG), both chaired by Professor Gerald Collee. The Veterinary Products Committee (VPC), chaired by Professor Sir James Armour, advised on all types of veterinary products.

862 One source of relevant evidence was information on adverse reactions to licensed medicinal products, reported by the medical profession and the pharmaceutical industry on yellow cards, which gave their name to the system of reporting – the yellow card system.

863 Informal methods were often preferred to formal licensing action under the Medicines Act. 'Guidelines' and 'recommendations' were issued, with which manufacturers were expected to conform. They had the merit of offering some flexibility in the light of particular circumstances and avoiding contentious litigation. We were told that in practice they were a powerful tool.

864 By 1987 the licensing arrangements in both DH and MAFF had developed a number of weaknesses. Faced with EU deadlines for reviewing 'Product Licences of Right' (those granted as an interim measure to products already on the market at the time that the UK licensing system was first set up), Ministers commissioned management reports from Dr N J B Evans and Mr P W Cunliffe about how arrangements might be improved. They found that the basic system was sound, but a two-year backlog in handling applications was mainly associated with understaffing, antediluvian data-holding systems and blurred management lines. The subsequent restructuring into Executive Agencies was intended to rectify some of the defects but itself caused some transitional turmoil.

Medical devices

865 Devices such as heart valves and pericardium patches were not covered by the Medicines Act. When BSE emerged, they were the responsibility of the Procurement Directorate (PD) of the National Health Service (NHS), which operated a voluntary registration scheme for manufacturers. The purchasing power of the NHS gave it considerable leverage over manufacturers. The need to consider this type of product in relation to BSE was not recognised until February 1989. Thereafter officials in PD lost no time in issuing guidelines that paralleled those issued to manufacturers of human and veterinary medicines (see below). Volume 7 recounts the actions they took on the products thought to carry risk. The last two such products were dealt with in early 1990 – one company had come into line with the guidelines by January 1990, while the other, after unsuccessfully attempting to find alternative material, ceased production of its device in April and recalled stocks. The response of PD was prompt and adequate.

Phase 1: the initial response on veterinary medicines

866 MAFF was quick to recognise in 1987 that veterinary medicines using bovine material might carry a risk, in particular where, as in cattle medication, there was no species barrier. Mr Wilesmith's initial investigations of BSE cases had included medications as a potential transmission agent, but by the end of 1987 he had ruled this out as not fitting the pattern of cases.

867 However, Dr Little, the CVL Deputy Director responsible for veterinary medicines, had meanwhile been giving the implications for these medicines some thought. He went out of his way to attend a meeting on 9 September 1987 of the BSC (the section 4 subcommittee of the CSM referred to above) in order to see how it handled a licence application in which possible transmission of CJD was a concern. We have already noted in Chapter 3 that differing perceptions about what happened at that meeting were to create an unfortunate misunderstanding between MAFF and DH about how much thought the latter was giving to BSE. We return to this below when we look at initial action taken by DH.

868 Within MAFF, Dr Little carried matters forward by commissioning a paper in November 1987 from a member of his staff, Mr Peter Luff. The paper was impressive as an initial overview of what was known about BSE in relation to safety of veterinary medicines. It reviewed options for action. Unfortunately, those responsible for human medicines were not sent Mr Luff's paper.

869 The paper was discussed twice in early 1988 by the Biologicals Committee, a working group of MAFF officials who handled routine biological product applications. They decided to leave the matter in abeyance for the time being.

870 It was resurrected in June, soon after a special discussion on BSE organised by Dr Philip Minor of the National Institute for Biological Standards and Control (NIBSC), and after Ministers' decision to introduce a ruminant feed ban. Dr Little and his staff acted swiftly. By 6 July Mr G W Wood of the CVL had prepared a set of draft guidelines for producers of veterinary medicines using bovine material.

871 These draft guidelines were given in July to NOAH, the trade association representing veterinary medicines producers, and were discussed with them on several further occasions.

872 Meanwhile MAFF provided letters of warning both to the *Veterinary Record* and to individual practitioners about the dangers of pituitary hormone material prepared outside the ambit of Medicines Act licensing. The concerns about BSE coincided with a review of hormone-based products that had Product Licences of Right. A warning about BSE was issued in general guidance produced in November and approved by the VPC on completion of the review. By the end of 1988 MAFF officials were also ready to seek the endorsement of the VPC for the proposed general guidelines on BSE.

873 All these were admirable initiatives so far as veterinary medicines were concerned. The problem was that the parallel interest of those dealing with human medicines had been neglected. Apart from a copy of the MAFF draft guidelines sent to Dr Harris, the Deputy Chief Medical Officer at DH, in July 1988, at the suggestion of Dr Minor of the NIBSC, we could find no trace of any significant contact between the two licensing authorities about BSE and medicines throughout this period.

874 In December, Dr Paul Adams of DH, who was following up recommendations by the CSM on human medicines, had some discussion with Mr Bradley at the CVL, and the penny began to drop that MAFF and DH should work together on advice about the same biological material forming the basis of both animal and human medicines.

Phase 1: the initial response on human medicines

875 We have looked at what was happening during the same 18 months within DH.

The period up to March 1988

876 As we have already seen, up to March 1988 DH had been neither informed nor consulted by MAFF about BSE. We looked at two occasions during the period when this might have happened.

877 The first was the BSC meeting on 9 September 1987, which Dr Little attended. Also present was a DH pharmacist, Mr John Sloggem, who had been researching an application for a Clinical Trial Certificate (CTC) for a product containing bovine brain extract. Fortuitously he had learned of BSE in August from Dr David Taylor at the Neuropathogenesis Unit (NPU) in Edinburgh, whom he had asked about the risk from 'slow viruses'. Dr Little told us that he mentioned BSE at the BSC meeting, although others present could not remember this. We think it unlikely that Dr Little referred to BSE in the course of the formal proceedings in such a way as to register with any of those present. Equally, however, we believe that there must have been some informal conversation about it between Dr Little and Mr Sloggem after the formal meeting was over. From this Dr Little gained the impression that DH was aware of BSE and was giving it some thought. He reported this to Dr Watson, Director of the CVL, who in turn told the CVO, Mr Rees.

878 However, matters were not as Dr Little thought. He did not appreciate that Mr Sloggem was pursuing his interest individually, on the narrow front of the particular application in front of him, and had learned of BSE quite by chance. More generally DH was still in the dark.

879 Had Dr Little taken steps subsequently to follow up his conversation with Mr Sloggem, the true state of affairs might have emerged. Although we do not think Dr Little is to be criticised for not doing more, once he thought that DH had taken the matter on board, we do think it regrettable that the opportunity was lost for joint consideration of BSE at an early stage by those responsible for the safety of human and veterinary medicines.

880 We also considered whether Mr Sloggem might have shared the information he was collecting more widely at that stage. However, DH had not been formally notified about BSE. Mr Sloggem had learned of it only by chance in the process of a particular investigation and thought it was a slow virus. It was not incumbent on him to inform Medicines Division or DH generally about what he had learned.

881 The second occasion on which DH might have been alerted was at a meeting of the BSC on 6 January 1988, when Mr Sloggem presented his paper about the product he had been reviewing. This was the first time that a number of those present had heard of the new disease. The CTC was turned down, partly with the 'slow virus' risk in mind. We do not think it unreasonable that the subcommittee and the officials of MD did not identify any wider considerations.

882 However, we think it was a pity that no system existed to capture information of the sort acquired by Mr Sloggem on a readily accessible form of working database. We see such a database about concerns and queries as being of value to both the licensing authorities.

March–December 1988

Initial action by the CMO and MD

883 We have seen already that DH was formally notified of the emergence of BSE in March 1988. When the CMO, Sir Donald Acheson, heard about the disease, he had an immediate concern about the safety of bovine insulin and of vaccines prepared using bovine serum. No doubt the unhappy story of human growth hormone was fresh in his mind. He asked his deputy, Dr Harris, who had long experience of medicines licensing, to seek advice from the NIBSC.

884 It was also agreed that the safety of biological-based medicines was a priority question for the proposed group of experts – set up shortly thereafter as the Southwood Working Party.

885 During April officials in MD saw a submission from the CMO to DH Ministers alerting them to the disease, and minuted one another about its implications. We were told they knew 'virtually zero' at that time about TSEs. They decided to await the outcome of the Southwood Working Party's deliberations. Although some preliminary steps might usefully have been taken in the meantime, such as searching their database of licensed products, we thought the decision to await the views of the Working Party was a reasonable response by MD at this juncture.

The NIBSC discussion

886 On 16 May 1988 the NIBSC organised a discussion about BSE to consider what the disease might mean for medicines using biological material. The meeting was attended by Mr Wilesmith, the CVL epidemiologist, Dr Kimberlin from the NPU, Dr Rosalind Ridley and Dr Harry Baker from the MRC's Clinical Research Centre, and Dr A J Beale and Dr A J M Garland from Wellcome. Surprisingly, no one from MD attended. It has not been possible now to unravel why. Dr David Jefferys, the obvious candidate as head of the new drugs and biologicals branch of MD, believes he did not receive an invitation. Among the outcomes of the discussion was a recommendation that tests of the infectivity of calf serum should be undertaken. We return to this later.

Galvanising MD

887 In May Dr Pickles, the newly appointed DH joint secretary of the Southwood Working Party, moved into action. She summoned up some information from the existing database and suggested to Dr Jefferys that a number of questions should be put to the BSC. He was not in favour of doing so, noting that the BSC had already discussed BSE informally in January. He did, however, respond with some preliminary thoughts and suggested that others in MD should also be involved in any further discussions.

888 Dr Pickles returned to the charge on 21 June immediately after the first meeting of the Working Party. In a forthright minute intended to 'galvanise Medicines Division into action', she listed further questions needing answers and pressed for these to go to the BSC. Dr Gerald Jones, the senior medical officer in MD, told us that by now it had become clear that they had 'a serious problem'. They decided to refer the issue of BSE to the BSC and during July Dr Frances Rotblat, a Senior Medical Officer working for Dr Jefferys, and Dr John Purves, Pharmaceutical Assessor to the CSM and the BSC, were commissioned to write a joint paper for the BSC's November meeting.

889 We were concerned whether the matter was put to the section 4 committees sufficiently promptly, and whose responsibility this was. One of the defects identified by the Evans/Cunliffe report was the divided responsibility in MD and lack of clear management lines on many matters. BSE was inherently an awkward topic for MD to handle. It had implications across the different administrative, medical and pharmaceutical branches and potentially affected both new, and as yet unlicensed drugs, and drugs already on the market.

890 We accept that responsibility for BSE did not naturally fall to a single branch within these arrangements, but consider that good management pointed to a lead responsibility being assigned. We consider it fell to Dr Gerald Jones, having discussed the matter with senior staff, to decide the priority to be accorded to BSE in relation to other work within MD and to set in hand appropriate action.

891 We also consider that he should have asked for the paper to be prepared for the September rather than the November meeting. It seemed from the evidence we received that, even allowing for the logistics of preparing and distributing papers in good time, this could have been achieved had Dr Jones assigned the matter a higher priority. The consequence was that two months were lost when progress might otherwise have been made.

The paper for the BSC

892 The paper prepared by Dr Rotblat and Dr Purves served its purpose. It elicited advice from the BSC in November. The subcommittee made a number of recommendations, which were to apply to all licences for new products, including:

    i. No immediate licensing action on oral products.

    ii. All bovine materials to come from appropriately certified healthy herds, not fed with animal protein. No brain or lymphoid tissue to be used in parenteral products.

iii. Manufacturing processes for parenteral products to be capable of eliminating scrapie-like agents.

iv. MAIL (Medicines Act Information Leaflet) article to request manufacturers to identify products in which bovine materials had been used. Serum to come from appropriately certified healthy herds.

893 These recommendations were subsequently endorsed by the CDSM, which among other things was responsible for sutures, the CRM, which was reviewing all the Product Licences of Right, and the subcommittee on Safety, Efficacy and Adverse Reactions (SEAR). They were then endorsed by the CSM itself on 17 November.

894 The Chairman of the CSM, Professor Sir William Asscher, told us that experience with human growth hormone and dura mater implants had made the Committee very wary of parenteral products. However, the fact that scrapie had not transmitted to man gave reassurance that BSE was unlikely to be acquired orally.

Sir Richard Southwood's concerns about biologicals

895 A copy of the recommendations was sent to Sir Richard Southwood. Sir Richard had been taking a continuing close interest in the question of the safety of biologicals. He had written to the CMO in August about this and had been reassured that the topic would shortly be coming before the CSM and other committees. He had written to Professor Asscher just before the CSM's November meeting pressing for any action to apply then to existing products and making a number of suggestions for the contents of informal advice to manufacturers. A round of further correspondence ensued, mainly consisting of Sir Richard's continuing concern that he was not getting his point across about existing products, and Professor Asscher's replies assuring him that he was. When he gave oral evidence Sir Richard told us that by existing products, he thought the Working Party meant products that were already licensed and stocks of those products. It is not at all clear whether Professor Asscher and the CSM appreciated that the second category was included.

896 Sir Richard Southwood also wrote in December to Dr Little about veterinary products, making similar points. It is plain from this letter that Sir Richard was unaware of the advanced preparation of MAFF guidance.

897 We have already noted that MAFF did not go out of its way to inform officials in MD, or involve them in the discussions about BSE in MAFF's Biologicals Committee. Equally, MD officials did not seek to find out the situation on veterinary medicines when the issue of BSE and human medicines arrived on their desks in April 1988, or when the MAFF draft guidelines were despatched to them in July 1988. The consequence was that DH had to catch up with several months' head start by MAFF before it could begin to address the problems.

Phase 2: preparing joint guidelines, January–March 1989

898 On 3 January 1989 MAFF and DH officials eventually sat down together to work out a joint policy towards medicinal products. They agreed it was essential to keep in step, especially as MAFF concerns about animal vaccines would cause DH great difficulties of supply if current stock – in some cases up to five years' supply – had to be lost. Joint guidelines should be published in MAIL together with a request for information. These conclusions were relayed by Dr Jefferys and Dr Adams to Dr Harris.

899 Within MAFF, Mr F J H Scollen, who handled the policy side of veterinary medicines licensing in Animal Health Division, minuted Mr Cruickshank with his views. He saw the issue as one to be addressed 'first and foremost in the human health context' because of the risks associated with maintaining or disrupting the supply of vaccines for human health purposes. He went on: 'Judgements about what is needed and feasible on the animal medicines front can be more readily taken afterwards.' This was the line that was subsequently taken.

900 A text for draft joint guidelines was agreed by an *ad hoc* working group of officials from DH, MAFF and the NIBSC, chaired by Professor Collee, which met on 1 February. The group decided that further action, especially on current stocks of affected products, should be determined once the scale of the problem had been more precisely identified with the help of the manufacturers. Any such action 'would need to be based on a human health risk/benefit assessment'.

The final draft of the *Southwood Report*

901 Licensing officials had been keen to know what the *Southwood Report* would say about medicines. They were looking to it to provide reasoned grounds for any action they might take. At the 1 February meeting those present were shown the currently proposed wording of this section by Dr Pickles, and reacted with dismay.

902 Mr Scollen, who had attended the meeting, gave a graphic account in a minute to Mr Cruickshank:

> There was general dismay at the drafting, which tends to highlight the (theoretical) risk via medicines and to relegate the qualification that the risk is remote.

903 After listing a number of criticisms the group had made of the draft, Mr Scollen continued:

> Even if the report is modified in the light of these reactions, its appearance seems likely to trigger a need for a major public relations job which takes full account of the medicines angle. Consistency between MAFF and DH will be essential and should be achievable. The guidelines themselves could subsequently generate similar pressures since they clearly do not address the issue of current stocks and they could prompt questions – for example – on

the standards applicable in the collection of animal material at slaughterhouses for biological medicinal purposes.

While I have no doubts about the Working Group's staged approach and the balance to be struck between risks and benefits to human health, this will not be the easiest position to present to a potentially critical public prone to seeing the influence of commercial interests.

904 Dr Pickles, too, got the message. The next day she wrote to Sir Richard Southwood reporting:

They have now realised that virtually none of the current essential human or animal vaccines could comply with the CSM guidelines as agreed by their November meeting and there may be several years of some vaccines in stock to make matters more difficult. Public confidence in the vaccination programme must not be put in jeopardy and yet supplies of some vaccines are very limited. After a late start, it now seems that both human and veterinary sides of the medicines business are working together and putting together a package of measures that seem sensible and workable (and indeed now incorporate all the points you raised with Professor Asscher in your earlier letters, and which I had raised with them separately).

905 She went on to suggest a revised passage for the Report on the grounds that:

This treats CSM/VPC like HSE ie the problem has been referred to the body with the statutory responsibility in that area and it is then for them to take appropriate action.

906 The Southwood Working Party went along with this line of reasoning at its final meeting on 3 February and adopted the revised wording suggested. The report as finally published said on medicines:

5.3.3 The greatest risk, in theory, would be from parenteral injection of material derived from bovine brain or lymphoid tissue. Medicinal products for injection or surgical implantation which are prepared from bovine tissues, or which utilise bovine serum albumin or similar agents in their manufacture, might also be capable of transmitting infectious agents. All medicinal products are licensed under the Medicines Act by the Licensing Authority following guidance, for example from the Committee on Safety of Medicines (CSM), the Committee on Dental and Surgical Materials (CDSM) and their subcommittees. The Licensing Authority have been alerted to potential concern about BSE in medicinal products and will ensure that scrutiny of source materials and manufacturing processes now takes account of BSE agent . . .

5.3.5 In these, as in other circumstances, the risk of transmission of BSE to humans appears remote.

The continuing concern on vaccines

907 Shortly after the final version of the *Southwood Report* was agreed, Dr Pickles sent a copy to the CMO with a draft submission to Ministers. This draft alerted Sir Donald Acheson to the continuing concerns about vaccines. He decided to take a personal hand in matters and asked Dr Harris on 9 February to look into the matter urgently with Medicines Division. He told us that this intervention was quite contrary to his normal practice; he was trying to 'stir up more activity in the Medicines Division'.

908 Stir up activity he did. On 13 February MD officials met and agreed to carry out a telephone survey of all manufacturers of children's vaccines. They mooted a working group of officials and experts to follow matters through, and this suggestion led eventually to the setting up of the BSE Working Group.

909 Twenty-four hours later, MD had collected a useful body of information from those manufacturers identifying what they knew about vaccines that contained bovine material or which might have used it during manufacture, and about the stocks held. This suggested that in some cases considerable stocks were held, described variously as 'large', five years, and 63,000 litres.

910 An *ad hoc* group of experts and officials met again on 22 February. This meeting was a key precursor to discussion and advice from the CSM the following day. For this meeting the group added to its number several outside experts – Professor Asscher, Chair of the CSM, Sir John Badenoch, Chair of the Joint Committee on Vaccination and Immunisation (JCVI), Dr Kimberlin of the NPU, Dr William Martin (Southwood Working Party member) and Professor M D Rawlins, Chair of the CSM subcommittee on Safety, Efficacy and Adverse Reactions (SEAR).

911 Those present at that meeting were told of the information on vaccines collected at Sir Donald's instigation. They considered the *Southwood Report*, the proposed guidelines, a draft questionnaire seeking information from licence holders and a draft letter to licence holders. There clearly remained a number of concerns about the content of the guidelines and whether they ought to be going out at all. It was agreed that the guidelines should be seen as 'gold standard' and that this should be made clear.

CSM and VPC approval and the issue of the guidelines

912 The CSM met the next day and approved the various drafts, including a covering letter and also a position statement of its own. This said that the Committee had considered the safety of human medicines in the light of the *Southwood Report* and agreed that the risk to humans of infection via medicinal products was remote. It said the CSM and the VPC had agreed joint guidelines 'as a precautionary measure, and for the sole aim of seeking to guard against what is no more than a theoretical risk to man'. The VPC had approved the guidelines a few days earlier.

913 The main points covered in the guidelines were:

- they applied to all licensed products for injection, application to the eye or to open wounds;
- no brain, neural tissue, thymus or other lymphoid tissue, placental tissue or cell cultures of bovine material should be used in manufacture;
- collection techniques to avoid contamination should include no brain-penetrative stunning, the use of sterile and disposable equipment, calves to be under 6 months, all cellular components to be removed from serum;
- sterilisation advice; and
- the guidelines applied also to material from sheep, goats, deer and other animals susceptible to TSEs.

914 An MCA paper for the Committee drew attention to products already produced and awaiting distribution. It noted that the questionnaire asked companies about their stocks and said: 'The Committee's advice on this issue will be sought at a later date.'

915 Ministers were told on 23 February that the CSM and VPC had concluded that the risk of transmission of BSE through vaccines was remote. To ensure the safety of medicines, however, guidelines would be going out to producers in March. The Cabinet took this into account when they discussed the *Southwood Report* later that day.

916 The guidelines and questionnaire were issued on 9/10 March by DH. The covering letter took the wording a stage further by referring to the guidance as 'a purely precautionary measure' and said that it represented 'a standard that is deemed to be best practice for the future, and steps should be taken to implement it. However, it is realised that this guidance may not be fully applicable in all circumstances.' MAFF issued parallel documents for manufacturers of veterinary products on 15 March 1989.

Was the action taken adequate?

917 The guidelines were the single most important step taken to secure the safety of medicines. They were the only specific protection put in place to guard against BSE infection via medicines, since the SBO Regulations of November 1989 expressly excluded from staining and sterilisation the material going for pharmaceutical use. Here we consider how matters were handled between January and March 1989, looking at:

   i. the Southwood message and how it was interpreted;
   ii. whether non-binding guidelines were appropriate;
   iii. the scope of the guidelines; and
   iv. treatment of existing stocks.

The Southwood message and how it was interpreted

918 When discussing the *Southwood Report* earlier in this volume, we noted that the wording the members of the Working Party finally adopted to describe the risks from bovine material in vaccines and other injected products failed to convey their true concerns.

919 The potential risks from parenteral injection had been one of the Working Party's most serious worries. They were concerned about existing products and existing stocks. Their identification of risk as remote was predicated on action being taken to address these matters.

920 Those preparing the guidelines, on the other hand, believed that the risk even before taking any precautions was theoretical and remote. Dr Martin observed to the Inquiry that his impression on attending the meeting on 22 February was that those on the human medicine side regarded BSE as an animal problem, and considered that the Southwood Working Party were being excessively apprehensive.

921 The Working Party were anxious to avoid a vaccine scare. Nevertheless, as discussed earlier in this volume, they should not have allowed their Report to give a false impression of their assessment of the risk posed by medicinal products. The message that flowed from it was that risk was remote even if no remedial measures were taken. This interpretation became the conventional wisdom both inside Departments and among medicines manufacturers and others outside government.

Were non-binding guidelines appropriate?

922 It could be argued that suspect material could have been cut off promptly and decisively had formal licensing action been initiated at once on individual items of high risk. We were, however, persuaded by the arguments put to us that guidelines were a more appropriate approach. In essence these arguments were that this approach was quicker and cheaper, and as effective. We agree that had regulatory action been attempted based on an unproven risk, a shoal of legal challenges might have resulted.

Was the scope of the guidelines adequate?

923 The question here was whether covering parenteral products and those applied to open wounds or to the eye was enough: should orally administered and all topical products – such as creams and ointments – also have been included in the guidance?

924 Oral products were carefully considered by the experts who sat on the section 4 committees. Nothing in the *Southwood Report* pointed to the need to alter the assessments made by them in November and sent to Sir Richard at that time. No recommendations were made by the Southwood Working Party regarding subclinically infected cattle entering the food chain. We felt that it was not unreasonable for the section 4 committees to assume that if it was safe to eat meat, it must be safe for humans to eat the minimal amount of bovine material contained in oral medicines such as gelatine in capsules.

925 As for topical applications, the guidance covered the two most obvious risks, application to open wounds and to the eye. The decision not to include other topical material at this stage seemed reasonable.

Were existing stocks of injected products treated appropriately?

926 The issues that exercised us most were whether suspect stocks of injected products should have been immediately withdrawn and how this should have been handled and presented.

Keeping them in use

927 There were two principal arguments against immediate withdrawal of stocks. The first was the difficulty of procuring sufficient guaranteed 'clean' stocks to maintain the vaccination programme or provide life-preserving medication. Many of the contemporary documents and the statements we saw emphasised the difficulty of replacing stocks overnight. In particular, 'growing' batches of vaccines was a lengthy process. For this reason, stocks tended to be built up and kept for a number of years ahead.

928 The second argument was that such action risked causing a general panic that would deter parents from having their children vaccinated, as had happened on previous occasions over other 'scares'. Discussing his later concern that the proposed ban on bovine offal should not raise alarm about pharmaceuticals, Sir Donald Acheson told us:

> I had in mind a marked and extended previous reduction in the acceptance of whooping cough vaccine which had followed incorrect public allegations by a scientist that the administration of the vaccine carried a significant risk of encephalitis. On the one hand I was aware that during the period 1980–1988, due to incomplete vaccination of our population of children, there had been 123 deaths from measles and 50 from whooping cough in England, together with a many times larger burden of illness and some long-term complications. Against this I had to balance a remote risk of a fatal disease.

929 Professor Asscher told us he saw the risk-benefit analysis of existing stocks as comparatively easy because the risk according to the *Southwood Report* was remote, and because vaccines were very important in protecting human health:

> The CSM's judgement was that the risks associated with interruption of the UK vaccination programme were far greater than the potential risk of BSE being transmitted.

930 We weighed carefully all the evidence provided to us. It is clear that the overwhelming opinion of the medical professionals at this time was that existing stocks should not be immediately withdrawn. Officials in MD accepted this advice and in our view it was reasonable for them to do so. Experience had shown that incomplete vaccination of children led to significant numbers of deaths that would otherwise have been prevented.

FINDINGS AND CONCLUSIONS

Handling and presentation

931 The decision not to withdraw existing stocks immediately gave rise to a separate but related dilemma: the question of what information should be given to the public about the risks associated with BSE and the continued use of existing stocks.

932 The message in the various Q&A briefs prepared at the time of publication of the *Southwood Report* was that the CSM and the Southwood Working Party were agreed that the risk of transmission of BSE via medicinal products was remote, and that there was no reason to question the safety of existing stocks.

933 There was concern that publicity about the steps being taken would create the very situation that it was desired to avoid. This raised ethical as well as practical considerations, calling for judgement rather than scientific expertise. We believe that vaccine scares, like food scares, are likely to be fostered by a belief on the part of the public that the full picture is not being disclosed. A decision in an individual case not to disclose the full picture in order not to alarm the public is likely to perpetuate, in the long term, the distrust that leads to alarmist reaction. We can appreciate the short-term attraction, in the case of BSE, of not telling the public that there was a degree of concern about vaccines. Taking a long-term view, however, we believe that a policy of giving the public full information about risk is, on pragmatic grounds alone, the correct one, whether the subject matter is food, vaccines, or any other area of potential hazard. If we are correct, the ethical requirement must also be one of openness.

934 We were unable to establish in precisely what terms the decision to go on using existing stocks was brought to Ministers' attention and what express consideration they gave to it. It seems to us that it must have been at least implicitly understood, if not expressly discussed, at a ministerial level, that there was an issue regarding existing stocks of vaccines, and that a decision had been taken that they were not to be immediately withdrawn while the guidance worked its way through. However, there is no doubt that the decision was not taken at a ministerial level.

935 When we put to various Ministers the question of whether they would have expected to be consulted or informed, we received various answers. Mr Clarke, who was Secretary of State at the time, thought that if the experts were agreed, they probably need not refer it to Ministers. Mrs Virginia Bottomley and Mrs Edwina Currie, who had also served as Ministers in DH, took a different view. Mrs Currie added that she would not dream of overruling people who were on the various senior medical committees. However, she went on to say: 'If it was an issue that was likely to arouse public concern, for example a dodgy batch of vaccine, then Ministers would be alerted very quickly.'

936 Had the decision in February 1989 about the continued use of stocks of potentially infected vaccines and its sensitivity in relation to the vaccination programme been explicitly put to Ministers, we believe they would have accepted the overwhelming advice of the expert committees, CMO and other DH officials. However, we also believe they would have taken a lively interest in how soon the doubtful material would be phased out and the steps to encourage this. Such interest would have influenced the subsequent pace of events.

Phase 3: implementing the guidelines after March 1989

937 We look now at the third phase of action and one that has attracted great public interest. When they put the guidelines to the CSM for approval in February, officials had emphasised that they were practicable and capable of being implemented over as short a time period as possible. They now had to ensure this happened. They also had to deal with the matter of existing stocks, on which they had undertaken to come back to the CSM. Were these tasks carried out adequately for both human and veterinary medicines?

The context for handling matters

938 Before we trace the way in which DH and MAFF respectively carried out these tasks over the years that followed, we draw attention to two significant changes that took place in the context in which they were acting.

939 The first was the reorganisation of the administrative arrangements for handling licensing that we have already touched on, in order to create Executive Agencies. Preparatory changes were made in 1989 with the redesignation of MD as the MCA and the appointment of a new head from outside the public sector, Dr Keith Jones. This was paralleled by the appointment of Dr James Rutter as the head of the newly constituted VMD. After a 'shadow' period, during which reporting lines remained much the same, the two Executive Agencies came into formal existence in 1991 and 1990 respectively. The Medical Devices Agency followed in 1994.

940 Although these new arrangements did not alter the way the medicines licensing system worked, they affected how officials were organised, their accounting lines and the performance standards they were expected to meet.

941 The second major change was increasing EU involvement in medicines matters and the handling of BSE risk. European guidelines on BSE and human medicines came into operation in May 1992 and closely similar ones on veterinary products a year later. In addition, the World Health Organisation offered a formal view in November 1991 that the careful sourcing of material was the best way of securing safety from the remote risk in medicinal products. The international dimension to medicines dominated the later years covered by this Report.

Collecting and analysing the information

942 The first step for both Departments was to collect the information asked for in the questionnaires issued in March. The date set for questionnaire returns was 1 May 1989, with a view to discussion at the first meeting of the newly constituted BSEWG in July. Six weeks proved far too short a deadline. It was to take many months of chasing to get in all the responses. The delay in getting returns collected and analysed meant that the first meeting of the BSEWG had to be postponed until September.

943 Meanwhile work continued within the MCA on analysing the responses. The different products were ranked according to risk, and MCA officials were asked to prepare papers on those falling in the three highest risk categories for consideration by the BSEWG when it met. We thought this was a sound approach. The ranking, which was influenced by Dr Kimberlin's views, and was subsequently adopted by the BSEWG, was as follows:

 i. Injected products with bovine brain/lymphoid tissue as ingredients.

 ii. Injected products with bovine ingredients other than the above.

 iii. Tissue implants, open wound dressings, surgical materials, dental and ophthalmic products with bovine ingredients.

 iv. Topically administered products with bovine ingredients.

 v. Orally administered products with bovine ingredients.

 vi. Products with other animal/insect/bird ingredients.

 vii. Products with materials produced from animal material by chemical processes, eg stearic acid, gelatine and lanolin.

The SBO ban and pharmaceuticals

944 Meanwhile, as we described earlier in this volume, action in MAFF was developing on another front. Mr MacGregor's decision to introduce an SBO ban had initially made DH nervous that this would awaken public concerns about pharmaceutical safety and thus threaten the vaccination programme. However, Sir Donald Acheson told us that, apart from this anxiety, DH welcomed the proposed measure as a step to protect human health. When MAFF set about defining the scope of the ban, DH became involved in the process. This was handled mainly by Dr Metters, who was Dr Harris's successor as Deputy Chief Medical Officer, and by Dr Pickles.

945 Dr Pickles quickly spotted that the list of risk tissues included some used for medicines and medical devices, such as intestines, spinal cord and thymus. However, the approach being adopted was that the SBO ban could not and should not apply to material used for pharmaceutical purposes. At a definitive MAFF meeting on 27 September 1989 about the scope of the ban, it was agreed that the Regulations 'were not the correct vehicle' for a ban on non-food items. This was consistent with the existing exemption for unfit meat sent to a manufacturing chemist, in the 1982 Meat (Sterilisation and Staining) Regulations. In November Ministers agreed with the advice put to them that the CSM/VPC guidelines already in place were the appropriate safeguard in relation to the use of SBO in medicines. Manufacturing chemists should therefore continue to be allowed to receive the unsterilised and unstained material.

946 We noted that when the question of this exemption came up again in March 1991, there was a further debate and the position changed. Mr Lawrence saw the exemption as 'rather anomalous' and argued that it should be removed. MAFF Ministers agreed with the proposal and the new Regulations in March 1992 removed the specific exemption for 'manufacturing chemists'. However, bovine material for pharmaceutical use may have continued to fall within the general exemption for premises used for the manufacture of products other than food.

947 This sequence of events highlighted the differences between the legislative frameworks for ensuring the safety of food and medicines.

948 We consider the legislative framework in Chapter 14, and examine there the extent of general statutory powers to ban the use of potentially hazardous bovine tissues for any purpose which might involve a risk to health, or even to destroy them. Differing legislative powers made it difficult to adopt a consistent approach to preventing the use of SBO in food, animal feed, medicines, medical devices and cosmetics.

949 We recognise that there are different considerations in play, and that much is dictated by relevant European legislation. However, the different frameworks make it more difficult to achieve a consistent approach. The most glaringly anomalous outcome in the case of BSE was the ban on the use of intestines for food purposes while they might still be used for sutures – thought to be a higher-risk route of infection.

How the BSEWG operated

950 The BSEWG was set up specifically to advise on the implications of BSE for human medicinal products. Its membership was high-powered. Chaired by Professor Collee, it included the chairmen of the section 4 committees it was advising, together with Dr Tyrrell, Dr Will and Dr Kimberlin of the Spongiform Encephalopathy Advisory Committee (SEAC) and Dr David Taylor of the NPU. Any conclusions it reached were therefore going to have great authority. However, it was purely advisory. It depended on the problematical cases and information about them being brought to its attention by officials, and on officials' subsequent action to follow matters up. Dr A Lee, an official in the VMD, was given the role of MAFF representative on the Working Group to maintain a link with the parallel action by the VMD. Altogether the BSEWG met five times between September 1989 and July 1992. These meetings provide convenient milestones, which we follow below.

First meeting of the BSEWG on 6 September 1989

951 At its first meeting the Working Group considered a list of products identified by officials from questionnaire returns and other data held. It agreed the ranking of risk categories proposed by the MCA and considered that the last four gave no cause for immediate concern. In respect of the first three it made four general recommendations to the effect that:

i. no action was needed where raw materials were sourced outside the British Isles in suitable conditions;

ii. the guidelines should apply to material from the British Isles, and companies should be encouraged to comply as soon as possible. The timescale should be agreed for each individual product;

iii. no licensing action should be taken at present on non-bovine materials; and

iv. the licensing authority should follow scientific progress on BSE so as to be in a position to take future licensing action when necessary.

952 The second of these recommendations depended on officials offering the encouragement and deciding any timescales. One of the papers put to the BSEWG at this meeting gave some indication of their line of thinking about the way the exercise should be handled. It suggested that considerations to be taken into account included 'the findings of the Southwood report in which it was stated that "the risk to man of infection via medicinal products was remote". It is important not to undermine this considered advice by demanding unnecessary assurances and information from manufacturers.'

953 Officials in the VMD appear to have taken a similar view of the Southwood findings. Mr Alastair Kidd told us that manufacturers were advised to change sources of bovine materials as quickly as possible, where necessary, but were allowed to exhaust existing stocks, as the *Southwood Report* and the VPC and CVL specialists in BSE had considered that the risk of BSE transmission by medicinal products appeared remote. The VMD told us that this advice was not given generally – the use of existing stocks was considered on a case-by-case basis.

954 At the BSEWG meeting two types of product were identified as needing special consideration. On the first – some homeopathic medicines with Product Licences of Right – it was agreed that more information was needed. The CRM carried this matter forward and decided in November that no action was necessary.

955 On the second, surgical sutures, there was a difference of view within the Working Group. They had a substantial paper prepared by MCA officials before them. Discussions had been taking place for some months with the major UK manufacturer about interim measures that might be adopted while a switch was made to non-UK material. This was not a simple operation as 25 million metres of intestines were used annually. This represented 10 per cent of the annual cattle kill in Australia and nearly a quarter of the New Zealand kill. The upshot of the BSEWG discussion was that, although the company's plans for a general switchover (in the event begun in February 1990 and completed by the summer) were acceptable, a minority thought that the sutures should be excluded forthwith from neurosurgery, on which the company itself had envisaged offering a warning. Professor Collee was one of these.

The follow-up to the first meeting

956 The CDSM opted for the majority view on sutures at its meeting on 20 September, and the CSM at its 28 September meeting endorsed the BSEWG's general recommendations.

957 On 10 October Mr Murray Love, an administrator working in Mr David Hagger's division in MCA, minuted Dr Jefferys and others suggesting a way forward following the BSEWG meeting. The matters he raised were highly pertinent. They included telling firms what the BSEWG had said, timescales for the three high-risk categories, dealing with stockpiled products, and the need for a coordinated licensing authority approach with clear allocation of responsibility. This minute received a lukewarm response from Dr Jefferys, who had discussed it with Mr Hagger, Dr Adams and Dr Purves. Their view was that a meeting of the BSEWG should be arranged for January, and that an in-house procedure for writing to individual companies about products and setting timetables should be agreed. Dr Jefferys told us that the follow-up with companies lay with Mr Hagger's division. Mr Hagger's division, however, was already in the process of being deconstructed as part of the MCA reorganisation.

Second meeting of the BSEWG on 10 January 1990

958 The key issues on this second agenda were the state of play on the 1989 questionnaire and how to deal with products not complying with the guidelines, particularly the remaining four vaccines which by that stage did not comply.

959 Apart from these vaccines, the only products using high-risk materials were some allergens using bovine brain in their preparation, not as an ingredient. The Working Group wanted a tough line on these allergens. The licensing authority should insist on a changeover to Australasian material within a reasonable timescale. It was reported that discussions were still continuing at the time of the next BSEWG meeting in July 1990. In October 1990 officials reported that satisfactory progress had been made. We were unable to ascertain when a final outcome was obtained.

960 On vaccines, Dr Rotblat now had more concrete information than that obtained from her ring-around 11 months earlier. She identified four products, the first three of which were produced by Evans Medical and the fourth by Wellcome:

> i. MMR (measles, mumps and rubella) vaccine with stocks to December 1990 – not yet licensed
>
> ii. Measles vaccine with stocks to September 1990 – not used much now
>
> iii. Tuberculin PPD with stocks to September 1991 – no other source available
>
> iv. DTP vaccines (diphtheria, tetanus, pertussis) with unadsorbed stocks to May 1991 and adsorbed to June 1990 – adsorbed used in preference to unadsorbed (not used much now).

961 The meeting decided that 'the benefits accruing from continuance of the vaccine programme outweighed the very remote risk to the population from the use of bovine material in these products'. The minutes go on to say:

> It was considered after some discussion that negotiations should take place to ensure that sources are changed as soon as possible and to replace existing stocks with new material whenever feasible. Replacement of Wellcome unadsorbed DTP vaccine, by Wellcome adsorbed vaccine should ensure that the former, which is not much used, is replaced earlier than 1991. In the case of the Tuberculin PPD, no other source is available at present, but the company (Evans) should be asked to move over to the new product and replace stocks as soon as this is feasible.

The follow-up to the second meeting

962 The CDSM, at its meeting on 17 January, praised the speed with which the company making sutures had responded to the BSEWG recommendation: it was to begin the changeover in February to Australasian sources.

963 Concerns about BSE in bovine insulin were raised that spring by the British Diabetic Association. Dr Jefferys told the Association in April 1990 that none was being sourced from the British Isles. Although 42 licensed bovine insulin products had been originally identified for the CSM in 1988, none figured among the items put to the BSEWG in the light of the questionnaire. We infer that they were by then sourced outside the UK.

Third meeting of the BSEWG on 4 July 1990

964 Professor Collee told us that at this meeting the Working Group discussed the safety of foetal calf serum at length. He had sought the advice of Dr Taylor of the NPU and others before the meeting. The Working Group reiterated its view that the risk relating to serum was low. Taken together with the fact that the risk of transmission of BSE was theoretical and the view that the benefit of availability of vaccines outweighed any potential risk from their use, the use of foetal calf serum in the process of manufacture was accepted.

965 The Working Group returned to the issue of the non-complying vaccines. Correspondence with the two companies concerned had produced updated information.

966 The Working Group decided that a licence should not be given to the first product (unlicensed MMR vaccine) unless it complied with the guidelines, and that existing trial batches should not be used.

967 There was still no alternative to the third product (Tuberculin PPD), which used glycerol beef broth during the process of manufacture. Stocks were available up to September 1991. These would be changed over 'as appropriate' as the new supplies, which were peptone-based, came on stream. The Working Group thought that the replacement of stocks should take place as quickly as practicable, but meanwhile, given the low risk from glycerol broth, the danger of having no stocks outweighed the risk from the product.

968 The source of the measles vaccine was being changed to New Zealand and present stocks would be depleted in three months.

969 The company preparing DTP vaccines had changed the source of its bovine media, but meanwhile was still using non-complying material. The Working Group recommended a meeting with the company to discuss bringing forward the time when there was compliance with the guidelines.

970 The safety of topical products was also reviewed at this meeting, in the light of action taken earlier that year on cosmetics. The only two products using bovine material sourced it from West Germany, and it was decided that no further action was needed on licensed topical products.

Fourth meeting of the BSEWG on 31 October 1990

971 This turned out to be the main 'wash-up' meeting of the Working Group. They unanimously decided that the special circumstances of the experimental transmission of BSE to a pig did not warrant a fresh look at porcine material. On allergens, they were told that progress with the company concerned was satisfactory.

972 By now the last of the replies to the questionnaire had been received, some 18 months after they had been sent out, and gave no cause for concern. On the outstanding issue of the stocks of the DTP vaccine, the Working Group was beginning to take a more hawkish line. The stock-out dates for the adsorbed vaccines were now between June and December 1991. Those for the unadsorbed vaccine ran beyond 1991. The Working Group asked its secretariat to explore with the licence holder whether the stocks of the latter could be replaced sooner.

Veterinary products

973 On the veterinary side assurances were still awaited from some companies that appropriate action had been carried through. The BSEWG had received progress reports from Dr Lee at each of its meetings, although this item appears to have been treated as purely for information. The difficulties and delays experienced by the VMD over collecting returns, clarifying obscurities and phasing out certain products had broadly mirrored those on human products. We note that when the VPC had its second and final discussion about the exercise in December 1990, there were at least two companies with considerable stocks of vaccines expected to last another four years. The VMD provided us with a table outlining the 143 products that did not initially comply with the CSM/VPC guidelines and the outcome of compliance measures taken. This indicated that apart from one fish vaccine, all manufacturers had complied with the guidelines, so far as their manufacturing processes were concerned, by 1992.

Final meeting of the BSEWG in July 1992

974 After its meeting in October 1990, the BSEWG lay fallow for almost two years. One or two proposals for a meeting came to nothing. BSE did not figure on either the CSM or BSC agenda. However, in July 1992 what proved to be the final meeting of the BSEWG was held. The Working Group considered the implications of the emergence of BSE overseas for medicines, in particular sutures from France. By now there were European guidelines in place for human medicines. These were in some respects a little looser than the UK guidelines, though based on the same principles. They did not, for example, cover sutures. The BSEWG view was that the UK should treat sutures as if they were covered by the guidelines even though other countries did not do so.

975 Once again, concerns about foetal calf serum were raised, with Professor Collee stressing that continued vigilance was necessary. Besides the unanswered question of whether it could in itself transmit infectivity, there were also concerns about collection methods. These concerns were similar to those raised by Dr Pickles some three years earlier and referred to by Mr Scollen in his report to Mr Cruickshank in February 1989.

976 One item that does not appear to have been raised at the meeting was the safety of gelatine. Dr Minor had suggested shortly before that it might be discussed there. He had been disturbed to learn at a meeting in Heidelberg about the 'shockingly mild' German manufacturing process after 'any old cow bone went into the production vat including spine and skull'. There was a pharmaceutical interest in gelatine because it was used for capsules as well as in some other forms. The matter was in the event followed up by a written opinion being commissioned from Professor Collee. His advice was that the BSE guidelines on sourcing should apply to gelatine. Dr Purves told us that this was taken into account in dealing with product licences subsequently. Problems over gelatine rumbled on thereafter, with British suppliers taking steps to exclude UK material in order to meet increasingly rigorous demands from their overseas customers.

Overview of the way the guidelines were implemented

977 We discuss at some length in Chapter 6 of Volume 7 some features of the way in which phasing out existing products was handled and the reasons for this. We note in particular three factors that directly influenced the response:

i. Uncertainty about the risk. Officials and expert committees had to operate mainly on the basis of value judgements, unable as they were to assess and cite proven adverse reactions.

ii. The management situation. The heavy task of conducting a case-by-case approach was superimposed on a creaking system that was overloaded and understaffed. Meanwhile the licensing divisions were undergoing restructuring and had new management preoccupied with other pressing tasks.

iii. Mixed messages about the urgency. The general perception after February 1989 was that although the measures were in themselves quite drastic, they did not have to be treated as an emergency given that Southwood assessed the risk as remote. The low-key presentation of risk, carefully crafted to avert public alarm about the vaccination programme while remedial action was being taken, had the unfortunate result of being taken as the message itself. This must also have influenced manufacturers' attitudes.

Veterinary medicines

978 In the case of veterinary products, a decision was taken that the VMD should pace and match its action to that of the MCA. Although we thought this was a reasonable approach, it seemed, unfortunately, that playing second fiddle was one of the factors that led to a less urgent and decisive approach than was originally envisaged. We are in no doubt that a further factor was that, like the MCA, the VMD read the Southwood message as basically reassuring. Whether the decisions on veterinary medicines had an impact on the numbers of BSE cases may never be known. It is impossible to say today whether continued use of bovine-based medication may have added to the total number of BABs.

Human medicines

979 In the case of human products, the problems in tackling the exercise were greater and the organisational arrangements more complex. The lack of an obvious lead branch in MD continued in the MCA. While there was a team effort, this lacked leadership to prescribe what it was expected to achieve overall and who was to do what by when. Matters were not helped along by changing responsibilities during the process of integrating the administrative and professional branches.

980 The BSEWG was a useful means of achieving speedy advice from the key experts. But the Working Group relied on the MCA to refer matters to it and to act appropriately after receiving its advice. It did not itself lay down any imperatives, such as deadlines for action to be completed, other than to urge that things be done 'as soon as possible' in some cases. Officials were not accountable to it. However, once the BSEWG ceased to meet, the impetus for officials to prepare progress reports appeared to disappear.

981 The three most sensitive groups of products used for humans were (i) those containing brain and other high-risk tissues as an ingredient, (ii) sutures and (iii) vaccines. We concluded on these as follows:

- *Products directly containing high-risk tissues, eg brain and glands:* the small number of products concerned were identified and dealt with reasonably promptly.

- *Sutures:* discussions were promptly and effectively conducted with the major UK producer, safeguards introduced and use of UK materials phased out as speedily as practicable. The experts' recommendations on sutures for general use were reasonable. On the specific question of continuing use in neurosurgery, we think with hindsight that it would have been preferable if the minority view among the experts that this should not continue had prevailed. We note, however, that as yet no cases of vCJD appear to be associated with their use.

- *Vaccines:* bovine material was not an ingredient in the finished product. What was unclear was whether its use as a growth medium for cells allowed infection to transmit. Results of studies on serum carried out by the NPU in which no infectivity was detected were not available until 1993. The general view before then was that this was a very low-risk material and that there was in any event only a remote risk of the BSE agent passing to humans via medicines. Given this, and the dangers of interruption to the vaccination programme, we think it was not unreasonable to conclude that the balance of risk to benefit favoured using the existing vaccines until alternative supplies became available.

982 The corollary, it seemed to us, was that the replacement process needed to be as speedy as possible. While the individual decisions taken by DH about each of the products concerned were reasonable, it can be seen with the benefit of hindsight that they contributed overall to a protracted process of achieving compliance with the guidelines. Parallel delays were incurred in the treatment of veterinary products. It seems highly unlikely that so long a period of grace was envisaged by those taking decisions on vaccines in February 1989. Knowing what is now known, a harder line might have been taken to reduce the length of time that both people and animals continued to be exposed to suspect products. Although this is in part attributable to the false impression on risk, there was undoubted room for improvement in the way the guidelines were followed up. In particular we think it would have been better if:

i. there had been a handling plan with well-defined leadership that 'managed' the whole process to specific deadlines; and

ii. there had been clear expectations about reporting to top management and Ministers. We believe Ministers should take a lively interest in what is being done in their name, and that there should be clear presentation to them of important policy decisions.

983 We have noted that, once medical devices were identified as a concern, action to ensure their safety was handled purposefully. The PD style of administrative approach (see paragraph 865 above) might with advantage have been mirrored elsewhere and have led to a brisker momentum in phasing out suspect products.

984 Taking animal and human medicines as a whole, matters that were handled well included the heroic venture of a questionnaire to all licence holders to make good the faults in the database. Despite believing that action was purely precautionary, officials worked diligently to carry the follow-up action to its conclusion. The most urgent items were identified and dealt with promptly. A voluntary total switch of sourcing was secured, despite there being no firm evidence to offer of human risk. All this was achieved while struggling with the legacy of serious past failings in the running of the licensing system that were still being addressed.

Research into pharmaceuticals

985 As the story of the way medicines, and in particular vaccines, were handled has shown, there was a pressing need to establish whether bovine serum was infective. The only way to do this was by research. In Chapter 7 of Volume 7 we look at what happened to proposals for research into this.

986 The need for this research had been identified at the NIBSC discussion in May 1988, though it appears that no studies into the infectivity of serum were carried out as a result of this meeting.

987 However, the subject was not forgotten. When the Tyrrell Committee prepared its Report on research in spring 1989, one of the items it identified as a top priority was research into which bovine tissues were infective. Given the limitations on the numbers of animals, staff and suitable housing to carry out this research, the Committee agonised over which items should be done first. In its Report it said: 'Nowhere else has the decision on priorities been more difficult.'

988 The decision it reached included ranking work on foetal calf serum and bovine serum albumin as a three-star (ie, top) priority.

989 In Chapter 7 of Volume 7 we trace the events that followed after the *Tyrrell Report* was presented to MAFF and DH. The proposal had a chequered history. In August 1989 Mr Gummer proposed and Mr Freeman agreed that it should be jointly sponsored and funded by both Departments, reflecting their joint responsibilities under the Medicines Act. Money was earmarked. However, following the first BSEWG meeting in September, Dr Pickles indicated to Mr Hagger that the MCA might want to consider whether the work was still needed, given that the action agreed by the BSEWG should ensure that contaminated material would not be entering pharmaceutical processing. She pointed out the need to secure Dr Tyrrell's support for such an approach. In January Dr Pickles informed Ministers, at the time the *Tyrrell Report* was being published, that the MCA was acting on the recommendation together with its experts.

990 When Mr Lawrence circulated a chart showing progress on the Tyrrell recommendations in April 1990, he noted that work on serum research was being carried out at the NPU with industry funding, adding that trade restrictions and industry sourcing from outside the UK had lowered the priority on research into serum.

991 It is plain now that MAFF and DH had to an extent been operating at cross-purposes. DH had been concentrating solely on the proposal allocated to it, namely to secure research on serum. The *Tyrrell Report* had identified this item as just one element in the general programme of tissue testing. That other general work was being taken forward by MAFF and the NPU.

992 Mr Bradley of the CVL had reached the judgement in December 1989 that foetal calf serum was one of the top priority items for the limited animal resources available. The CVO agreed with him and it was included in the quota of tissues for transmission studies in the first year of the project with the instruction that it was important to get these studies under way as soon as possible. MAFF emerges with credit for its purposeful handling of the matter.

993 The work was done by the NPU and the results were made available in 1993. No infectivity was shown in these tests of foetal calf serum.

994 Thus despite its apparent downgrading by DH, the work was actually done. However, it seemed to us that this outcome was in some respects achieved despite inconsistencies in approach and a degree of mutual misunderstanding. Four features struck us as having complicated the process:

- The notion that industry might voluntarily sponsor and share the results of the work.

- The compartmentalising of the serum and other tissue study items, first by the Tyrrell Committee and then by MAFF, in how they allocated responsibilities. This led to confusion about how the work was carried out thereafter and who was calling the shots.

- The detached attitude of the medicines licensing divisions, which had an interest in the outcome.

- The divergent perceptions of MAFF, DH and SEAC about what was actually happening on the Tyrrell proposals.

Cosmetics and toiletries

995 We have grouped our material about the risk of transmission of BSE from cosmetics and toiletries in the same volume (Volume 7) as medicines because these products had much in common. In particular, both might apply animal materials to the skin, the eye or to mucous membranes. But, as we shall see, they were covered by a very different set of safety provisions.

The main products

996 Cosmetics using bovine materials fell into three categories. Those most likely to present a risk of BSE contamination were some 'exotica'. They included anti-ageing and anti-wrinkle creams and 'cellular extracts' such as premium face creams. They might contain only lightly processed brain extracts, placental material, spleen and thymus. This was the most urgent category to tackle.

997 The second category consisted of 'High Street' topically applied products such as creams and toiletries applied to the skin, lips and eyelids. It also included items like soaps, shaving sticks and stick deodorants. The bovine materials used were heavily processed. Although questions were asked about ensuring the safety of this group of products, they were never considered a serious risk.

998 The third category of concern was bovine collagen used in implants. Dr Pickles was concerned initially about their use in unlicensed clinics as beauty preparations. We looked into their status. DH told us that in practice this material was used under medical supervision and thus treated as 'prescription only medicines'. We concluded that we need not explore their cosmetic use separately.

Regulation

999 The Department of Trade and Industry (DTI) had regulatory responsibility for the cosmetics industry. At the time BSE emerged, Mr Richard Roscoe, who was a Grade 7 officer, headed the branch in charge of the safety of cosmetics sold in the UK. DTI looked to DH, and in particular to Dr R J Fielder, for advice about toxicity of products that were causing concern.

1000 The legislation governing safety was the EU Cosmetics Directive and Regulations made under the Consumer Protection Act 1987. We set out details of these provisions in Volume 7. Although cosmetics had to meet various safety requirements, they did not require a licence. Enforcement lay with local authority Trading Standards Departments, which would require some evidence of harm before seeking to intervene. The Secretary of State for Trade and Industry also had certain intervention powers. In practice the regulation of the industry operated very much on an informal and voluntary basis, relying on the industry to cooperate.

1001 Although identified in the *Tyrrell Report* in June 1989 as needing consideration, the cosmetics industry received no advice or guidance until February 1990. We deal briefly first with how this happened. We then look at what happened thereafter.

The Tyrrell recommendation on cosmetics

1002 The *Tyrrell Report* submitted in June 1989 had this to say about cosmetics:

> Some uncertainty remains as to whether all the possible routes of transmission from bovine (and ovine) tissues to other species have been considered and appropriate action taken. Small scale users of bovine products such as the cosmetic industry, may not be covered by the present regulations and guidelines.

1003 Coupled with a wider proposition about investigating the fate of bovine products passing through as yet unrecognised routes, this item was given a three-star recommendation for further work. We return later to what happened to this wider proposal for an audit of bovine tissues.

1004 Despite what the Report said, no steps were taken by MAFF or DH to contact DTI about cosmetics. By good fortune, Mr Roscoe at DTI learned of the possible risk from BSE and independently decided to ask DH about it in January 1990. After he had consulted medicines licensing officials and Dr Pickles, Dr Fielder provided advice to Mr Roscoe. The gist of it was that DTI should warn the cosmetics industry via its trade association, the Cosmetic, Toiletry and Perfumery Association (CTPA), that it should reformulate products so as to exclude bovine offal or source it from outside the UK.

1005 This Mr Roscoe promptly did. The CTPA in turn relayed this advice in full, first to those of its members that made 'premium skincare products' (the ones most likely to contain offal extracts), and second to members generally. Ms Marion Kelly of the CTPA told us she was confident from members' replies at the time about premium face creams that no products were using UK material. Replies to a request for information from the wider membership had not been retained.

Was the initial action adequate?

1006 We considered first the failure to alert DTI in 1989 to the need to consider cosmetic products in relation to BSE. We think that Dr Pickles, who had the lead on BSE in DH, should have done so. We were not impressed with her argument that the risk had been 'so slight that effectively it could be disregarded'. This ignored the need to inform DTI as the regulatory Department and the fact that she could not have known which products were involved.

1007 Throughout the BSE story, Dr Pickles took many prompt and commendable initiatives to alert those concerned and to carry action forward. Sadly, in this case, Dr Pickles fell short of her normal high standards. She acknowledged to us that had she informed DTI, it could have addressed the issues six months earlier than it did.

She should have done so; but this lapse is minor in comparison with the commendable action taken by her in many other respects.

1008 Within MAFF, we considered that responsibility for informing DTI lay with Mr Lowson, the head of Animal Health Division. We were not persuaded by Mr Lowson's argument that he had only a hazy notion of DTI involvement in the cosmetics industry and that this was a human health matter so 'something where one would expect other Departments to take the lead, particularly the Department of Health'. In our view Mr Lowson shared responsibility with Dr Pickles for ensuring the recommendations were properly assessed and followed up. We consider that, jointly with Dr Pickles, Mr Lowson should have promptly ensured that what the *Tyrrell Report* said on cosmetics was drawn to the attention of DTI. The failure to do so contributed to several months' delay in initiating action to secure the safety of cosmetic products.

Was DTI action adequate?

1009 Mr Roscoe deserves credit for registering that BSE might pose problems for the cosmetics industry, and for acting promptly in seeking advice from DH, and passing it on to the CTPA. We agree that the Department's statutory powers to intervene were not appropriate in these circumstances and that the only realistic course open to DTI was to persuade the industry to take voluntary action. Mr Roscoe's letter and the response by the CTPA were together the most significant single action taken to address the risk from cosmetics.

1010 However, we think it is unfortunate that Mr Roscoe did not make efforts to contact firms which were not members of the CTPA. It was indeed, as he said, a 'flaw in the system . . . that we could not reach all manufacturers'.

Action taken thereafter

1011 We turn now to the way matters were handled after the CTPA had distributed the DTI warning. Initially, everything went quiet. Dr Pickles had included a question about the adequacy of the action taken on cosmetics in a draft paper for the first meeting of SEAC in May 1990, but Mr Meldrum raised some concerns about the paper, and it did not go forward.

1012 Three members of SEAC, Dr Tyrrell, Dr Kimberlin and Dr Will, attended the meeting of the BSEWG in July 1990, at which the DTI action on cosmetics was noted, and topical medicinal products were again given the all-clear.

1013 However, SEAC itself did not turn to cosmetics until March 1991, when it asked for a paper on the topic. This task fell to Mr Murray, who had taken over from Dr Pickles as the DH secretary to SEAC. Mr Murray asked one of his staff to make enquiries of the CTPA into the use of bovine material in cosmetics. It was unusual not to approach DTI as the Department responsible for cosmetics safety.
Mr Murray's paper identified the uncertainties about the use of bovine material in cosmetics, and about small-scale producers that were not members of the CTPA.

1014 SEAC discussed Mr Murray's paper in July 1991, along with a paper from Dr Pickles about non-food uses of bovine material more generally. The Committee thought that in general no problems arose, but asked that DTI be reminded of the need to update the guidance to cosmetics manufacturers in the light of the emergence of BSE in other countries. After the meeting, Mr Murray asked Dr Pickles for her view on updated guidance, and she queried whether 'fringe' cosmetics companies were being kept informed by DTI, and advised Mr Murray, when writing to DTI with the guidance, to ask to be told about what happened thereafter.

1015 Although Mrs Diane Whyte in DH drafted a letter to Mr Roscoe, it appears not to have been sent. Work continued somewhat slowly on the text of a draft letter to revise the guidance, but no contact was made with DTI. Meanwhile Mr Bradley of the CVL had told Mr Lawrence with some perspicacity that 'contacts via DH/DTI do not inspire me with confidence'. He felt that MAFF needed either to go out to the industry to assess what kind of bovine material was really used in cosmetics and for what, or to have closer contact with the trade association. He observed:

> I am not satisfied yet that the industry is in the clear and it is us that may shoulder some blame if it is later found ladies are rubbing cow brain or placenta on to their faces.

1016 DH, as it happened, shared Mr Bradley's view that they needed hard facts about the situation, and matters now took a different turn. DH had drawn attention to the lack of knowledge in its paper for SEAC. This led the Department in early 1992 to decide to put a series of detailed questions to the industry to clarify the situation and what action was being taken. The plan now was that, depending on the outcome, a meeting with the CTPA might be arranged, and, if need be, guidance considered later. DH officials did not consult DTI about these ideas. Although the object was sound, the exercise proved abortive. It was simply impracticable for the CTPA to provide answers from its members within three weeks to a list of 20 detailed questions asked out of the blue. There was no obligation on the industry to provide such information.

1017 However, the CPTA did put a note in the May edition of its scientific newsletter to say that an enquiry had been received from DH about the use of bovine and ovine materials, and asked any of its members using these to contact the Association urgently. There was no positive response from CTPA members. Ms Kelly told us she read this as meaning the members were not using such materials.

1018 The CTPA's response led DH to press ahead instead with efforts to draft the guidance letter originally called for by SEAC a year earlier. In July, Dr Fielder who, besides being the toxicological adviser to DH, was a UK member of the EU expert committee on cosmetics, took a hand. He pointed out that there was a risk of getting into deep water with the European Commission if they sought a voluntary ban. He suggested a meeting involving DTI before the CTPA was contacted again. This was a timely proposal. Among other things, it brought DTI back into the frame.

1019 A meeting was held in September 1992 between officials from DH, DTI and MAFF and CTPA staff and members. There was a useful exchange of information. The outcome was agreement that DH would provide advice to the CTPA on gelatine; the CTPA would list products using risk materials; the CVL would offer advice about suppliers of material; and the CTPA would consider further what guidance might be prepared. Dr Wight had called this meeting to bring the parties together as suggested by Dr Fielder, but it was not clear whose call it was next. The initiative on preparing guidance had now been passed from government, whose job it was to ensure the safety of cosmetics, to the trade association which would be disseminating it.

1020 However, there was some follow-up contact by telephone and letter. The CTPA subsequently wrote to Dr Wight at DH to say that it had contacted a company using cerebrosides and that this material would be phased out by early 1993.

1021 From this point on, action moved to the European arena, with DTI in the lead. Before long the EU Working Party on Cosmetics became involved, with a view to preparing guidance at the European level. The DH reaction was that this was welcome as it helped to avoid the impression that the problem was solely one for the UK cosmetics industry. However, Dr Fielder flagged up the danger that the exercise might drag on, when in fact guidance needed to go out as soon as possible.

1022 Dr Fielder's fears were realised – the exercise did indeed drag on. Preparation of European guidance became embroiled in slow procedures, infrequent meetings and national differences of view. COLIPA, the European trade association, played an active role providing reassurance that voluntary action had been taken.

1023 In March 1994, at the EU Health Council, all Member States except Germany supported the view that existing measures to contain BSE and protect public health were sufficient. It was eventually decided that the Cosmetics Directive need not be amended to ban the use of bovine material. It was later amended, after the period covered by this Inquiry and the emergence of vCJD.

1024 Meanwhile the CTPA had told DTI that it would prepare UK guidelines jointly with the French industry. The CTPA guidance to UK manufacturers was eventually issued in March 1994. It followed closely guidance from the World Health Organisation that had been issued in 1991 on inactivating TSEs and categorising tissues into four categories of infectivity. It is difficult to see how much, if any, value was added by the long delay.

The adequacy of the response

1025 A problem in assessing the adequacy of the response is the lack of knowledge that persists today about what cosmetics that contained bovine ingredients were on offer at the time and what precisely they were used for. With hindsight, we agree with Mr Bradley's view that first-hand knowledge needed to be sought. We revert to this matter in Chapter 9.

1026 We recognise the handling problem created by the limited powers available to deal with an unproven threat like BSE which affected raw materials. We have commented elsewhere on the desirability of statutory powers to destroy dangerous material at source.

1027 Given these considerations it can be seen with hindsight that two things were needed.

1028 The first was purposeful leadership. There was continuing vagueness about who was in the lead. This confusion operated both between Departments and within DH. We are in no doubt that the lead should have lain with DTI, with professional advice from DH. Dr Pickles's instinct that DTI should be asked to carry forward the guidance and required to report progress was sound.

1029 The second was a sense of urgency. This was patently lacking. DH thought the risk was remote. Dr Wight told us that when she arrived in DH in 1991 to take over from Dr Pickles, she understood that all the significant action on BSE had by now been taken and her role was principally a watching brief. The perception that revised guidance for cosmetics was urgently needed and that certain matters needed to be vigorously followed up had faded away. Manufacturers were left to use up stocks, and checks were not made to ensure that they had reformulated their products.

1030 Taken together, the effect was to leave large gaps in knowledge and to delay inordinately the issue of further advice. As with medicines, this has left unanswered questions about the products affected, how long production continued and on what scale. It seems to us undesirable that so little is known about products which offer a potential pathway to infection. This is a matter we believe DTI should review.

8. Occupational risk

1031 We turn now to the important matter of occupational risk from BSE. This largely escaped the limelight, save briefly when it seemed that farmers might be particularly vulnerable to CJD. Contact with live animals and with their tissues was a well-known disease hazard. One of the early steps taken by MAFF was to issue detailed advice to its staff on precautions to take if they were if in contact with bovine material. The Health and Safety Executive (HSE) warned farmers and hauliers about the risk of aggressive behaviour in any BSE-affected animals they were handling.

Those at risk

1032 However, these early warnings during the period up to December 1989 reached only some of those handling risk material from cattle. Others were vets in private practice, waste tip and incinerator operatives, slaughtermen and butchers, knackermen, hunt kennel and maggot bait farm workers, renderers and animal feed handlers. Laboratory workers, teachers and students were handling cattle glands and tissues. Workers in zoological parks needed guidance. Later on, medical and healthcare professionals, mortuary workers and undertakers needed to take special precautions in respect of human victims of vCJD. There was also a wide spectrum of occupations handling bovine material being processed for food and other uses, such as fertiliser and collagen.

1033 Ultimately the main occupations at risk were identified and advice given. But this was a long-drawn-out process. It took over three years to complete the task of issuing simple warnings and basic advice to the most obvious high-risk trades. A further two years passed before full guidance went out to those handling risk tissues in laboratories, hospitals and mortuaries.

1034 The following, heavily condensed chronology of the events traced in Volume 6, Chapter 8 shows when advice was issued to the main at-risk groups. It illustrates how protracted the process was, even where it was agreed that a particular group of workers needed to be speedily alerted.

Chronology of occupational safety advice

May 1988	HSE issues guidance for **cattle handlers** about aggressive BSE cases.
July 1988	MAFF issues guidance to its **veterinary and laboratory staff**.
November 1988	Further MAFF guidance to its **staff handling tissues**.
February 1989	*Southwood Report* says HSE is considering appropriate action.

9 March 1989	MAFF asks HSE for meeting on guidance to **farmers, knackermen and workers at disposal sites**.
29 March 1989	HSE identifies **slaughterhouses** as a possible risk.
April 1989	MAFF puts draft interim advice to HSE about **carcass handling**.
9 June 1989	Meeting between HSE, MAFF and DH (first of series). Brainstorming identifies **farmers, vets, slaughterers, knackers, butchers, stockmen, market handlers, fell mongers, renderers, lab workers, those working at incinerators, artifical inseminators, local authority inspectors**. MAFF considers the first four need urgent advice.
25 July 1989	MAFF drafts advice to **vets**. British Veterinary Association (BVA) agrees to draft own guidance.
8 August 1989	HSE issues news release on advice to **carcass handlers** mooted in April.
11 September 1989	HSE undertakes to redraft MAFF draft guidance note to **abattoirs**.
December 1989	HSE issues general information sheet on **handling zoonoses in agriculture**. Passing mention of BSE says no evidence that it is transmissible to humans.
January 1990	MAFF issues guidance agreed with BVA for **veterinary surgeons**. HSE considers no immediate guidance for **farmers and farm workers** is needed. MAFF disagrees.
February 1990	HSE publishes pocket carry cards on BSE and **carcass disposal**.
March 1990	HSE publishes Guidance Note 5 on occupational risks of BSE for **workers in abattoirs and meat trade**.
10 May 1990	When a TSE is diagnosed in a cat, Dr Pickles suggests **neurophysiologists** and others might need advice.
31 May 1990	Agreement that guidance is needed for **renderers**. More meat trade advice desirable on deep cuts, use of bandsaws and inhalation of material.
June 1990	MAFF advisory note for **farmers** on handling BSE suspects, and breeding.
24 August 1990	MAFF guidance to **zoo workers**.
6 September 1990	HSE/MAFF/DH decide against further advice for **meat trade**, which is opposed to it as drafted, and to leave aside for the time being guidance to **renderers**.
February 1991	Working Group of Advisory Committee on Dangerous Pathogens (ACDP) set up to prepare **health and safety advice on handling human and animal TSEs**.
15 March 1991	HSE/MAFF/DH decide guidance for **renderers** is needed.
7 October 1991	HSE/MAFF/DH identify dangers of cuts from splitting cattle heads and spines. Agree guidance needed for **knackers, hunt kennels and maggot bait farms**.

FINDINGS AND CONCLUSIONS

October 1991	Draft fast-track letter for **medical professionals** circulated in ACDP Working Group (ACDPWG).
28 November 1991	Food National Interest Group (NIG) advice note to HSE Inspectors about forthcoming comprehensive advice on precautions for **knackers, renderers and slaughterhouses**. Emphasis on employer surveillance, hygiene of pithing rods and risks of hand-scooping of brains.
9 June 1992	HSE issues comprehensive advice for **knackers, renderers and maggot bait farms**.
8 December 1992	Fast-track letter issued to **medical professionals**.
September 1994	ACDPWG **guidance on TSEs** published.
April 1995	Guidance issued in 'Communicable Disease Report Review' for all those **handling human cadavers**.
December 1995	HSE/MAFF/DH working group meets for first time since October 1992. Agrees to reinforce present guidance.
January 1996	Update of HSE's **Guidance Note 5 for slaughterhouse/meat trades** issued.
June 1996	ACDP guidance issued for **all workers in contact with BSE**.

1035 As we built up this reconstruction of events from documents made available to us and witness statements, we were dismayed by the delays that occurred in advising workers at risk from contact with the BSE agent. Time was not available to explore this large field of evidence in depth at our oral hearings and with further witnesses. Our Report therefore does not attempt to pinpoint the actions of individuals but rather to look at weaknesses in the system that caused us concern. Two illustrative examples are described below.

1036 The first was the issue of advice from the Advisory Committee on Dangerous Pathogens (ACDP) to laboratories, medical workers and undertakers. Fuller details are in vol. 6: *Human Health, 1989–96*, Chapter 8. The second was the issue of advice from the Department of Education and Science (as it then was) to schools about dissecting bovine eyeballs. Fuller details are in vol. 6: *Human Health, 1989–96*, Chapter 9. We conclude by drawing attention to some general points that struck us on the handling of occupational safety advice.

ACDP advice to laboratories, medical workers and undertakers

1037 The HSE had an established role on national guidance about handling dangerous pathogens. It also looked for expert outside advice to the ACDP, which reported jointly to it and DH. The ACDP Chairman was Dr Tyrrell.

1038 The ACDP had been closely involved in the categorisation of levels of risk from pathogens and advising on appropriate precautions. It had reviewed procedures for handling CJD and it was natural for the HSE and DH to look to it for advice on handling other TSEs. A Working Group of the ACDP (ACDPWG) was set up in February 1991 to:

> ... report to the ACDP on the need for additional guidance on health and safety aspects of work with animals or humans, their tissues or *in vitro* systems infected or potentially infected with spongiform encephalopathy agent, and to draw up guidance.

1039 Professor Peter Biggs was asked to chair the Working Group, but Dr Pickles stood in for him at the first couple of meetings. Displaying the same energy and purposefulness as on other matters, she launched the work with her own paper. This not only provided draft outlines of the scope of the document that might be prepared but suggested a handling plan and timetable to enable the guidance to appear at the earliest possible date. Unfortunately, that timetable soon faltered and sank into a drafting morass. The following chronology illustrates this. The 14-month history of a so-called 'fast-track' professional letter for neurosurgeons is distinguished by italics. The 'Guidance Document' had a gestation period of over three years.

Chronology of drafting of ACDPWG advice

4 December 1990	ACDP agrees to set up ACDPWG.
January 1991	Dr Pickles circulates draft paper.
28 March 1991	**First meeting of ACDPWG** discusses Dr Pickles's paper.
13 May 1991	**Second meeting** adopts Dr Pickles's paper for internal use as the 'Reference Document'. This is to be the basis for briefer practical guidance for wider circulation – the 'Guidance Document'.
6 August 1991	**Third meeting** reviews second draft of Reference Document.
	First draft of Guidance Document.
October 1991	*ACDPWG secretariat circulates draft 'fast-track' professional letter (PL) for neurosurgery and ophthalmic staff.*
24 October 1991	**Fourth meeting** discusses third draft of Reference Document and agrees extensive redrafting needed for wider circulation.
	Guidance Document needs recasting.
	Professional letter (PL) to be issued quickly. DH to take forward.
30 October 1991	ACDP meeting told fourth draft of Reference Document is 'more or less the final draft'.
	ACDPWG welcomes any comments on second draft of Guidance Document 'as soon as possible'.
28 November 1991	SEAC discusses draft Reference Document. Final draft promised for new year.
December 1991	Secretariat circulates third draft of Guidance Document.

14 January 1992	**Fifth meeting:** Reference Document to be available on request; Guidance Document to be widely distributed. Both need redrafting.
	First draft of PL considered.
27 January 1992	*Second draft of PL circulated.*
June 1992	Fourth draft of Guidance Document circulated.
15 June 1992	**Sixth meeting:** no further work being done on Reference Document. Guidance Document has higher priority.
	Third draft of PL considered and amended.
7 August 1992	*Fourth draft of PL distributed.*
8 December 1992	*Fifth and final draft of PL issued to neurosurgery and ophthalmic staff.*
15 February 1993	Fifth draft of Guidance Document circulated.
February 1993	Sixth draft circulated.
5 March 1993	**Seventh meeting:** members asked to consider all aspects of Guidance Document.
24 May 1993	**Eighth meeting:** ACDP should aim to issue Guidance Document only. Final comments sought on Guidance Document.
May 1993	Guidance Document agreed.
14 June 1993	ACDP accepts Guidance Document.
June 1993 to September 1994	Correspondence about publication details.
24 September 1994	Publication of 'Precautions for work with human and animal TSEs' (The Guidance Document).

1040 Witnesses suggested a variety of reasons for this sorry tale. They included uncertainty about appropriate decontamination procedures and about blood products; pressure of other work on the secretariat; and being side-tracked into protracted drafting time on the professional letter of warning on neuro- and ophthalmic surgery procedures. Once the Working Group had become caught up in a cycle of widely spaced meetings to consider substantial redrafting, they were constantly overtaken by emerging new information. Professor Biggs described it graphically:

> In a way, the Working Group was on a treadmill in the sense that any delay arising from the time needed to address a subject, or any other reason, was time during which new information became available requiring re-addressing subjects already dealt with.

1041 A background factor influencing the handling of the exercise was the controversy over whether human TSEs were a category 2 risk or, as some argued, should be in category 3, requiring more rigorous safeguards. There was also debate over whether BSE should be categorised at all, it being open to question whether it was a human pathogen.

1042 Further delays were then incurred until September 1994, after the document had been agreed by the ACDP in mid-1993. We were told that this was while DH finalised advice on at-risk patient groups inadvertently treated with CJD-infected medicines or tissue grafts.

1043 While each of these reasons was no doubt thought to be valid justification at the time for taking a measured pace, collectively they produced what seems to us a quite unacceptable delay. The workers concerned were in occupations that potentially exposed them to particularly high risks, yet they were among the last to receive guidance. The best was allowed to become the enemy of the good.

The issue of guidance to schools about dissecting bovine eyeballs

1044 We turn now to our other cautionary tale, this time involving a different part of Whitehall, the Department of Education and Science (DES), as it was then known. We deal with this topic at some length in vol. 6: *Human Health, 1989–96*, Chapter 9. The dissection of bovine eyeballs in biology lessons was one of the 'unusual pathways' for possible disease transmission – to teachers and pupils – and needed to be addressed, since the eye is closely associated with the brain structure. There was no basic disagreement among officials about that. What went wrong was that the relatively simple task of agreeing the text of a brief warning note about it turned into a two-year saga.

Chronology of guidance on bovine eyeball dissection

27 September 1989	MAFF discusses the issue in context of the SBO ban. Agreed effect would be minimal due to availability of sheep and pigs' eyes as alternatives for dissection. However, MAFF suggests amending Regulations to remove eyes before staining.
February 1990	Scottish Education Department consults and issues advice against using bovine eyeballs in Scottish schools.
20 February 1990	Dr Pickles raises issue of theoretical risk with MAFF and with Dr Diana Ernaelsteen, Medical Adviser to DES.
June 1990	SEAC advises that eyes of cattle more than 6 months old should not be used for dissection in schools.
July 1990	Dr Pickles informs Dr Ernaelsteen of SEAC advice and about the advice issued in Scotland. Dr Ernaelsteen to discuss within DES whether there is a need for promulgation of general advice within England.
July 1990	Welsh Office officials ask Dr Pickles whether guidance has been issued following SEAC advice. Dr Pickles refers them to Dr Ernaelsteen.
	DES Schools Branch 3 accepts responsibility for issuing guidance.

28 August 1990	First draft of submission to the Minister recommending the discontinuance of eyeball dissection.
21 September 1990	DES is reluctant to ban all bovine eyeball dissection and asks about ovine and pig eyeball dissection.
4 October 1990	Dr Ernaelsteen says ovine dissection is unsuitable, but pig or horse eyeball dissection and using bovine eyeballs from calves under 6 months old is acceptable.
5 October 1990	Mr Ron Jacobs (DES) undertakes to revise the first draft of submission to Ministers.
8 January 1991	DES prepares second draft submission to Ministers.
25 February 1991	Dr Ernaelsteen expresses concern to DES at delay in issuing advice.
19 April 1991	HMI queries whether advice issued yet.
25 April 1991	DES circulates revised draft of proposed advice to be cleared with Ministers.
9 May 1991	MAFF tells DES it is content with advice.
February 1992	Mr Jacobs leaves post and passes third and final draft to Mr M B Baker (DES).
March/April 1992	DES seeks cross-departmental views on guidance.
	MAFF queries why the procedure is taking so long but is content with advice. DH is content with advice.
	HSE doubts there are any problems but will contact DES soon.
16 April 1992	Mr J Creedy of HMI draws attention to articles in medical journals which state that risk is minuscule.
21 May 1992	Dr Ernaelsteen advises DES that guidance is not timely now.
June 1992	DES draft submission to the Minister stating it is wise not to take advice further. This is sent to Mr Baker.
	DH queries progress on advice.
August 1992	Welsh Office queries progress on advice.
7 September 1992	DH again queries progress on advice.
30 September 1992	Mr Baker states he is not willing to give this high priority due to Dr Ernaelsteen's advice.
14 October 1992	DH stresses that advice should be issued and that DES should not reject SEAC advice.
28 October 1992	DES responds stating that it will put submission to Ministers.
29 October 1992	DES sends a submission to the Minister of State on bovine eyeball dissection.
15–21 December 1992	Guidance issued and sent to education establishments in England.
7 January 1993	Guidance issued and sent to educational establishments in Wales.

1045 We know this episode has rightly been investigated by DES itself. Mr Baker had identified the issue as a matter for his branch, Schools Branch 3, in July 1990. We have concluded that steps should have been taken to avoid the delay that occurred from May 1991 to December 1992. As he himself acknowledged, responsibility for this delay fell in considerable measure to Mr Baker. Mr Jacobs, who had day-to-day responsibility for the issue within Mr Baker's branch until February 1992, also shared some of the responsibility. Mr Baker and, to a lesser degree, Mr Jacobs should have ensured that this matter was promptly and properly addressed. Mr Baker and Mr Jacobs faced a heavy workload of competing priorities at that time and this is something we have borne in mind.

1046 Unfortunately, it seemed to us that some delay was also caused by Dr Ernaelsteen's advice in May 1992 that guidance was no longer timely. Having commendably stood her ground up to then, we consider it regrettable that, in the absence of any new medical facts, Dr Ernaelsteen countenanced any further delay in issuing advice on stopping the practice of bovine eyeball dissection.

1047 The story seemed to us to offer salutary lessons. The people handling the matter were far from the scene of action on BSE. That was other Departments' business. Their own Minister was not involved. No framework of overall action was in place through which they were accountable. All in all there seemed to be no hurry. Meanwhile other work was more pressing. Safety of pupils and teachers was outside DES's normal remit and many people had to be consulted. As with all civil service documents, there was an urge to refine and polish wording. As time went by, the delay itself made the issue of guidance less appealing.

1048 Here as in other areas, excessively reassuring language about the risk from BSE sedated those who needed to act. Insofar as they had a perception of the situation, it was that the risk was remote. There was no strong sense of 'ownership' of the topic to overcome the difficulty of working across normal boundaries in unfamiliar territory. There was no overall frame of reference and accountability.

Overview of occupational health

1049 The factor that most influenced the pace of action in both these case studies, and in reviewing occupational safety generally, was the belief that the risk from BSE was remote. We discuss elsewhere the reverberations of the wording used in the *Southwood Report*. In particular, the recommendation to the HSE in the Report about the issue of further advice to at-risk groups was scarcely a clarion call to action. The HSE attributed MAFF's eagerness to get advice issued to political and media pressures. It saw no reason to depart from its normal number-based risk assessment approach and measured processes for evolving guidance. These were sound but slow.

1050 A second factor was the absence of a comprehensive review of pathways of transmission to ensure that all the critical points had been identified. As discussed in Chapter 9 of Volume 7, Dr Matthews of MAFF, immediately after his meeting with the HSE on 9 June 1989 to discuss the issue of advice, had commissioned a list of slaughterhouse products and their destinations. This was intended to assist thinking about high-risk occupations that should be given early consideration.

Unfortunately, this exercise was not taken much further. Had the audit of possible pathways of infection proceeded, it might have helped to pinpoint where the issue of urgent advice could not wait.

1051 The third factor was the inherently slow metabolism of the consultative and drafting arrangements on occupational safety. While polished and carefully agreed detailed guidance was to be desired, it ought not to have been at the expense of prompt and straightforward interim warnings.

9. Potential pathways of infection

Consideration of an audit of the uses of cattle tissues

1052 The last part of vol. 7: *Medicines and Cosmetics* deals with a topic that concerned not just medicines and cosmetics, but also many other industries and activities where BSE posed a threat. This was the need to establish all the ways in which cattle tissues were used, in order to ensure that BSE infection was not spread by unrecognised routes.

1053 We consider it was a top priority to prepare an overview of this kind. A proper understanding of all the ways in which cattle tissues were used was fundamental to the planning of suitable measures to stop the disease from spreading. Those responsible for action in each area of concern needed to be contacted and the risk assessed. The industries and groups of workers involved stretched far beyond the ambit of MAFF. Coordination of measures and ensuring they covered all the ground was going to be important. Various pieces of safety legislation might have to be deployed. Many Departments and public bodies would be involved in enforcing and monitoring individual activities. Once action had been taken, the map of identified pathways could be used to monitor the situation, and ensure new information was relayed to those who needed to know it.

1054 However, a comprehensive overview exercise was not carried out. Gaps in knowledge were still causing problems seven years after the need for an overview was identified. This led to new proposals within MAFF for a research study. The term 'audit trail' was applied to it, a convenient description we have used here.

1055 What follows is a condensed account of what happened. The fuller story and analysis can be found in vol. 7: *Medicines and Cosmetics,* Chapter 9.

The Tyrrell recommendation

1056 The Southwood Working Party in 1988 had agreed with Dr Pickles that it would be useful to have an epidemiological flowchart to determine what bovine material was used for. They followed up the most pressing issues they had identified, but did not themselves prepare an overview of all uses.

1057 The matter was picked up in the *Tyrrell Report* on research into TSEs a year later, in June 1989. This had an item as follows:

> Item A1d More detailed investigation into the fate of bovine (and ovine) tissues and products that could lead to infection being spread by as-yet-unrecognised routes.

FINDINGS AND CONCLUSIONS

> Some uncertainty remains as to whether all the possible routes of transmission from bovine (and ovine) tissues to other species have been considered and appropriate action taken. Small scale users of bovine products, such as the cosmetic industry, may not be covered by the present regulations and guidelines. There are no formal proposals for work of this sort and consideration should be given as to whether such a study should be commissioned *** [ie, three-star, top priority]

1058 Along with the other three-starred items, A1d was approved by Ministers for immediate action. MAFF had divided all the recommended projects into two tables. Table 1 contained items to be wholly funded by MAFF; Table 2 listed the remainder, to be jointly funded or to fall entirely to others. The audit and cosmetics item, captioned 'Spread of infection by unrecognised routes', was listed in Table 2. The wording was vague: 'Those routes currently considered important are being pursued. Scientific progress may reveal the need for further action. This issue is of importance also to DH.'

1059 There had meanwhile been a meeting with the HSE in June to follow up the Southwood recommendation on occupational risk. This was attended by Dr Matthews, a veterinarian in MAFF. After the meeting he asked Mr Hutchins, a Senior Veterinary Officer in MAFF's Meat Hygiene Veterinary Section, for a background paper listing the destinations of slaughterhouse products. The object was to help identify workers at risk. Mr Hutchins promptly produced a businesslike list of raw by-products, processed by-products and their use. MAFF officials were at this stage heavily engaged in deciding which tissues needed to be covered by the proposed SBO ban. They did not seek to trace any further the fate of the items identified in the list or to contact other Departments that might have an interest in them.

1060 In March 1990 Mr Lawrence of MAFF's Animal Health Divisiom drew up a progress chart of where each of the Tyrrell proposals now stood. This revealed that nothing had been done about the audit since Ministers had agreed it the previous August. Cosmetics, as we have seen, were being tackled, thanks to Mr Roscoe at DTI.

1061 Faced with this awkward situation, Mr Lawrence turned to the MAFF Meat Trade Adviser, Mr Chris Rogers, for advice about outlets for slaughterhouse material. Mr Lawrence appears to have been unaware of the list prepared by Mr Hutchins. Mr Rogers identified many of the same items, adding his own observations. One of these concerned a different sort of by-product, namely slaughtering and rendering waste. We return to that later in this chapter.

1062 In May, while being briefed for a Parliamentary Debate on BSE, the MAFF Minister, Mr Gummer, learned that the audit had not yet been set in hand. He instructed that it should go ahead forthwith and that MAFF should fund it. Some confusion and misunderstandings then ensued about whose job it was to draft the protocol for the work. The details appear in Volume 7.

1063 The upshot was several more months of inaction. SEAC was told on 2 July 1990 that the project had not been followed up, but that MAFF was seeking information from slaughterers about where bovine tissues went so as to provide the basis for a comprehensive picture of the products in which they might be used. It appears that MAFF officials were taking a narrow view of what was required. A few days later, Mr Lowson told Dr Kenneth MacOwan, who managed the MAFF research budget, that MAFF had kicked this off through an enquiry at slaughterhouses to establish what happened to the whole range of bovine tissues, and that pending the results of this enquiry he would not see a need to direct resources to the item.

1064 Thereafter matters gathered dust until March 1991, when SEAC called for a paper on non-food uses of bovine materials and MAFF set about updating its progress chart. Dr Pickles queried the assertion in the MAFF chart that DTI, MAFF and industry had the item in hand. The dust was blown away with a vengeance. It was now revealed that nothing had been done. Mr Maslin told Dr Pickles:

> From our papers it would seem that there has been no 'study' initiated. The references to 'DTI, MAFF, Industry' was I assume included in the summary chart in the early days and has simply been perpetuated in later charts. Alan Lawrence recalls that this was a matter raised with Mr Gummer before the BSE Parliamentary debate last year. It seems however that this area has fallen through the cracks.

1065 The chart had confused the follow-up on cosmetics – where DTI had been in touch with the industry – with the wider audit.

1066 By way of response to the situation, Mr Bradley of the CVL provided an 'off the top of my head' set of suggestions about non-food uses. Mr Maslin suggested to Dr Pickles that these and Mr Rogers's list of the previous year might be amalgamated to form the paper sought by SEAC. There ensued a spirited exchange between MAFF and DH about who was to blame for the item falling through the cracks.

1067 In a minute to Dr Pickles, Mr Lowson conceded this ought not to have happened:

> I entirely agree that it is not satisfactory that this item on the Tyrrell shopping list should not have received the attention it deserved.

1068 However, he did not accept that the blame lay with MAFF. He had understood Dr Pickles was drafting the protocol, though:

> . . . it was a hot afternoon, a long meeting and nobody produced a note so I would not want to be too critical of the fact that nothing seems to have happened as a result. No doubt for our part we should have been more assiduous in trying to find out what was going on.

1069 In response to Mr Maslin's suggestion about the list, Dr Pickles observed:

> Of course I could make a start at a 'list' but the purpose of a research study was to investigate more formally as to what actually happens, not what some of us think might happen.

1070 We entirely agree with Dr Pickles's observation. What was needed was a full and accurate picture tracing products through their various handling and processing stages. This was going to extend well beyond the boundaries of MAFF's knowledge.

1071 What in fact happened was that, with a few additions, Mr Hutchins's original list of uses from June 1989 was annexed to a paper by Dr Pickles for the SEAC meeting on 28 June 1991. SEAC was asked to consider if the list was complete and if these uses presented any risks to the public or to workers. SEAC thought that in general no problems arose but was still concerned about some matters. One of these was the risk that unstained and unsterilised SBO might end up in products that could come into contact with humans.

1072 Mr Lawrence prepared a paper reviewing the controls and the guidance on pharmaceuticals. It took an optimistic view that these covered the situation but suggested that a further check could be made through the abattoir owners on the destination of by-products. This suggestion does not appear to have been discussed by SEAC when the paper was tabled in September, nor does it appear to have been followed up. The SEAC interim report on research published in April 1992 said that the fate of bovine tissues had been examined in-house by MAFF and was not progressing as a formally commissioned piece of work.

1073 Thereafter the need for an audit of this kind did not resurface until 1995, when it emerged in the context of a review of MAFF-funded TSE research. The proposed audit was slow in getting off the ground. In February 1996 SEAC advised that it was high priority to carry out the audit and that sheep tissues should be included in the study. The work was commissioned from outside consultants in June 1996 and completed in May 1997.

Reasons for this outcome

1074 Why did the matter turn out this way? Various factors were at work. MAFF thought that what was required could be done in-house by existing staff. No association appears to have been recognised between the risk for workers in identified industries and the risks that might be continuing to be carried in the material itself. There had been some confusion from the start about the status of the study which the *Tyrrell Report* had identified. Was it truly research or simply a fact-finding exercise? The indeterminate wording of the initial allocation in Table 2 provided no impetus to anyone to move matters forward. Subsequently the compressed reporting in the progress chart of the coupled cosmetics and audit proposals gave a misleading impression about whether action was in hand and who was in the lead.

1075 However, given the importance of doing the work, all these difficulties could undoubtedly have been overcome had the project had a champion. None emerged to press for the work to be done and secure action. This lack of ownership of the project spelt its doom.

Where responsibility lay

1076 We have no doubt that whether or not the *Tyrrell Report* had listed it as an item, an exercise of this sort was a necessary precursor to an effective government response to BSE. Within MAFF, Animal Health Division, headed by Mr Lowson, was responsible for developing policy on BSE. The role of working up policy proposals and submissions for Ministers, and setting up the arrangements to carry them out, was generally a Head of Division responsibility. It seems to us that Mr Lowson had a responsibility to ensure as far as possible that the development of policy on BSE was properly informed by data from appropriate scientific research and field studies.

1077 The work done by Mr Hutchins and Mr Rogers to compile lists of uses was a good start but no more than that. They did not seek to trace through what happened to the products and what risks might be associated with them. Yet these lists appear to constitute the sum total of the 'in-house work' that SEAC was assured made a full audit premature for the time being. This was scarcely a systematic investigation, nor was it of value without policy action to follow up the clues it offered.

1078 We consider that the need for the work on an overview to be done should have been obvious at the time. Mr Lowson agreed that he needed no special advice from scientists about whether or how to carry out a fact-finding exercise to map all the ways in which cattle products might be used. New though he was in his post, in our view he should have ensured that this matter was promptly and properly addressed.

1079 We considered whether Dr Pickles shared responsibility for this. On reviewing her actions, it seems to us that at each stage she pushed hard for the audit to be carried out. She took independent action in an effort to secure DH funds to break the financing deadlock; and she drew the failure to carry out the project to Mr Gummer's attention, which led directly to his instruction that the work should go ahead. Thereafter she made efforts to get the protocol drafting under way at MAFF. We do not think she could have done more than she did.

1080 We have been at pains to explore what happened to the audit. We see the failure to carry it out as a serious shortcoming in the response to the emergence of BSE. Time and again the story we have explored has shown that in the main the right action was taken, but often more belatedly than it could have been. Some matters, such as the safety of gelatine and tallow which were used for a wide range of different purposes, were dealt with only late in the day. Others, such as waste disposal from slaughterhouses and rendering plants dealing with SBO, were barely identified at all. Where work was put in hand there were often no deadlines. Urgent warnings were delayed while drafts were refined. Some of this could have been avoided if all had been working within a recognised overview and timetable as a framework for tackling matters, under a firm guiding hand.

10. Pollution and waste control

1081 When slaughter and compensation measures were introduced, the carcasses of the cattle in question became the property of MAFF. The Ministry had already established disposal procedures to apply to the handling of BSE carcasses. Instructions were swiftly distributed to field staff. The preferred option was incineration at MAFF premises followed, 'in order of decreasing desirability', by off-farm burning on waste ground or at a local authority site, incineration on farm, burial at a local authority tip, and burial on farm by a contractor.

1082 The problem that arose with the BSE cases was their sheer volume. As numbers rocketed in 1989 and 1990, constantly outstripping forecasts, MAFF was forced to adopt various expedients while new incinerator capacity was being sought. This was not a simple undertaking. There was local hostility both to the emergency measures of open burning and tipping that had to be adopted and to the issue of planning permissions and other licences for new incinerator capacity. As fast as new provision was made, the number of reported BSE cases grew yet greater. Only in 1992, a year in which 43,449 carcasses had to be destroyed, did MAFF get the disposal situation fully under control. Thereafter carcasses were no longer buried and virtually all incineration was at designated premises.

1083 Volume 6: *Human Health, 1989–96*, Chapter 10 describes the steps that MAFF took and the difficulties it encountered. Ministers took a close interest in what was happening, both because they wished to be assured that the policies adopted were not creating health risks, and because of the continuing public sensitivity about some of the measures adopted. The potential impact on overstrained waste disposal facilities was not unnaturally a consideration in some of the policy issues that arose. By their nature carcasses had to be disposed of promptly if they were not to constitute a threat to public health. Moreover this was far from cost-free.

1084 Overall MAFF handled this difficult and unpopular task of carcass disposal both energetically and competently.

1085 In the process, however, they had to deal with various objections from those with responsibilities for environmental protection. There was growing public concern about the nature and persistence of the BSE agent in waste whether burnt, used as landfill or discharged as effluent. We shall return to this point. But first we review what happened with a different sort of waste, the Specified Bovine Offal.

1086 Here matters were not so straightforward. Responsibility for disposal did not rest with MAFF but with the owner of the material. Initially there was no requirement to distinguish SBO from other meat unfit for human consumption. Most of this unfit meat was not regarded as waste but rendered to produce MBM and tallow. As one renderer put it, 'We were very much a by-product industry. We cleared up the mess from the slaughtering industry trades.' In 1991, after the introduction of the animal SBO ban and SEAC's advice that the protein product of SBO should not be used as fertiliser, MBM could only be disposed of at a licensed destination. It had become controlled waste.

1087 The disposal of some other sorts of BSE waste was given much less attention than SBO. These were the side products of slaughtering cattle, and destroying, treating or processing cattle material.

1088 They took various forms. Effluent passed down drains to sewers and rivers. Blood, slaughterhouse or rendering plant waste, including that from plants that rendered SBO, and sewage sludge from works handling their effluents might lawfully be spread as fertilisers on land where animals subsequently grazed or crops were grown.

1089 While emissions from plants required formal consents from water authorities and others, in practice none of the usual precautions or conditions which applied to discharges would have inactivated the BSE agent. It appears to have been assumed that it was sufficiently diluted to pose no risk. This was a matter that was not thoroughly investigated until work was commissioned by the Environment Agency in 1996 to trace all the environmental pathways along which BSE material might travel, and to assess the degree of risk and appropriate precautions for each.

1090 Although much of the evidence offered to us about the BSE risk from effluent from the Thruxted Mill rendering plant in Kent related to a time outside the period covered by this Inquiry, the concerns expressed and the action taken in response to them illustrated some of the difficulties posed by BSE for those responsible for dealing with secondary wastes.

1091 The environmental regulation regime had been found wanting in many respects towards the end of the 1980s. Discussing the disposal of solid waste the Select Committee on the Environment observed in 1989:

> Never, in any of our enquiries into environmental problems, have we encountered such consistent and universal criticism of existing legislation and of central and local government as we have during the course of this enquiry.

1092 The system was at the same time having to be adapted to meet EU requirements designed to ensure that waste was recovered or disposed of without endangering human health or harming the environment. The principle of 'producer pays' for disposal costs was being introduced. Major reorganisation of responsibilities was undertaken and new powers brought in under the Environment Act 1990.

1093 Thus the task of disposing safely of BSE carcasses and SBO took place within a regulatory system that was in trouble and in transition. Chapter 8 in vol. 14: *Responsibilities for Human and Animal Health* describes the main features of the system and the major changes introduced to rearrange responsibilities and to regulate waste and sewerage, waste tips, waste spreading and air quality.

1094 These were wide-ranging matters. We could not attempt to add detailed exploration of them to the many other topics our Inquiry has had to cover. It is clear, however, that as a potential transmission pathway for BSE, general waste disposal systems received scant attention prior to 1996. This matter was not specifically referred to or addressed by the Southwood Working Party, the Tyrrell Committee or SEAC. Yet all of them advocated a systematic review of the destination of all bovine materials. Had this been carried out as discussed in the earlier section of this chapter, it might have been expected to identify many of the matters touched on above, and to indicate where more research or development of new techniques would be valuable.

11. Wales, Scotland and Northern Ireland

1095 BSE was a UK-wide threat needing a UK-wide response. That was speedily and sensibly agreed by all concerned once it was apparent that BSE extended throughout the United Kingdom. By common consent, MAFF and DH took the lead role. In order to simplify our exposition of a highly complex and extended series of events we have in our Report mainly concentrated on the actions of MAFF and DH in England and the legislative measures that they introduced. These applied to or were copied by the other three parts of the United Kingdom. We in turn have copied the terminology that they often used in describing themselves collectively as the Territories.

1096 In vol. 9: *Wales, Scotland and Northern Ireland*, we have been concerned to see how the links between central government in London and government in the Territories functioned in relation to BSE. We have been particularly interested in identifying the extent to which the Territories sought to play an independent role or to make an independent contribution in relation to the handling of the disease. In this chapter we shall set out a summary of our main findings about the role of the Territories.

1097 We found no fundamental differences in the nature of the response to BSE throughout the UK. Like their colleagues in Whitehall, Ministers and officials in the Territorial Departments worked closely together. Decisions were taken on the basis of submissions and discussions. Where there were minor or temporary variations from the general UK line in their actions, these did not in our view bear on the course of the disease or expose animals and humans to a significantly greater or lesser degree of risk.

1098 It was plain from all the evidence that the Territorial Departments were strongly influenced at first by the MAFF perception of BSE as purely an animal disease. They then found this perception confirmed by the *Southwood Report*. The risk to humans was remote. The Report gave 'quite a comforting message'. It is difficult not to infer that this perception, coupled with the Government's drive towards 'lifting the burden' of regulation from industry must, as elsewhere, have tempered enforcement zeal.

1099 Nonetheless, officials pressed ahead diligently with the agreed precautions.

1100 Inevitably with a canvas covering ten years, and a vast complex of administrative actions, there were things that could with advantage have been done a little differently and perhaps a little better. However, we were not looking for perfection. We were interested in the light thrown by some of the failings we noted on the way collective government works among Departments with different geographical responsibilities, rather than different functional ones.

1101 We note first some features of what happened in Wales, Scotland and Northern Ireland, and then set out some more general findings.

Wales

1102 Welsh legislation and administrative arrangements closely resembled those of England. This simplified the task of coordinating action. We were struck by the quality of independent thinking that the Welsh Office medical team led by the CMO for Wales, Dr Deirdre Hine, applied to the issues raised by BSE. The team's attitude reflected its effective combination of medical and epidemiological skills with first-hand knowledge of the realities of slaughterhouse operation. A similar working combination of skills at national level in Whitehall Departments could well have been fruitful.

1103 There were no special features of the Welsh situation that dictated a different approach. However, the Welsh Office team had valuable insights to offer for national policy development and did their best to register them. Dr Hine wished to get closer to the thinking of SEAC. We applaud the alternative strategy she adopted towards its chairman, Dr Tyrrell, of successfully inviting him to Cardiff. Her interest in exploring the issues was natural in the context of her responsibilities to the people of Wales. It seemed to us that the various information blockages that she and her colleages encountered could have been overcome had there been a wish in Whitehall to involve the Territorial Departments more closely in the policy-making process.

Scotland

1104 Here there was not the same happy combination of skills and knowledge in place to bring together the animal and human health implications of BSE. Matters were very much left in the hands of the Agriculture Department. However, in 1990 Dr Gerald Forbes, a former member of the Scottish Home and Health Department, expressed concerns about the risk that BSE posed to humans, which appear initially to have sounded a cautionary note with the CMO, Dr Kenneth Calman, and with Mr Graham Hart, who headed the Health Department. Dr Robert Kendell on the other hand, who took over as CMO in 1991, did not seek Dr Forbes's views, regarding the Environmental Health (Scotland) Unit which Dr Forbes now headed as a 'one man band'. Dr Kendell looked mainly to Mr James Scudamore, the Assistant Chief Veterinary Officer, Scotland, for advice about BSE. Mr Scudamore seems to have fulfilled his role admirably, both towards the CMO and in working closely with the Animal Health branch in the Department of Agriculture and Fisheries for Scotland (DAFS). However, as he told us, he had expected that his contributions from the veterinary and general MAFF perspective would have formed no more than one element in any Scottish Office assessment of an issue. We agree. But no such wider assessment appears to have been made by DAFS officials in relation to BSE.

1105 We thought that this shortcoming could be attributed to weak links and lack of shared perceptions in the Scottish Office between those responsible for animal and human health. Dr Kendell told us that he simply assumed that it was his job to keep careful tabs on the human disease, and it was the job of DAFS to ensure that everything was right and proper on farms and in abattoirs. We saw little sign of joint working on BSE between the administrators in the Health and Agriculture Departments. One manifestation of this was the pigeonholing of the hard-won

SEAC papers by DAFS administrators as scientific, technical and 'all Greek'. These were never discussed and assessed jointly with Health officials, or indeed at all, nor brought to the attention of the CMO, who later thought they would have been 'enormously helpful'.

1106 It seems to us that those dealing with animal and human health could profitably have shared knowledge about and discussed slaughterhouse practices, the food chain implications if enforcement of Regulations was inadequate, and any impact that this might have on handling BSE in Scotland. We also think that it was desirable that a working competence in understanding the papers of a key advisory committee such as SEAC should have been available in the Scottish Office.

1107 Happily the poor liaison did not create delays in the action taken by DAFS to introduce Scottish legislation and apply the various precautionary measures agreed on BSE. We have no criticisms of this. The House of Commons Agriculture Committee had, in 1990, censured the delay in introducing the Scottish human SBO ban to mirror the England and Wales Regulations of November 1989. However, given the last-minute addition of sausage casings, which had a bearing on haggis manufacture, and the troubles that immediately arose over the lawfulness and adequacy of the 1989 SBO Regulations, we thought it not unreasonable that those producing the Scottish equivalent should take the time necessary to avoid these pitfalls.

1108 That said, the border between Scotland and England, and indeed between England and Wales, is meaningless so far as the movement of people, animals and goods is concerned. In these circumstances, human and animal health threats need a common approach. As a general principle, it seems to us highly desirable that when animal and human health safeguards are urgently needed, there should be available powers to bring those into effect simultaneously across the whole of Great Britain.

Northern Ireland

1109 Here there was indeed a significant physical border. Besides differing more markedly in terms of its legislation and administrative arrangements, Northern Ireland was separated from Great Britain by a wide sea crossing. It was reasonable that Ministers and officials there should have given careful thought to whether to follow the policy lead from London on making BSE a notifiable disease, and on the ruminant feed ban. They decided not to do so at first.

1110 We did not think the delay in formalising notification made any difference. However, we were concerned about the decision not to take immediate action on a feed ban. Recycled infective material might already have been in local MBM, and cattle eating it might already have become infected, thus prolonging any epidemic in Northern Ireland. We noted that the decision to delay the ban was taken only after outside consultation and analysis of various options. It was put to us that it was justified by the absence of BSE outside Great Britain and by the beliefs held at the time about the cause and distributing mechanism of the disease. Moreover, import controls were put in place for MBM and live cattle. We concluded that the decision was not unreasonable at the time, though with hindsight it would have been preferable not to delay. However, immediate precautionary introduction of a ruminant feed ban would probably have reduced the cases of BSE in the Province by only a small number. Northern Ireland was in any case far less affected by BSE that the rest of the UK.

1111 After the first case in Northern Ireland was confirmed in November 1988, the NI administration closely followed the UK line on all matters, despite a hankering for independent health status for its cattle, with a view to restoring beef exports. We think they were right to keep in step with the rest of the UK.

1112 We heard differing accounts of the usefulness of the NI cattle-tracking system in alleviating the effects of the BSE crisis in the Province. It does not appear to have been a significant factor during the period with which we were concerned, although it may have helped since in allowing the earlier resumption of exports than in the rest of the UK.

Collective government and working relationships

1113 Tackling BSE entailed a huge exercise in public administration. It required close working between Ministers and officials, consultation and cooperation between Departments and efficient follow-up action. Our Inquiry has been a review of all these matters and of how far collective government rose to the challenge.

1114 Collective government across the different parts of the UK required its own set of working relationships. By and large the machine worked reasonably well, but there were many recognised endemic difficulties. Unsurprisingly these sometimes gave decision-making on BSE a bumpy ride. We were told with some vigour of frustrations about failures and delays in communication between Whitehall and the Territories.

1115 In some respects this mirrored communication failings between Whitehall Departments, and between the cadres of administrative and professional advisers. For the Territories, travelling times to and from London exacerbated the problems. Typical examples of these difficulties, where BSE was concerned, included MAFF delays in telling Scottish administrators about the disease, DH disinterest in views from Scotland and Wales, and the absence of territorial officials from formative meetings.

1116 Communication problems were particularly significant in relation to the Territories' reliance on Whitehall for scientific expertise and risk analysis. It made sense that such work was not duplicated. But if the material passed on was meagre and late, consultation was purely token. Moreover, without access to the basic information, the Territorial Departments had to rely on the judgements already made in Whitehall and on Q&A briefing that might itself slide over the underlying issues. The handling of BSE cast some of these difficulties into strong relief. The lessons they offer for the future are described in Chapter 14.

12. Science and research

1117 Although only one member of our Committee is a scientist, our terms of reference have required us to review, at second hand, a substantial body of scientific learning and research. We are required to establish the history of the emergence of BSE. In order to attempt to answer the questions of where BSE came from and why it emerged in this country we have had to consider, among other things:

- epidemiological research;
- evidence on the technical aspects of rendering and the inactivating effect of rendering processes on TSE agents;
- transmission properties of BSE compared with those of scrapie; and
- strain-typing of the BSE agent after transmission to mice.

1118 More fundamentally, we have had to consider the complex research on the very nature of TSEs. This is critical to the theory, now widely accepted, that BSE has been transmitted as a result of recycling bovine protein that included infective prion protein.

1119 In the course of our Inquiry we have received evidence from scientists who espouse alternative theories, for example:

- the organophosphate theory; and
- the autoimmune theory.

We have had to consider whether these were viable alternatives to the prion protein theory.

1120 More generally, our requirement to review the adequacy of the response to BSE, taking into account 'the state of knowledge at the time', has required us to follow the development of scientific knowledge about BSE between 1986 and 1996, paying particular attention to those aspects which had a bearing on the likelihood that BSE might be transmissible to man.

1121 We are also required to establish the history of the emergence of vCJD. This has required us to consider the scientific research, both before and after 20 March 1996, which has focused on the question of whether the link between BSE and vCJD is clearly established.

Scientific conclusions about BSE

1122 Our analysis of the scientific knowledge occupies the major part of vol. 2: *Science*. We shall not attempt a summary in this volume. We shall simply set out the conclusions that we have drawn from the scientific response to BSE:

i. The vector responsible for the epidemic of BSE in cattle was MBM

The spread of BSE in cattle to the point where it became an epidemic came about from the use of meat and bone meal (MBM) in cattle feed. The MBM in question was infective because it had been made by rendering infective offal from cattle suffering from, or merely incubating, the disease. As little as 1 gram (or less) of this material could cause death if ingested by other cattle. It was so infective that accidental contamination of cattle feed with pig or poultry feed containing MBM was a significant factor in continuing to spread BSE after the ban on the use of MBM in cattle feed. Apart from MBM in feed, transmission from mother to calf is likely to have played a part. We cannot yet say whether contamination of pastures played a part. The suggestion has been made that the BSE agent may have been spread in the early stages in hormones used in veterinary preparations. This possibility cannot be discounted. But the overwhelming vector of the epidemic was MBM in cattle feed.

ii. The unmodified scrapie agents were not the agents responsible for BSE

While it was reasonable in February 1989 to accept the hypothesis that the cases of BSE being reported had come about through the rendering of carcasses of sheep infected with extant strains of scrapie established in the national flock, this theory is no longer plausible. We think it likely that the passive surveillance system failed to detect several earlier cycles of BSE in the South West of England in the 1970s and early 1980s. Each cycle was followed by more extensive contamination of MBM. Much of the recycling could not be detected because tissues from animals incubating the disease but not showing signs were involved; but it is likely that there were isolated animals which did develop signs and were slaughtered or died of the disease. BSE was unknown at the time and it seems possible that the disease in such cattle might have been ascribed to known disorders such as hypomagnesaemia or simply not explored. These early cycles began because a novel TSE agent originated in the early 1970s. The cause of this novel agent is likely to have been a new prion mutation in cattle, or possibly sheep. Moreover, other mammalian species whose carcass waste was included in MBM cannot be excluded. It is conceivable that the conversion of normal prion protein into its infective form was initiated not by a gene mutation, but by an environmental agent, such as a toxic chemical; this has not yet been achieved experimentally. Current knowledge suggests that the original agent was not the unmodified scrapie agent or agents. We have also noted a number of pointers which could have led to the conclusion by mid-1990, and certainly well before 20 March 1996, that the agent fuelling the BSE epidemic was not then (if it ever had been) the unmodified scrapie agent or agents. It is now not possible to be sure which of the hypotheses as to the origin of the novel agent is correct.

iii. Changes in rendering

It is a common misconception that reduction in temperature or a failure to prescribe minimum holding times in the rendering of carcass waste led to failure of inactivation of the scrapie agent and transmission across the species barrier to cattle. Changes in the rendering process in the late 1970s and early 1980s, namely the switch from batch to continuous processing and the abandonment of solvent extraction of tallow, might have led to reduction in inactivation of the agent in MBM, but it is now known that the processes used previously were also incapable of completely inactivating TSE agents. No commercial rendering procedure has been designed capable of completely inactivating BSE in MBM before or since.

iv. Confirmation of the central role of prion protein

All evidence points to the specific association of an abnormal form of the prion protein and TSEs. In its normal shape, the prion protein (PrP^C) does not cause harm. In its abnormal shape (signified by PrP^{Sc} – a generic term for the agents causing TSEs), it is resistant to the normal cellular processes of degradation. Contact between normally shaped and abnormally shaped proteins induces the normal to convert to the abnormal. This leads to a build-up of the abnormal form of the protein, which accumulates in, and eventually causes the death of, nerve cells. Nerve cells are particularly susceptible to PrP^{Sc} because they cannot regenerate. The presence of PrP^{Sc} can be demonstrated in the brain and spinal cord of all humans and animals affected with TSEs. Incubation times in experimental animals correlate with the infective dose of the agent, and these times are increased by treatment with agents (β-sheet breaker peptides) which reverse the conformational change leading to PrP^{Sc}. These observations virtually eliminate other hypotheses as to the direct cause of TSEs, such as autoimmune disease of the central nervous system, because those hypotheses do not incriminate the prion protein. In both scrapie and vCJD, susceptibility and resistance to disease is associated with polymorphisms within the prion protein gene (though no such genetic susceptibility factors have yet been identified for BSE). It remains possible that environmental factors, including toxic chemicals, may additionally be implicated in susceptibility to prion disease.

v. BSE is caused by a single strain of agent

Strain-typing in mice has shown that all sources of the BSE agent so far examined produce the same lesion profile and incubation times in experimental mice. The same strain has been identified in cats, which have developed FSE since 1990, and in exotic ungulates and carnivores from zoological parks.

vi. Variant CJD is caused by the BSE agent

Strain-typing studies in mice reveal that the disease patterns produced by the agents causing BSE and vCJD are identical. The glycosylation patterns of the prion protein associated with each condition are also identical and different from other TSE strains. In transgenic mice in which the mouse prion gene has been replaced by the bovine prion gene, inoculation with the BSE agent from cattle brain produces the same disease pattern and incubation period as agent derived from patients with vCJD. Following inoculation with the scrapie agent, the incubation period and disease patterns in the transgenic mice are markedly different from those produced by BSE and vCJD. In the absence of any other plausible factor, the evidence that BSE caused vCJD is so strong that all other hypotheses are now excluded.

Alternative theories

The organophosphate theory

1123 The theory that BSE was caused by a reaction to the use of organophosphorus compounds (OPs) poured on cattle as systemic pesticides cannot be reconciled with the epidemiology and is not supported by research. One experiment has, however, given some limited support to the possibility that the OP phosmet might modify the susceptibility of cells to the prion disease agent.

The autoimmune theory

1124 There are a number of reasons why this theory does not seem viable, including:

- the fact that mouse-adapted BSE can be transmitted by intracerebral inoculation to mice lacking a functional immune system; and
- the fact that the theory is incompatible with what has been established about the central role of the prion protein in TSEs.

Research

1125 An important aspect of the response to BSE was the research that was undertaken in order to learn more about the disease. Before 20 March 1996 MAFF had funded over 120 research projects in relation to different aspects of BSE. Research work into TSEs, and more particularly BSE, was also funded by the Research Councils. We have not interpreted our terms of reference as requiring us to review the adequacy of all these projects. What we have explored are the broader questions of the funding, planning and coordination of BSE research. Our consideration of these topics is to be found in vol. 2: *Science* and vol. 11: *Scientists after Southwood*. Here we propose to do no more than set out a brief summary of our conclusions.

1126 BSE did not emerge at a propitious time so far as research was concerned. In 1985 Ministers had accepted a recommendation from the Priorities Board for Research and Development in Agriculture and Food that expenditure on research into animal diseases was disproportionate and should be reduced by 20 per cent. Implementation of this policy was resulting in staffing cuts at research establishments.

1127 The Neuropathogenesis Unit (NPU) in Edinburgh had been set up jointly by the Agricultural and Food Research Council (AFRC) and the Medical Research Council (MRC) in 1981 as an independent unit to study scrapie and the similar human diseases of the central nervous system such as CJD. The need to relocate staff and facilities and to build up suitable mouse colonies, coupled with financial constraints on the appointment of necessary new staff, meant that it had not yet been able fully to address this remit, although it had brought together a wide range of expertise in genetics, strain characterisation and transmission of scrapie. In 1986, however, it had been brought within the framework of the Institute for Research on Animal Diseases, later to become the Institute for Animal Health. Shortage of funding and the loss of independence had resulted in the disillusionment of its Director, Dr Alan Dickinson, who resigned in 1987, and for whom for a long time it proved impossible to find a suitable replacement. There was also uncertainty about where the various parts of the new Institute should be located. Thus the emergence of BSE found the NPU in a state of some disarray and with its future in doubt.

1128 Despite these problems, both at the NPU and more generally, research into BSE was not significantly impeded through lack of funding, although some research projects got off to a slow start. An application for additional funds from the Treasury Reserve was laboriously put together, finally presented in August 1989 and rejected. Alternative sources of funding were then identified, which involved the diversion to BSE of funding earmarked for other projects.

1129 Between 1987 and 1996 the Government spent over £60 million on research into BSE and other TSEs. Of this, £37.9 million was spent by MAFF and £27.4 million funded by the Research Councils. DH's expenditure was £1.6 million, largely spent on funding the CJD Surveillance Unit (CJDSU).

1130 Almost all the research funded by MAFF was carried out either at the CVL or at the NPU, with CJD research being carried out by the CJDSU. The BSE research programme was developed within the CVL by the BSE Group, headed by Mr Bradley, in consultation with the NPU. One project involved collaborative work between the two laboratories. Priorities were allocated by the Tyrrell Committee. The research that was carried out was extensive and wide-ranging, for example:

- It identified that BSE had the histopathology of a TSE.
- It quickly identified that BSE was transmissible to mice, both by inoculation and in feed.
- It identified that BSE was similarly transmissible to sheep and to goats.
- It confirmed the infectivity of brain and spinal cord and identified the infectivity of the distal ileum of calves.
- It identified that ½ gram would suffice to transmit BSE orally to a sheep and 1 gram to a calf.
- It identified the fact that BSE was a single and distinctive strain of TSE agent.
- It swiftly identified the emergence of a new variant of CJD.
- It identified the link between vCJD and BSE.

1131 In 1990 Sir Donald Acheson set in train an initiative to place the AFRC/MAFF/MRC research effort on BSE under the coordination of a single 'director'.

This met with resistance on the part of the Research Councils, which saw it as a threat to their independence, and was supported by MAFF only on condition that the director would report to the MAFF Minister. The proposal foundered. Instead it was agreed that SEAC would perform a limited role in facilitating interchange between the various bodies responsible for research. The demands on SEAC for advice were so onerous that members did not have the time to carry out a review of the adequacy of the research effort and to identify gaps in the research programme. The most that they were able to do was to check that the projects recommended by the Tyrrell Committee as having high priority were under way. In June 1992 they published a paper that recorded that they were 'content with the progress of implementing the recommendations overall'.

1132 We have concluded that it might have been advantageous to have had an individual or committee with a remit to coordinate research and to draw attention to research needs. As it was, these were largely identified by the CVL, which then played the role of contractor in supplying much of the research identified. Thus most of the projects were awarded without competition and were not peer-reviewed. We have identified, with hindsight, areas where research could profitably have been started earlier or been pursued with more vigour. Also, an attempt might have been made with advantage to recruit expertise from the wider scientific community. It is at least possible that had an overview been kept of all BSE research, some of these issues would have been identified and addressed at the time:

1133 *Scrapie-into-cattle transmission* – Experiments to see if and how scrapie would transmit to cattle were begun in 1997. It would have been valuable to test the theory that BSE was caused by the scrapie agent or agents ten years earlier, although we accept that there were difficulties in the way of doing this.

1134 *BSE in sheep* – The possibility that BSE might have been transmitted to sheep was recognised as early as 1987. So too was the risk that, if it had done so, it, like scrapie, might become endemic in sheep. Research to check whether this has happened is now being carried out. It is perhaps the most important unanswered question about the BSE epidemic.

1135 *Minimum infective dose* – The NPU experiment to transmit BSE to sheep and goats, which was initiated in 1988, was, incidentally, a valuable test of whether a dose as small as that contained in ½ gram of material would transmit in feed across the species barrier. It was not, however, designed or used for the purpose of providing this information. The 1992 attack rate experiment was the first occasion on which MAFF sought to see how much infective material was needed to transmit BSE in feed, and even this was not designed to identify the minimum quantity. The results of the attack rate study were of great practical importance.

1136 *Sensitivity of the mouse bioassay* – the infectivity of different tissues in BSE-infected cattle was tested by bioassay in mice. Tests begun in 1993 have demonstrated that mice are at least 1,000 times less susceptible to BSE than cattle. It would have been advantageous if the extent of this species barrier had been identified earlier.

1137 *Ante- and post-mortem tests for BSE* – Simple ante- and post-mortem tests for BSE would have been of the greatest practical value. These are areas which could have been developed with greater vigour and in which a research 'supremo' might have stimulated open competition.

1138 *ELISA test for ruminant protein in compound feed* – Research was carried on 'in house' at a leisurely pace. This was in part because the importance of developing such a test was not appreciated until the significance of cross-contamination of feed was brought home in 1994. A research director might have identified external sources that would have advanced this area of research more rapidly.

1139 *Epidemiology* – One of the remarkable features of BSE research is that the epidemiology was left largely to Mr Wilesmith and the members of his small epidemiology department at the CVL. This perhaps reflected the lack of veterinarian epidemiologists in this country. There was, however, scope for human epidemiologists to address questions such as the cause of the BABs, the pattern of the epidemic and the number of subclinical cases going into the human food chain.

13. What went right and what went wrong?

1140 In previous chapters we have described the BSE story. Here we review certain aspects of the story, discussing what went right, what went wrong, and why. We begin with the practice which ensured that BSE spread so widely – the use of meat and bone meal (MBM) in cattle feed. Then we look at the identification of BSE, and the major policy decisions, before considering what may have been the major causes of shortcomings. We conclude with general comments on the criticisms of individuals found elsewhere in our Report. An Inquiry inevitably focuses on shortcomings, and these comments are designed to redress the balance.

A recipe for disaster

1141 There is a body of opinion that believes that farmers had only themselves to blame for the epidemic of BSE. Cows are ruminants. They do not naturally eat animal protein. They were fed animal protein in order to boost their milk yield or fatten them up. Some say that it offended against nature to feed animal protein to ruminants. Some say that it was doubly offensive to turn grass-eaters into cannibals. Some say that it was not surprising that a plague was visited upon those that tampered with nature in this way.

1142 Objection can be taken to many intensive farming practices on ethical or aesthetic grounds. We have resisted the considerable temptation to enter into this debate, which would take us well beyond our Terms of Reference. Of relevance to our Inquiry is the narrower question of why those responsible for the practice of using MBM in cattle feed did not foresee that this might be a recipe for disaster?

1143 The MBM used in cattle feed was produced by rendering. This involved pooling and then processing material from hundreds, perhaps thousands, of animal carcasses at a time. As with other processes where ingredients are pooled, there is a risk of contaminating the pool if any single source is infective. It is thus of crucial importance to make sure that the rendering process will destroy any potentially harmful organisms or other agents in animal carcasses. This is particularly important if animal protein is being recycled within the same species, so that there is no species barrier to infection.

1144 The suggestion has been made to us that the 1979 Royal Commission on Environmental Pollution warned against the risk of recycling animal waste. The risk to which the Commission drew attention was that of recycling poultry litter by including it as a protein supplement in ruminant feed. But the Committee went on to encourage this practice as an environmentally sound re-use of materials 'given that care is taken to avoid health hazards'. An Agricultural Research Council report on 'The Nutrient Requirements of Ruminant Livestock' in 1980 drew attention to the value of undigested protein, of which MBM is a prime example, in ruminant rations to promote milk and flesh production. This authoritative report by leading animal nutritionists, including the Agricultural Devolpment and Advisory Service (ADAS), gave a boost to the use of MBM by feed manufacturers.

1145 The practice in the UK of recycling animal protein as an ingredient of animal feed dates back to at least 1926. In the 1970s attention was directed within MAFF to the danger that this practice would result in the spread of infectious diseases. The diseases considered were those caused by conventional viral and bacterial organisms. No consideration appears to have been given to the risk that scrapie might be recycled in sheep, or even transmitted to other farm animals. This may seem surprising. The answer probably lies in the fact that half a century had elapsed without any indication that animal feed containing ovine protein was infecting sheep or any other animal.

1146 The measure that MAFF introduced to address the risk of the spread of infectious diseases as a consequence of incorporating MBM in feed was the Diseases of Animals (Protein Processing) Order 1981. This laid down a mandatory sampling regime designed to ensure that the rendering process inactivated all conventional viral and bacterial pathogens. The measure was not designed to ensure that the rendering process would inactivate Transmissible Spongiform Encephalopathies (TSEs). No rendering process has yet been devised that will guarantee to inactivate BSE.

1147 What went wrong was that no one foresaw the possibility of the entry into the animal feed cycle of a lethal agent far more virulent than the conventional viral and bacterial pathogens, and one which would be capable of infecting cattle despite passing through the rendering process. When regard is had to the experience of what, by 1981, was over 50 years of recycling of animal protein, we can understand why the risk of a disease such as BSE was one which was not anticipated or addressed by farmers, renderers, feed compounders, animal nutritionists or government.

The identification of the disease and its cause

1148 Identification of the emergence of BSE was always going to pose a challenge:

- It had a long incubation period.
- It tended to strike down a single animal in a herd.
- It produced clinical signs which resembled those of other conditions.
- It could only be identified as a TSE by histopathology.

1149 It is to the credit of the system of passive veterinary surveillance and the skill of the Central Veterinary Laboratory (CVL) pathologists that the disease was identified at a relatively early stage of the epidemic.

1150 Great credit is due to Mr Wilesmith for his rapid identification of MBM in feed as the immediate source of infection. His individual contribution to the response to the challenge of BSE was of the highest value. His deduction as to the probable reasons why MBM was infectious was reasonable, but wrong. It was unfortunate that his explanation – the scrapie theory – was one that provided unwarranted reassurance that BSE was likely to behave like scrapie and would thus not be transmissible to humans.

1151 It was also unfortunate that, although problems with Mr Wilesmith's theories became increasingly apparent to the scientists as more was learned about BSE, no reappraisal ever received publicity. When our Inquiry began, most members of the public remained under the impression that BSE was scrapie in cattle and that the reason why cattle feed had become infectious was that renderers had altered their methods of production to the detriment of safety standards.

The Government's response

1152 In earlier chapters we have seen how the emergence of BSE confronted government with three challenges:

- how to eradicate BSE in cattle;
- how to address the possibility that BSE might be transmissible through animal feed or otherwise to other animals; and
- how to address the possibility that BSE might be transmissible through human food or otherwise to humans.

1153 Those chapters summarise our discussion in Volumes 3 to 9 and 11 of the adequacy of the response to those challenges, having regard to the state of knowledge at the time. In the remainder of this chapter we draw attention to the major policy decisions in relation to these matters, which we have concluded were appropriate. We have shown that shortcomings attended the introduction, implementation, enforcement and monitoring of the measures pursuant to these decisions, and we identify some underlying features which led to shortcomings.

Eradication of BSE

1154 Banning the incorporation of ruminant protein in ruminant feed was the correct policy to adopt in order to eradicate BSE. Had it been fully implemented it would probably, by today, have achieved its object. As it is, it brought about a massive reduction in the number of new cases of infection so that, by 1996, it was apparent that the epidemic had been brought under control.

1155 Precautionary measures could have been taken to address the possibility that BSE would prove to be maternally transmissible. Maternal transmission, of itself, might prolong but could not perpetuate the disease. It was reasonable to refrain from culling the offspring of BSE dams unless and until it was shown that maternal transmission was taking place on a scale that justified this. There was room for argument as to whether or not breeding from the offspring of BSE dams should be discouraged or forbidden, but this was not a major policy issue.

1156 The possibility that BSE might be horizontally transmissible was addressed by:

- guidance to farmers on preventing other cattle from coming into contact with the placenta of a calving dam; and
- a ban on the use of protein derived from Specified Bovine Offal (SBO) as fertiliser.

Possible transmissibility to other animals

1157 Although the primary motivation for the compulsory slaughter and destruction of cattle showing signs of BSE was the protection of human health, it had the added benefit that the carcasses of these animals could not be rendered for animal feed. Thus this measure was in part a response to the possibility that BSE would transmit to other animals.

1158 The measure specifically adopted to address this possibility was the animal SBO ban. The object of the ban was to prevent the inclusion in animal feed of protein derived from SBO. The leading pet food companies and the bulk of the animal feed industry had previously adopted this ban on a voluntary basis. MAFF made it compulsory after experimental transmission to a pig by inoculation had been achieved. This ban affected predominantly the content of pig and poultry feed. Although no express application of the ALARP principle was involved in this decision, we consider that, if effective, it would have reduced the risk of transmission of BSE to other animals through feed as low as was *reasonably* practicable, having regard to:

   i.   the reasonable belief that BSE was unlikely to be zoonotic;

   ii. the fact that there was no history of transmission of TSEs to, or experience of TSEs in, either pigs or poultry; and

   iii. the economics and waste disposal consequences of going further and imposing a total ban on including any animal protein in animal feed.

1159 Measures were also taken to reduce the risk of transmission of BSE to other animals through veterinary medicines. Guidelines were issued to manufacturers of both human (see below) and veterinary medicines, which advised that certain bovine products should not be used in the manufacture of certain medicines, suggested that action should be taken to reduce contamination in the collection and production processes, and advised on sterilisation or discarding of the equipment used.

Possible transmissibility to humans

1160 The principal policy decisions which addressed the possibility of transmission of BSE to humans through food were those to introduce:

   i. compulsory slaughter and destruction of cattle with symptoms of BSE; and

ii. the human SBO ban.

1161 These were two vital measures for the protection of human health. Each was introduced at a time when the possibility that BSE might be transmissible to humans in food was considered remote. On that basis we consider that they constituted a proportionate response that satisfied the ALARP principle, albeit that the policy decisions did not result from the application of that principle. It is necessary, however, to go on to consider the decisions about which tissues should be proscribed as SBO.

1162 For the reasons that we have set out earlier, we consider that the decisions about what should and what should not constitute SBO were reasonable, having regard to what was known at the time. It should be recognised that in drawing the line so as to exclude the abomasum, from which tripe and rennet were derived, and offal from calves aged less than 6 months, commercial considerations are likely to have weighed in the balance.

1163 The possibility that BSE might be transmissible to humans through non-food products was addressed by issuing guidance to a number of relevant industries about the potential risk, including occupational risk, from the use of bovine products. Perhaps the most important was that issued to manufacturers of medical products, which, as we have noted, applied equally to medicines for human use and veterinary medicines.

1164 The potential risk from occupational contact with bovine materials was also addressed by advice and guidance to many of those whose jobs brought them into contact with such materials. This advice was developed and issued over a period of time.

1165 The final policy decisions with which we are concerned were those reached on 20 March 1996:

    i. a requirement that carcasses from cattle aged over 30 months be deboned in licensed plants supervised by the Meat Hygiene Service (MHS) and the trimmings classified as SBO; and

    ii. a prohibition on the use of mammalian MBM in feed for all farm animals.

1166 If there had been no need to consider practicality or public perception, a case could have been made for saying that the deboning scheme satisfied the ALARP principle. In the event it was not viable. On this occasion the wrong policy option was selected.

1167 The prohibition on the use of mammalian MBM in feed for all farm animals was we consider an appropriate response under the ALARP principle to the change in knowledge of the risk posed by BSE to humans, consequent upon the conclusion of the Spongiform Encephalopathy Advisory Committee (SEAC) that the cases of vCJD were probably linked to exposure to BSE.

Shortcomings and possible reasons for them

1168 Putting hindsight aside, we have no doubt that the policy decisions that there should be a ruminant feed ban, that clinically affected cattle should be destroyed and that SBO should be kept out of human food and animal feed, were right. Because the right policy decisions were taken, BSE is today within reach of eradication and millions have received a high degree of protection from the risk of ingestion of potentially infective products or by-products of the cow. This reflects credit on our system of government and, in particular, on the State Veterinary Service (SVS), which bore the brunt of the demands made on this country by BSE.

1169 Plaudits must, however, be muted. Not all went well. All too often the correct policy decision was marred by:

- the time that had been taken to reach it;
- lack of rigour in considering how to give effect to it;
- lack of rigour in implementing, enforcing and monitoring the Regulations introduced to give effect to it.

1170 In order to see what lessons can be learned from the BSE story it is necessary first to consider what may have been the major causes of the shortcomings that we have identified.

Was there a conflict of interest in MAFF?

1171 We begin with a criticism that has been widely made of MAFF's position in relation to BSE. This starts with the complaint that MAFF had a conflict of interest between the aim, on the one hand, of supporting producers of agricultural produce (as 'sponsor department' for the industry) and, on the other, of protecting consumers of agricultural produce. The criticism continues that in resolving that conflict MAFF was more concerned to protect the interests of the producers.

1172 We discussed the question of conflict of interest with a number of witnesses, including Sir Michael Franklin, who served as Permanent Secretary at MAFF up to the end of September 1987. He accepted the potential for conflict of interest, but commented:

> . . . you have to ask yourself whether it makes sense, and this is a great philosophical discussion on the machinery of Government, whether it is better to have these potentially conflicting interests in a separate department so that the tension is between the two departments, or whether it is better to have a single department with a single minister who can resolve those tensions within his own command. I have said earlier why I think, in terms of the food chain going from the farmer to the food industry and through to the rest of the food chain, there is in fact a positive advantage in having it all under one minister, and where tensions arise resolve them within the department.

1173 We do not propose to be diverted at this point into a great philosophical discussion. At the general level, it should be recorded that Mr Gummer initiated measures which addressed this conflict by creating a separate Food Safety Directorate within MAFF and a Consumer Panel to advise the Ministry. We are concerned, though, to deal with the criticism that in the course of the BSE story MAFF leaned in favour of the agricultural producer to the detriment of the consumer. So far as the policy decisions are concerned we are satisfied that this criticism is without foundation. The ALARP principle does not aim to achieve zero risk. It involves an exercise in proportionality. For the reasons given earlier, we are satisfied that the consideration given to the details of the human SBO ban was a fair application of that principle. Once SEAC had been set up, MAFF's approach was always to consult it on whether the risk BSE posed to humans called for further precautionary measures. Whether that was the best way to use SEAC, we shall discuss in due course. The fact is that MAFF never did less, and on occasion did more, than SEAC recommended. MAFF officials and Ministers were, in our judgement, as concerned as anyone else that if there was a possible risk to human health, appropriate measures should be taken in response to it. Concern for the industry meant, however, that officials and Ministers were particularly concerned about how the public would perceive the risk from BSE.

Other conflicts of interest

1174 Many Departments have potential conflicts of interest between responsibility for regulating an industry and being custodians of its interests within general government business. Examples in the BSE story included the dual role of the Department of Trade and Industry (DTI) on cosmetics and toiletries, and the multiple role of DH in fostering the pharmaceutical industry, looking after the interests of the NHS as a large-scale purchaser, and licensing individual products to safeguard consumers. Commonly, Departments seek to operate internal arrangements that keep the different roles separate. In the case of medicines, the ring-fencing arrangements, as we have seen, included heavy reliance on advice from statutory advisory committees of outside experts. This itself can create problems because many such experts may have their own financial links with companies for whom they are carrying out research or acting as advisers. There has been increasing emphasis on the need for all such interests to be declared when relevant to particular items under discussion.

1175 We have seen no indication that vested interests were allowed to influence the approach to safety in these areas.

Perception of risk

1176 We have identified three types of challenge posed by BSE: the need to eradicate the disease, the potential threat to other animals and the potential threat to humans. The rigour with which each of these challenges was addressed was bound to be affected by the subjective belief of those involved as to whether BSE was, in fact, a potential threat to human life. We have formed the view that the vast majority of those who were involved in this country's response to BSE believed, subjectively, that it was not a threat to human health. In their heart of hearts they felt that it would never happen – BSE was not, potentially, a matter of life and death for humans – and this belief was shared by many who could see, objectively, that the potential risk was there.

1177 This view is based largely on impression as a consequence of having heard oral evidence from those who were principally concerned. It is also supported by a small statistical survey that we carried out. We asked more than 270 witnesses, including those who were involved in the response to BSE either as Ministers, officials or scientists advising government whether they had changed their diet as a result of learning about BSE. All but a handful said BSE had had no relevant effect on their diet.

1178 Although most of those concerned with handling BSE believed that BSE posed no risk to humans and understood the available science as indicating that the likelihood that BSE posed a risk was remote, they did not trust the public to adopt as sanguine an attitude. Ministers, officials and scientific advisory committees alike were all apprehensive that the public would react irrationally to BSE. As each additional piece of data about the disease became available, the fear was that it would cause disproportionate alarm, would be seized on by the media and by dissident scientists as demonstrating that BSE was a danger to humans, and would lead to a food scare or, even more serious, a vaccine scare.

1179 From the moment in December 1986 when Mr Bradley classified his first minute about BSE as 'Confidential', to the Chief Medical Officer's (CMO's) reassuring recorded message of 20 March 1996, ending with the statement 'I myself will continue to eat beef as part of a varied and balanced diet', officials and Ministers followed an approach whose object was sedation. In the first half of 1987 there were restraints on the release of information about BSE. After this there was no attempt to conceal facts from the public. The approach did not set out to deceive. It set out simply to redress the balance that it was feared would otherwise remain tilted as a consequence of alarmist media cover. One witness described it nicely as 'leaning into the wind'.

1180 Examples of this approach are legion. Here is a selection:

- The repeated statements that 'there is no evidence that BSE is transmissible to humans', which did not explain that such evidence would take many years to emerge.

- The repeated invocation of the assessment in the *Southwood Report* that 'the risk to humans is remote', which continued long after the assumptions made by the Southwood Working Party had been shown not to be valid.

- The agreed presentation of the human SBO ban as being a convenient means of giving effect to the baby food recommendation.

- Presentation of oral transmission of BSE to mice and transmission to a marmoset as demonstrating that BSE behaved like scrapie.

- Statements that the cat did not increase the likelihood of BSE transmission to humans.

- Dr Metters's statement that: 'Every effort has thus far been made to underline the Government's position, based on advice from the Southwood and Tyrrell Committees, that the disease is not a risk to humans.'

- The attempt to get SEAC to produce publicity soundbites.

- The publicity documents submitted by MAFF officials to their Ministers on the very day in March when the balloon went up.

- The public presentation of the medicines guidelines as if they had secured the situation without indicating that products were not required to be withdrawn.

1181 The campaign of reassurance focused particularly on the safety of beef. Successive DH CMOs, and a CMO for Scotland, made unqualified statements that it was safe to eat beef. They did so, not on the basis that they were satisfied that BSE was not transmissible in food, but on the basis that they were satisfied that the portions of the cow which might infect were not permitted to enter the food chain. This was not made clear to the public, who equated statements that it was safe to eat beef with statements that BSE posed no risk to humans.

1182 The official line that the risk of transmissibility was remote and that beef was safe did not recognise the possible validity of any other view. Dissident scientists tended to be treated with derision, and driven into the arms of the media and to exaggerated statements of risk. Thus views expressed on risk became polarised. Dispute displaced debate.

1183 The need to provide a reassuring message also featured strongly in the presentation of measures to ensure the safety of medicines. Concerns that the public might boycott vaccines if their safety was called into question were considered paramount.

1184 The anxiety of Ministers and officials not to provoke alarm was shared by the scientific advisory committees. The Southwood Working Party told us that they did not wish to raise needless alarm in those who might have been infected with BSE before any precautionary measures were taken. They accommodated the concern of those responsible for advising on the safety of medicines that their Report should not suggest that vaccines posed any risk. Their Report gave the impression that *in all circumstances* the risk of transmission of BSE appeared remote. It had the caveats that they had had little evidence to go on and that, if their assessment were proved wrong, the implications would be extremely serious. These caveats were, however, quickly lost sight of. So that, for instance, the Committee on Safety of Medicines (CSM) in a position statement said: 'The CSM agrees with the Southwood Working Party that the risk to man of infection via medicinal products is remote. As a precautionary measure, and for the sole aim of seeking to guard against what is no more than a theoretical risk to man, the CSM and the Veterinary Products Committee (VPC) have agreed joint guidelines on good manufacturing practice for the manufacturers of human and veterinary medicines who use bovine, or other animal materials either as an ingredient or in the production process.'

1185 SEAC's 1994 'Summary of Present Knowledge and Research' on TSEs could have been the occasion for a public reassessment of the risk of transmissibility of BSE to humans in the light of all that had been learned since Southwood. It should have replaced the *Southwood Report* as the document to which anyone seeking an up-to-date and authoritative assessment of risk referred. But the message that it gave as to the reassessment of risk was muted and, so far as the public were concerned, it seems to have vanished without trace.

1186 What was the effect of the campaign of reassurance? The precautionary measures that the Government introduced against the possibility that BSE might be transmissible to humans called for care and diligence in their implementation and enforcement. This was to be expected from those involved only if they were persuaded that such a possibility was a real one and that the precautionary measures were therefore important safeguards of human health. We have noted evidence from those responsible for enforcing the SBO Regulations in slaughterhouses that BSE was not regarded as a risk to human health. Local authorities told of the confusion among their staff about the line to take. We have also identified areas where the bureaucratic process ground on very slowly in responding to BSE – the preparation of guidance on operational risks and dissecting bovine eyeballs are examples. In the case of medicines and cosmetics, a relaxed attitude was taken to using up stocks. We believe that lack of diligence in implementing Regulations and lack of urgency in other areas of response to BSE were attributable, in part, to the success of continuous efforts to make sure that news about BSE did not give rise to public concern.

1187 We do not suggest that all were sedated by the official presentation of risk. Some were sceptical and the media were not slow to point out that, while MAFF persisted in maintaining that the risk to humans was remote, this message was accompanied by a series of measures aimed at reducing risk still further.

1188 Whether they were sedated or sceptical, the reaction of many members of the public to the announcement on 20 March 1996 was the same. They felt that the Government had not been telling the truth about the risk to humans from BSE; the public had been deceived.

1189 It is in the context of communication of risk that we feel that there is more force in the argument that it was unsatisfactory for a single Department to be concerned with protecting both consumers and producers. MAFF's dual role meant that their officials and Ministers were particularly apprehensive about the possibility of alarmist consumer reactions causing harm to the producers. We note, however, that DH officials who were not confronted with this potential clash of interests with regard to food, showed themselves as eager as MAFF to present information in a manner calculated to cause the least alarm.

1190 To an extent the Government's response to BSE was driven not by its own, and its advisers', assessment of risk, but by the public's perception of risk. The introduction of the human SBO ban is the most notable example. At times media response to BSE was exaggerated, but often media critique was pertinent and well informed. The media played a valuable role in reflecting, and stimulating, public concerns which proved well-founded and which had a beneficial influence on government policy.

Ignorance and failures of communication

1191 Some of the responses to BSE were inadequate because those responsible for them were not party to aspects of the state of knowledge at the time which should have informed their decisions.

1192 The earliest example of this was the delay in discovering the extent to which cattle were succumbing to BSE consequent upon restraints imposed at the CVL on dissemination of information about the disease in the first half of 1987.

1193 Another example was the delay in deciding to introduce a slaughter and compensation policy, which resulted from the failure of MAFF officials to inform DH of the disease and to get its input into the consideration of BSE's implications for human health.

1194 A third example was the delay in addressing the risk from bovine products in human medicines, which resulted from that same lack of communication between MAFF and DH.

Ignorance of views as to the minimum infective dose for cattle

1195 A further example is provided by the consequences of the failure to focus on the question of the minimum infective dose. At the end of 1990 interim results of the Neuropathogenis Unit (NPU) experiment to transmit BSE to sheep and goats had indicated that eating infective material weighing only ½ gram had sufficed to infect a sheep. Had scientists at the NPU, or Dr Kimberlin, or Mr Wilesmith been asked in 1988, they would have advised that it was at least possible that the minimum amount of material that would suffice for oral transmission to a calf would be very small. Yet the result of the CVL attack rate experiment, which showed, at the end of 1994, that a single gram had transmitted BSE orally to a calf, caused widespread surprise and concern.

1196 When the ruminant feed ban was introduced, some officials within MAFF were under the impression that a cow would have to eat a substantial quantity of infective material to contract BSE. This impression was shared by the UK Agricultural Supply Trade Association (UKASTA). It believed that the quantity of pig and poultry feed that might get into ruminant feed as a result of cross-contamination in feedmills was not a matter for concern. Mr Meldrum made the same assumption. The need to address the problem of cross-contamination of ruminant feed was not appreciated or tackled until 1994.

1197 When the animal SBO ban was introduced in 1990, none of those involved appreciated the extent to which contamination of MBM with SBO in rendering plants would give rise to infectivity in that MBM, let alone that this would be enough to pose a threat to cattle as a result of a second round of cross-contamination in the feedmills. Steps were taken to agree a rendering code to reduce contamination, but not with any urgency, and two years elapsed from the introduction of the ban before the code was in place. Even this was insufficient to prevent significant contamination. Only after the result of the attack rate experiment became known in 1994 was the decision taken that renderers would have to process SBO in separate facilities.

Ignorance of views as to the minimum infective dose for humans

1198 The question of the minimum amount that was capable of infecting was equally of importance in the context of the safety of human food. It was a vital element in any evaluation of the potential risk of contamination of human food by slaughterhouse practices, such as brain removal, and the production of mechanically recovered meat (MRM).

1199 In 1989, when the SBO Regulations were being prepared, the safety of MRM received consideration, which we have concluded was inadequate. Scientists were not asked for their views of the minimum amount which might infect. An assumption was made that any contamination of MRM with spinal cord was unlikely to be sufficient to be significant. In the following year MAFF officials adopted a similar approach to the question of contamination as a result of head-splitting and brain removal.

1200 SEAC's robust advice, that removing the brain before the head meat was not acceptable, gave some indication that the Committee considered that a small quantity of contaminant was cause for concern. SEAC never so stated expressly, and its paper on the safety of beef was capable of conveying the false impression that only a substantial quantity of infective material would pose a risk of transmission by the oral route.

Ignorance of pathways of infection

1201 One of the questions asked by Sir Richard Southwood before the first meeting of his Working Party was:

> What are the routes to man of parts/products of cattle, especially dairy cattle, before and after slaughter?

1202 MAFF and DH were unable to provide a detailed reply. He was told that there was a very low probability that spinal cord formed part of meat products, but that quantitative information on the fate of organs and tissues was unavailable. Since 1996 a survey has disclosed that, at the time that the Southwood Working Party were considering their recommendations, substantial quantities of spinal cord were going into human food as an ingredient of MRM. Had the Working Party known this we wonder whether they would have been content that it should be allowed to continue.

1203 The Tyrrell Committee advised, as a top priority item, that there should be a more detailed investigation into the fate of bovine (and ovine) tissues and products that could lead to infection being spread by as yet unrecognised routes. This recommendation was never implemented. Had it been, timely consideration might have been given to closing pathways of potential infection for humans or for animals that, at least initially, were overlooked.

1204 In June 1990 a survey of cutting procedures disclosed that lymph nodes removed in the course of dressing meat were used in meat products for human consumption.

1205 In 1994 members of SEAC were concerned to learn that the residues that settled in the tank bottoms in the course of refining tallow, including tallow derived from SBO, was still being incorporated in cattle feed.

1206 Until 1995 it was not appreciated by MAFF officials that gelatine derived from cattle was entering cattle feed in substantial quantities as an ingredient of recycled waste foods, in breach of the ruminant feed ban.

1207 Uncertainty prevailed throughout the period with which we are concerned as to the use of bovine products in cosmetics.

1208 Consideration was not given to the question of whether drainage waste from slaughterhouses or effluent from renderers of SBO might pose hazards of BSE contamination which called for review of their disposal.

Failures of communication

Between the Southwood Working Party, the Government and the public

1209 Many who read, or who were informed of the conclusions of, the *Southwood Report* failed to appreciate that:

- When describing the risk posed by BSE to humans as remote, the Working Party intended to indicate that such precautions as were reasonably practical should nonetheless be taken to address the risk.

- The description of the risk from medicinal products and occupational exposure as remote was predicated on the assumption that the responsible authorities had been alerted to the potential risk and were taking appropriate measures to address it.

- The Working Party's conclusions on risk were based on very limited data and were inferences drawn from knowledge of scrapie and CJD.

- The Working Party contemplated the possibility that their conclusions might be wrong, and that in that event the implications would be extremely serious.

Between SEAC, the Government and the public

1210 The breakdown of communication between MAFF officials and SEAC, when the latter considered slaughterhouse practices and MRM, resulted in the impression being given that the members of SEAC were not concerned by the degree of contamination described as 'inevitable' in MAFF's paper on the topic. That was not the position. Some, at least, of the members of SEAC were advising on the premise that there would be total removal of the spinal cord before MRM was extracted. They gave the same advice about MRM on the same basis in June 1995. Not until November 1995 was it brought home to SEAC that spinal cord was not always being cleanly removed, whereupon at last it advised against the practice of extracting MRM from the bovine vertebrae.

1211 We have commented above on the possibility that SEAC's 1990 paper on the safety of beef contributed to the erroneous belief that a substantial quantity of infective material would have to be eaten in order to transmit BSE.

1212 SEAC's 1994 paper on TSEs failed to spell out clearly that events since the *Southwood Report* had adversely altered the assessment of the likelihood that BSE was transmissible to humans, and this message was not conveyed to the general public.

Lack of rigorous consideration when giving effect to policy

1213 We have identified three occasions on which a lack of rigour when considering how to implement policy had adverse consequences. The first was at the time of the introduction of the ruminant feed ban. Because of the lengthy incubation period, years would necessarily elapse before any defects in the operation of the ban would become apparent. The technique of building a dam and then looking for leaks would not do. Rigorous consideration should have been given to ensuring that the dam was watertight in the first place. The question of whether cross-contamination in feedmills would be cause for concern should have been addressed. Advice should have been obtained on how much contaminant might suffice to infect. This would have led to UKASTA being advised that cross-contamination had to be prevented and focused attention on the urgency of developing a test that would detect ruminant protein in compound feed.

1214 The second occasion was when the question of the safety of MRM was raised in the course of the consultation exercise for the human SBO ban. The critical issues of the extent of likely contamination of MRM and the minimum amount of material that might infect were not addressed. It was simply assumed that any contamination would be too small to matter. Thus no guidance was given to local authorities or the Veterinary Field Service as to the importance of removal of all spinal cord.

1215 On these first two occasions the lack of rigour resulted in failure on the part of those considering the implementation of policy to obtain the information that was available and was needed in order to reach the correct decision.

1216 The third occasion was the preparation of the Order that was to give effect to the animal SBO ban. The terms of the Order were in a form that was unenforceable. Rigorous consideration would have led to the conclusion that this was not a ban where self-policing could be relied upon and that Regulations should be drawn up which could be enforced.

The best being the enemy of the good

1217 The production of written documents by officials and by advisory committees frequently entailed a process of wide consultation and drafting refinement. This was a 'Rolls-Royce' system, but one which tended to result in lengthy delays. Consultees would be tempted to suggest drafting improvements, which would then result in a further round of consultation. These were often not changes of sufficient substance to justify the delay that they caused.

1218 One area in which the effects of this were keenly felt was in the preparation of written guidance on precautionary measures and practices. On some occasions it took many months, or even years, after a decision was taken to issue written guidance, for that decision to be implemented. By way of example, it took two-and-a-half years for SEAC's advice on the dissection of bovine eyeballs to be passed on to schools and up to three years to issue simple occupational warnings and basic advice to some of the high-risk trades.

1219 When drafts were submitted to advisory committees for comment, delays could be particularly protracted. Again, by way simply of example, we can cite the comprehensive advice of the Advisory Committee on Dangerous Pathogens (ACDP) to those handling risk tissues in laboratories, hospitals and mortuaries. A further example is an excellent draft Advisory Note to farmers on the dangers of cross-contamination of cattle feed with pig or poultry feed that was initially drafted in November 1995, was considered and refined by, among others, both SEAC and Mrs Browning, and had not been issued by 20 March 1996. In all of these cases the desire to perfect a document was allowed to outweigh the need for speedy advice. The best became the enemy of the good.

Inappropriate use of advisory committees

1220 Advisory committees have a vital role to play in assisting government to formulate policy. However, if matters are referred to committees which only meet periodically, this can delay the process of taking decisions. We shall give detailed consideration to the lessons to be learned in relation to the use of committees at a later stage. For present purposes we would draw attention to the following principles:

- resort should be had to committees only where their expertise is needed;
- advice sought should be clearly targeted so as to fall within the expertise of the committee;
- advice given should be reviewed to ensure that it appears to be soundly based; and
- advice should be treated as such, and not as being determinative of policy.

1221 These principles were not always followed in the case of the BSE story. For example:

   i. In order to resolve the policy issue of whether cattle showing signs of BSE should be permitted to enter the human food chain, the essential question to answer was whether it was possible to be confident that this would involve no risk. There was no need to appoint the Southwood Working Party to resolve that question. MAFF officials had been able to reach a firm, and correct, conclusion on the limited available data that it was not. Had DH officials been involved with MAFF in considering the risk to human health from the outset, we believe that they would have concurred in that conclusion. The decision to refer the question to a Working Party resulted in a delay of over three months.

ii. The conclusions of the Southwood Working Party were not reviewed. Their recommendations were treated not as advice, but as definitive of the precautionary measures which did, and did not, require to be taken. It was left to public reaction, and the assistance of Dr Kimberlin through the good offices of Pedigree Masterfoods, the pet food manufacturer, to lead MAFF Ministers to conclude, over three months later, that an SBO ban should be imposed.

iii. The advice of the Southwood Working Party continued to be quoted as definitive of the precautionary action required by science long after some of the premises upon which the Working Party had advised were demonstrated to be unsound.

iv. SEAC was set up as a standing, part-time, committee to advise MAFF and DH on 'matters relating to spongiform encephalopathies'. The breadth of these terms of reference was reflected in the wide variety of matters on which SEAC was asked to advise.[103] It immediately became the practice to seek the advice of SEAC on any policy decision that had to be taken in relation to BSE, without identifying those aspects of the question on which SEAC was particularly qualified to advise or targeting the advice sought from it. Furthermore, once SEAC had advised, its recommendations tended to be treated as determinative of the action to be taken.

1222 On two occasions the intervention of SEAC proved positively unhelpful. The first was when slaughterhouse practices including MRM were referred to it. The untargeted request to SEAC to advise 'whether any action or guidance is necessary in relation to slaughterhouse practices' led to advice being given on the basis of SEAC's assessment of the efficacy of those practices. This assessment was unreliable and was one that MAFF officials were very much better placed to perform. The advice was not in clear terms and led MAFF officials wrongly to conclude that members of SEAC were not concerned about inevitable failure to remove all spinal cord before MRM was extracted from the vertebrae.

1223 The second occasion was when SEAC recommended the deboning scheme on 20 March 1996. This was unhelpful because the Government accepted it without time to review it to decide if it was practically and politically viable. In this context we would quote a pertinent observation made by the Agriculture Committee in its 1990 Report:

> Scientists do not automatically command public trust, particularly when they are in disagreement with each other, and when the issues concerned do not lend themselves to simple yes/no answers but involve computations of whether particular risks are acceptable or unacceptable to members of the public. Decision-making is not a purely scientific process.

1224 By the time that the 20 March policy decision came to be made, the reliance by government on SEAC to answer questions of policy had become so well established that officials and Ministers had been waiting to see what SEAC had to say rather than carrying out their own exploration of the policy options by way of contingency planning.

---

[103] The use of SEAC receives detailed consideration in vol. 11: *Scientists after Southwood*

Administrative structures

Interdepartmental structures

1225 Evaluation of whether, and in what respects, BSE posed a risk to humans was, in theory, primarily the responsibility of DH, but turned largely on questions that fell within veterinary expertise. Evaluation of whether, and in what respects, BSE posed a risk to other animals fell wholly within MAFF's responsibility and turned, to a large extent, on the same questions of veterinary expertise. Risk management in relation to both types of risk, as far as animals and food products were concerned, fell almost entirely within MAFF's area of responsibility, while DH took the lead on other areas, in particular human medicines. Occupational risk fell somewhere between these two Departments and the Health and Safety Executive (HSE). We have already observed, however, that it is difficult to draw the line between risk evaluation and risk management. It was important that MAFF and DH worked closely together. In particular, so far as food risks were concerned, DH needed to be satisfied that MAFF was taking appropriate action by way of risk management to ensure that potential food risks were satisfactorily addressed.

1226 So far as medicines were concerned, the licensing divisions of the two Departments were responsible for implementing the same legislation using the same assessment criteria – safety, quality and efficacy. A similar system of statutory advisory committees applied, and the Medicines Commission spanned both human and veterinary medicines, having an overview of the workings of the system as a whole. Moreover, veterinary and human medicines drew on similar raw materials, types of sterilisation and production processes. This called for a coordinated approach between the two.

1227 The need for such cooperation between MAFF and DH must exist in relation to other zoonoses, as well as BSE.

1228 Relations between MAFF and DH with regard to BSE did not fall within the framework of any formal interdepartmental structure for dealing with known zoonoses or potentially zoonotic animal diseases. If there had been satisfactory interdepartmental communication and collaboration on an informal basis, this would not have mattered. As we have pointed out, however, until Sir Donald Acheson was notified about BSE in March 1988, such communication and collaboration were absent. Had there been an effective interdepartmental body concerned with zoonoses and potential zoonoses, the BSE story might have got off to a better start. That does not, of itself, demonstrate the need for such a body – it raises the question of whether a formal structure may not be the best way of ensuring proper interdepartmental collaboration in this field.

1229 Matters were further complicated when other Departments were involved. The response on cosmetics called for effective communication and coordination between MAFF, DH and DTI, the industry's sponsor Department. Similarly, it was for DES to send out advice on the dissection of bovine eyeballs in schools, drawing on advice from DH, MAFF and the HSE. On waste disposal the Department of the Environment was involved. All this called for clear allocation of lead responsibility and efficient lines of communication between Departments. These were not always evident.

WHAT WENT RIGHT AND WHAT WENT WRONG?

DH role

1230 DH[104] had the lead in relation to human health surveillance, being the Department to which the CJD Surveillance Unit reported. On most other aspects of BSE, DH maintained a watching brief over MAFF's actions. As Sir Christopher France[105] told us, it was for the Chief Medical Officer and the professional staff who reported to him to take the lead on the DH response to BSE. In the early stages of the story Dr Pickles, who had the DH lead, played a notably proactive role in scrutinising and questioning MAFF's actions. As we discuss in Chapter 6, in the weeks leading up to March 1996 the DH role was passive, with the result that they did not raise with MAFF the need for contingency planning as soon as it became apparent that BSE might prove to be transmissible to humans.

1231 The other major area in which DH took the lead was in relation to human medicines. Veterinary medicines were to some extent treated as the poor relation of human medicines. MAFF from the beginning of 1989 took its cue from DH on the handling of existing products and stocks in relation to BSE. Within DH, medicines licensing was the province of Medicines Division (MD), which, as one witness put it, 'consumed its own smoke'. When reviewing products over which any questions arose, MD looked to advice from its 'section 4 committees' of eminent outside experts. During the period with which we are concerned there was a significant reorganisation of the arrangements for handling medicines licensing, in order to address structural and management problems identified in a management review. MD was reconfigured into an Executive Agency – the Medicines Control Agency – in 1991, and the Medical Devices Agency followed in 1994. This reorganisation itself led to some upheaval and confusion, which did not facilitate the management of the BSE measures.

1232 It seemed to us that clearer expectations about reporting to top management and to Ministers would have assisted in the handling of BSE and medicines. By way of example, had Ministers been asked explicitly to consider whether existing stocks of vaccines should continue to be used while guaranteed 'clean' replacements were procured, we believe they would have taken a keen interest in the follow-up. This in turn might have influenced the subsequent pace of events and perhaps led to the doubtful material being phased out rather more quickly than in fact happened.

Structure within MAFF

1233 During the period with which we are concerned, Mr Gummer sought to separate MAFF's sponsorship role from its role in protecting the consumer, by creating a new Food Safety Directorate. Within that Directorate there were what on the face of it appeared to be significant structural changes within MAFF whose aim was to improve the way administrators and veterinarians interrelated.[106] It had originally been suggested that the CVL and the Veterinary Investigation Service (VIS) should merge into a single Executive Agency. The SVS, however, had successfully made its case that it should retain the VIS within its structure. Administrators and veterinarians were, however, merged into the Animal Health and Veterinary Group in 1990, only to be sundered again in 1994. Most witnesses considered that neither change had much effect on how the two worked together in

---
[104] See vol. 15, Chapter 4 and Annex 1 for details of the interrelationship of professionals and administrators within DH
[105] DH Permanent Secretary to February 1992
[106] See Vol. 15: *Government and Public Administration* for details of the interrelationship of professionals and administrators within MAFF

practice. On a day-to-day basis the Chief Veterinary Officer had direct access to the Minister, and would assist him or her with professional advice in relation to policy decisions. Major issues of policy would be put before Ministers in formal submissions prepared by administrators with the assistance of professional advice from the veterinarians. A rather similar approach was followed in preparing papers for SEAC – the paper on slaughterhouse practices is a good example.

1234 So far as the quality of the advice was concerned, this system worked well. However, it was, as we have noted, a 'Rolls-Royce' system. Drafts were circulated and recirculated among a large number of officials, who might have input to contribute. Submissions were refined, polished and supplemented with minutes as they passed up the administrative hierarchy on their way to the Minister. The process could take a very long time.

1235 Where urgency was perceived, it was possible to cut through the red tape and reach a decision fast. This had its own dangers. Mr Gummer's insistence that the Government should announce its response to SEAC's advice on transmission of BSE to a pig simultaneously with announcing that advice, led to defective Regulations, prepared 'in secrecy and haste' without the normal consultation. Similar haste for a similar motive led to the announcement of a response which proved unviable in March 1996.

Chief Medical Officers and Chief Veterinary Officers

1236 The evidence we have heard about the parts played by the CMO and CVO in the BSE story suggests that consideration should be given to two aspects of their roles. It is not our function to define their roles in the abstract, as we have noted in Volume 6 in relation to the CVO.

1237 The first aspect calling for consideration is their ability to give independent advice to the public. Mr Meldrum, at least, considered that the CVO did not have the degree of independence afforded to the CMO in stating publicly his opinions. Indeed, Mr Meldrum has assumed that the CMO was required to advise the public, independent of government, even though this might cause difficulty for his Department, or other Departments. We are not aware of any secure basis for saying that the CMO can do this. This may be contrasted, by way of example, with the position of the Food Standards Agency, one of the functions of which is to advise the general public. The Agency also has a power to publish such advice. We think it desirable that the CMO and CVO should be in the same position.

1238 The second aspect relates to the effect of the relative status of the CVO and CMO. We note that the CVO is an official of high standing in the international arena, but we understand from the evidence we have heard that under civil service conventions the CVO ranks only with the deputy CMO. We feel it is important that this should pose no impediment to direct liaison between the CVO and CMO.

Central and local government

1239 The greatest impediment to the efficacy of the Government's response to the emergence of BSE was the structure laid down by statute for the enforcement of the Regulations that were designed to keep potentially infective tissues out of both human food and animal feed. The first and most critical control point was the slaughterhouse. In the slaughterhouse, the critical point for human health was the inspection and health-stamping of meat as fit for human consumption. For animal health, the critical control point was the gut room, where in practice, though not by any requirement of the Regulations, most of the SBO should have been kept segregated from material to be rendered to produce MBM for animal feed, and where, in accordance with the Regulations, the SBO should have been stained black.

1240 The statutory duty of enforcing the human SBO ban, together with many other Regulations relating to standards and practices in slaughterhouses, rested on the District Councils. In order to comply with European requirements, which were widely considered to be unnecessarily burdensome, District Councils had to employ an enforcement hierarchy, with the Official Veterinary Surgeon at the top. Local authorities faced severe budgeting constraints. Slaughterhouse supervision did not assert a strong claim in the competition for their limited funds, and in a climate of deregulation there was no encouragement from central government to accord priority to this issue. Most councils spent no more than was barely essential to cover enforcement duties in slaughterhouses. Some did not spend that much. When the MHS took over enforcement, it found that insufficient resources had been employed by at least some local authorities to ensure that the obligation imposed by the human SBO ban to remove all spinal cord from the carcass was universally enforced. It also found that familiarity with the Regulations, efficiency of line management and diligence on the part of local authorities in enforcing the Regulations were uneven across the country.

1241 Had the importance of the removal of spinal cord been emphasised in guidance to local authorities and to the Veterinary Field Service (VFS), which monitored performance, we believe that standards could have been improved, but only within limits. The limitations on the enforcement capability of local authorities could only have been remedied had they been persuaded to devote more resources to that task. We can see no way in which that goal could have been achieved.

1242 Turning to the animal SBO ban, the structural problems were that much greater. The County Councils responsible for enforcing that ban had no locus in the slaughterhouse. The District Councils were not in general enthusiastic about doing their job for them. The situation was exacerbated by the fact that the terms of the animal SBO ban imposed no obligations in the slaughterhouse, but we agree with Mr Meldrum and Mrs Attridge that, however well drafted the Regulations, the statutory structure of local authority enforcement would have prevented strict enforcement of the animal SBO ban.

1243 In this situation, monitoring by central government of the performance by local authorities of their enforcement obligations was desirable. MAFF Ministers thought the same, so far as concerned Regulations sponsored by their Department, and required the SVS to perform a monitoring role. The shortcomings in monitoring which we identify in vol. 5: *Animal Health, 1989–96* might well have been reduced if that monitoring had had a statutory foundation.

Central government and the Territorial Departments

1244 We have seen that the Territorial Departments were for the most part content to follow the lead of MAFF and DH with regard to BSE. Nonetheless, we have also seen that communication between Whitehall and the Territories was not always satisfactory. DH was not always interested in the views of the Territorial Departments. This was particularly unfortunate with regard to Wales, where the combination of skills and experience in the Welsh Office allowed its professionals and administrators to make some very useful and pertinent comments. It might well have been beneficial had these been taken on board by DH.

Individual criticisms: redressing the balance

1245 It is inevitable that an Inquiry such as ours focuses on what went wrong. The main point of having the Inquiry is to find out what went wrong and to see what lessons can be learned from this. This can be harsh for individuals. Their shortcomings are put under the spotlight. The overall value of the contributions that they have made is lost from view. We do not wish our Report to produce this result. Yet we cannot set out in detail the workload over the years of each of those who has received – at one point or another – a criticism in our Report. We must make some general comments.

1246 The more senior posts in the civil service are seldom sinecures. Ministerial office never is. We have limited our consideration of individual responsibility to those who occupied such positions. The shortcomings that we have criticised have not been the product of indolence; they have for the most part been mistakes made under pressure of work – pressure made the greater by the imposition on already busy lives of the considerable additional burdens of handling BSE.

1247 The day-to-day demands made by BSE on MAFF, and particularly on the State Veterinary Service, were considerable. By way simply of example, in the period with which we are concerned approximately 200,000 suspect cattle had to be inspected, slaughtered and autopsied by histopathology. The carcasses had to be collected and destroyed. Compensation had to be assessed and paid.

1248 Between 1988 and 1995 about 30 Statutory Instruments in Great Britain alone were brought into force making or amending Regulations dealing with BSE. Some of these involved a great deal of work, but more significantly they evidence the ongoing attention being focused on addressing the implications of BSE for both animal and human health during a period when it was considered unlikely that BSE was in fact a threat to humans. Thus the individual criticisms that we have made must be read in the context of participation in a positive response to BSE, which on the one hand brought the animal disease under control, and on the other resulted in the removal from human food and from medicines of a very high proportion of the material that might have had the capacity to infect.

1249 There are aspects of the response to BSE that stemmed from broader government policies, or from particular ways of handling the problem. Again, these may not be matters that give rise to individual criticism, but they may well highlight lessons for the future. For example, we have noted that Ministers often sought policy advice from SEAC during most of the period. A lesson we have drawn from this is that where the policy decision involves the balancing of considerations which fall outside the expertise of the committee, it will normally not be appropriate to ask the committee to advise which policy option to adopt. It is not our job to examine broad government policies, for example the deregulation initiative. Where relevant, we have examined their implications for the BSE story. For example, our consideration of the impact of the deregulation initiative for slaughterhouses is in Volume 6.

1250 Those who were most active in addressing the challenges of BSE are those who are most likely to have made mistakes. As was observed in the course of the Inquiry, 'if you do not put a foot forward you do not put a foot wrong'. In this context we think it right to single out for mention Mr Meldrum. Mr Meldrum was Chief Veterinary Officer in Great Britain for almost the whole of the period with which we are concerned. He involved himself personally in almost every aspect of the response to BSE. He placed himself at the front of the firing line so far as risk of criticism is concerned.

1251 Mr Meldrum impressed us as a particularly dedicated and hard-working civil servant. We are aware that many consider that he epitomises an approach on the part of MAFF that placed more weight on the interests of the farmer than on the safety of the consumer. We do not consider such an accusation to be fair.

1252 Mr Meldrum was at all times concerned that the livestock industry should not be damaged by a public reaction to BSE for which there was, in his opinion, no scientific justification. That is not an approach for which Mr Meldrum can be criticised. On the contrary, we consider that it was a proper approach for the Chief Veterinary Officer to adopt.

1253 In the BSE story there were a number of issues on which Mr Meldrum advanced the view that the possibility of risk to humans was too insignificant to warrant precautionary measures:

- Should offal of sheep be removed from human food?
- Should tripe and rennet from the abomasum be included in the SBO ban?
- Should tissues from calves under the age of 6 months be excluded from the SBO ban?
- Was MRM a risk to humans?

1254 We do not doubt that the views which Mr Meldrum advanced reflected his own beliefs.

1255 When Mr Meldrum had concerns about risks to humans, he acted on them. Thus:

- He recommended that there should be no exclusion from the SBO ban of intestines that had been procured to produce sausage skin.
- In 1990 he raised concerns in relation to peripheral nervous tissue going into MRM.

- In 1994 he raised the suggestion of banning recovery of MRM from the spinal column.

1256 We are satisfied that where Mr Meldrum perceived the possibility of a significant risk to human health he gave this precedence over consideration of the interests of the livestock industry.

1257 Pressures on busy people go some way to mitigate a number of other criticisms that we have made – for example, the failures to review the *Southwood Report*, and failures to give rigorous consideration to the form of the animal SBO ban.

1258 We have criticised the restrictions on dissemination of information about BSE in the early stages of the story, which were motivated in part by concern for the export market. We suspect that this may have reflected a culture of secrecy within MAFF, which Mr Gummer sought to end with his policy of openness. If those we have criticised were misguided, they were nonetheless acting in accordance with what they conceived to be the proper performance of their duties.

1259 For all these reasons, while we have identified a number of grounds for individual criticism, we suggest that any who have come to our Report hoping to find villains or scapegoats should go away disappointed.

14. Lessons to be learned

1260 We have reached the final chapter of this volume – consideration of the lessons to be learned from the events that we have been considering. First we summarise lessons from particular episodes of the story and then lessons to be learned about four topics which run right through the story: the use of advisory committees; dealing with uncertainty; legislative loopholes; and the experience of the victims of vCJD and their families.

1261 Aspects of this Inquiry make this an unusual and not entirely satisfactory exercise. The BSE story is an ongoing story. We have looked at a substantial section of the story, but one that ended over four-and-a-half years ago. We have conducted a particularly public Inquiry and believe that, while it has been proceeding, many lessons have already been learned from the BSE experience and acted upon. The scenery has shifted very considerably from that with which we have made ourselves familiar. The most significant changes have been the creation of the Food Standards Agency and the devolution of powers to a Scottish Parliament and Welsh Assembly. We have also been informed of the creation of a large number of interdepartmental bodies, covering areas which include zoonoses, animal disease surveillance and Transmissible Spongiform Encephalopathy (TSE) research.

1262 The Office of Science and Technology has addressed the questions of the Government's use of science, the Government's use of expert committees and the Government's approach to risk. These topics have also received consideration by a number of other institutions.

1263 It is not part of our remit to assess how well all these developments are now working. That is for others, including the Government, press and public. We propose to confine ourselves strictly to the lessons to be learned from the BSE experience up to 20 March 1996. If some of these lessons have already been learned, others may bear repeating.

Episodes in the BSE story

Lessons from the fact that BSE emerged

Commentary

1264 The fact that the origin of the BSE epidemic is unknown leaves many questions unanswered. In particular it raises the possibility that rare cases of autosomal genetic mutation may give rise to sporadic TSE in cattle, and possibly in other animals.

Lessons

- BSE is a novel and alarming zoonosis. There is much about it that is not yet understood. Precautionary measures need to be applied to reduce the potential risk to as low as is reasonably practicable.
- TSEs may occur in species in which they have previously been unknown.
- It is possible that TSEs develop sporadically in other animal species as they do in humans.
- If TSEs develop sporadically and rarely in farm animals, as they do in humans, they may well pass undetected. This is particularly the case where farm animals are slaughtered for consumption when young and thus before clinical signs normally develop.

Lessons from the transmissions of BSE

Commentary

1265 We have now learned much about the capacity of BSE to transmit to other animals, both naturally and experimentally. The lessons learned provide valuable data for risk management.

Lessons

- TSEs may be transmissible between the same species and between different species.
- TSEs may be transmissible within animal feed and human food.
- Tissues in an animal incubating a TSE may be infectious before the animal has developed clinical signs of the disease.
- It is possible to distinguish between the level of infectivity, or titre, likely to be found in the different tissues of an animal incubating a TSE. The brain and spinal cord, in the later stages of incubation, are the highest risk tissues.
- A very small quantity of infective material may be sufficient to transmit a TSE by the oral route.
- Risk of oral transmission of a TSE will be greatly reduced if high risk tissues are removed from the food chain.

Lessons from the spread of the BSE epidemic

Commentary

1266 What turned the initial case or cases of BSE from an incident into a catastrophe was the wide, and latent recycling consequent upon the practice of using meat and bone meal (MBM) as an ingredient of animal feed.

Lessons

- The process of rendering animal parts to produce MBM, which is then incorporated in animal feed, will result in the pooling of material from many animals and the wide dissemination of infection from a single infective animal.

- The rendering process cannot be relied upon to inactivate TSEs.

- Recycling animal protein carries a greater risk of spreading infection with a TSE when it is carried out within the same species.

- Recycling animal protein carries a greater risk of spreading infection with a TSE where the protein is derived from high-risk tissues.

- Where a TSE has a lengthy incubation period, recycling may spread the disease very widely before its emergence is detected.

Lessons from the identification of BSE

Commentary

1267 The identification of the emergence of a new animal TSE was of critical importance as the first step towards addressing the implications of the disease. The importance of a surveillance system that will identify the emergence of new animal diseases was demonstrated. The efficacy of the passive surveillance system depends upon farmers and their veterinarians drawing incidents of animal disease to the attention of the State Veterinary Service (SVS). When a new disease is identified, early publication of information about its characteristics will be desirable in order to encourage reporting of similar cases.

1268 We note with satisfaction the consideration currently being given to surveillance by the circulation of a consultation document: *Veterinary Surveillance in England and Wales – A Review, April 2000*. We emphasise the importance of pursuing this initiative.

Lessons

- An effective system of animal disease surveillance is a prerequisite to the effective control of animal diseases.

- An effective system of passive surveillance will depend upon farmers and their veterinarians having the incentive and the facility for drawing instances of animal disease to the attention of the SVS.

- Research into methods of diagnosis should form an integral part of an animal disease surveillance system.

- The proximity of the nearest veterinary centre of investigation to the farm where the disease occurs will be an important factor in determining whether or not a casualty is referred for pathological examination.

- The identification of BSE demonstrated the importance of the animal disease surveillance system of the SVS and of the close links that existed between the Veterinary Investigation Centres (VICs) and the Central Veterinary Laboratory (CVL).

- It is important that details of a new disease which may have implications for human and animal health should be disseminated within the State and private veterinary systems in order to encourage the reporting of similar cases.

Lessons from the consideration of the nature and implications of BSE

Commentary

1269 When BSE was identified as a new disease by the CVL in December 1986, it was at once appreciated that two important questions needed to be answered. Was it indeed a TSE? And did it have implications for human health? It was the greatest good fortune that, as a result of the joint initiatives of the Agricultural and Food Research Council (AFRC) and the Medical Research Council (MRC), there existed in the form of the Neuropathogenesis Unit (NPU) a world-renowned centre of expertise in TSEs. We have criticised the delay in seeking the collaboration of the NPU in answering the first important question. We have also criticised the more substantial delay in involving DH in the consideration of the second question.

Lessons

- Where animal or public health is at stake, resort should be had to the best source of scientific advice, wherever it is to be found, without delay.

- Collaboration between MAFF and DH, and between the Chief Veterinary Officer (CVO) and the Chief Medical Officer (CMO), must be maintained in considering the potential for animal diseases to threaten human health and the steps that should be taken in response to any potential zoonosis. Consideration should be given to whether a formal structure is the best means of achieving this.

- Advantage should be taken of the expertise and resources of the Public Health Laboratory Service (PHLS) whenever the possibility of a potential zoonosis exists.

- Lead responsibility must be clearly established for coordinating the scientific response to a new disease or a new outbreak of disease.

- Consideration should be given to combining in the same laboratory research on scientific issues that have common application to human and animal health by scientists practising in each field.

Lessons from the investigation of the cause of BSE

Commentary

1270 The investigation of the cause of the cases of BSE that were being reported in 1987 and 1988 was carried out by Mr Wilesmith. He was the only veterinarian on the staff of the SVS who had a postgraduate qualification in medical epidemiology. He told us of the dearth of veterinarians trained in epidemiology and of the absence of any training courses in veterinary epidemiology. Dr Tyrrell told us of the initial impossibility of finding a veterinary epidemiologist of high calibre to serve on the Spongiform Encephalopathy Advisory Committee (SEAC).

1271 The result was that the burden of the epidemiological investigation of BSE was shouldered by Mr Wilesmith, with the help of his subordinate staff, throughout the period with which we are concerned. The data on which he worked were not readily available to others interested in the epidemiology of the disease.

1272 Mr Wilesmith quickly and correctly identified MBM as the vector of BSE. His tentative conclusions as to why the MBM was infective proved to be erroneous. They were reasonable on the data available to him, but could profitably have been subject to epidemiological review as more data became available to which modelling could be applied.

Lessons

- Provision should be made for training veterinarians in epidemiology. Joint postgraduate training programmes in epidemiology for trainees in veterinary medicine and public health medicine should be encouraged.

- Epidemiologists, particularly those in the public sector, should make available the data upon which their conclusions are based.

Lessons from the introduction of the ruminant feed ban

Commentary

1273 When the ruminant feed ban was introduced, it was not appreciated that there was any need to be concerned about the amount of cross-contamination of cattle feed likely to occur in feedmills from the production of pig and poultry feed containing MBM. This was because it was assumed that the quantity involved would not be sufficient to result in transmission. There was a general impression that a large quantity of contaminated material had to be eaten in order to transmit this disease. There was no basis for this assumption, which should not have been made. Had rigorous thought been given to the matter, this would have involved seeking the views of the experts, who would have advised that a small quantity might suffice to infect.

Lessons

- When a precautionary measure is introduced, rigorous thought must be given to every aspect of its operation with a view to ensuring that it is watertight.

- Reliance on a trade association or other body to communicate the importance of a precautionary measure is not always appropriate.

Lessons from the introduction of slaughter with compensation

Commentary

1274 The decision that cattle showing clinical signs of BSE should be compulsorily destroyed was too long delayed. One reason was that DH was not involved until a very late stage. We have already referred to the need to maintain joint MAFF/DH involvement in dealing with potential zoonoses. Another cause of delay was the reference to a Working Party of the question of how to respond to BSE when the input of the Working Party was not essential to the decision on compulsory slaughter.

Lessons

- Where policy decisions turn on risks to human health, DH should be involved in the formulation of policy from the outset.

- Reference to outside expert committees involves delay. It should be avoided, where possible, in a situation of urgency.

- Uncertainty can justify action.

Lessons from the *Southwood Report*

Commentary

1275 We have drawn attention in Chapter 4 to certain aspects of the *Southwood Report* which detracted from its overall merit. We shall deal in due course in more detail with lessons to be learned in relation to the use of expert committees generally. We set out here those derived specifically from the *Southwood Report*.

Lessons

- An advisory committee should draw a clear distinction between any information provided by others, which it has not reviewed, and its own conclusions.

- An advisory committee should explain the reasoning on which its advice is based.

- When giving advice, an advisory committee should make it clear what principles, if any, of risk management are being applied.

- An advisory committee should not water down its formulated assessment of risk out of anxiety not to cause public alarm.

Lessons from the introduction of the animal SBO ban

Commentary

1276 The animal SBO Order suffered from fundamental defects which rendered it unenforceable. It was prepared in haste and without consultation. It was also prepared without the rigorous thought that should have been given to the need to introduce Regulations that were enforceable and the manner in which the Regulations should have achieved this.

Lessons

- Where a precautionary measure is introduced, rigorous thought must be given to every aspect of its operation with a view to ensuring that it is fully effective.

- If this cannot be done before the measure is introduced, it should be done as soon as possible afterwards.

Lessons from the implementation and enforcement of the animal SBO ban

Commentary

1277 The widespread disregard, both deliberate and accidental, of the animal SBO ban, was due in part to defects in the Regulations, in part to lack of enthusiasm among local authority inspectors and in part to lack of rigour by the Veterinary Field Service (VFS) in monitoring enforcement. We believe that the VFS's lack of rigour was in part a consequence of the fact that it had no statutory monitoring function and no right of access to slaughterhouses.

Lessons

- When Regulations that have implications for human or animal health fall to be enforced by local authorities:[107]

    - clear guidance should be given to the local authorities as to the importance of the Regulations and the manner of their enforcement;

    - there should be statutory provision enabling central government to monitor the standards of compliance and enforcement.

- Measures that depend on particular slaughterhouse procedures being followed need to be based on informed understanding of practical working conditions.

---

[107] This lesson is derived equally from the enforcement of the human SBO ban

Lessons from the introduction of the human SBO ban

Commentary

1278 We have been critical of the fact that the merit of the introduction of this precautionary measure was diminished by:

   i. the delay in appreciating that it was desirable to introduce a ban, consequent upon failure adequately to review the *Southwood Report*;

   ii. the public presentation of the reason for the ban, which suggested that it was not an important public health measure; and

   iii. the failure to identify that the practice of mechanical recovery of meat called for special consideration.

Lessons

- Government Departments must retain 'in house' sufficient scientific expertise to enable them to understand and review advice given by advisory committees.

- Government Departments must review advice given by advisory committees to ensure that the reasons for it are understood and appear to be sound.

- Where a precautionary measure is introduced, rigorous thought must be given to every aspect of its operation with a view to ensuring that it is fully effective and its purpose and application understood by those concerned.

- Government Departments should clearly tell both the public and those responsible for enforcement the reasons for, and the importance of, any precautionary measures that they introduce.

Lessons from the final months

Commentary

1279 The Government was taken by surprise and wrong-footed by the announcement by SEAC that a new variant of CJD had been identified which was probably linked to BSE. It should not have been. The growing apprehension that this might be the case had been expressed by Dr Will and other members of SEAC at its meetings on 5 January 1996 and, more forcibly, 1 February. Representatives of MAFF and DH present at those meetings did not put their colleagues on the alert that SEAC might be moving towards this conclusion. The possibility of this should nonetheless have been appreciated by those who received the reports of the SEAC meetings. They did not, however, consider any contingency plans. There were no interdepartmental discussions about the gathering storm. Everyone waited to see what SEAC had to say.

Lessons

- Departmental representatives attending meetings of advisory committees in the capacity of secretariat or observers should see that their Departments are promptly informed of any matters which may require a response from government.
- Contingency planning is a vital part of government. The existence of advisory committees is not an alternative to this. The advisory committees should, where their advice will be of value, be asked to assist in contingency planning.

Lessons in respect of Wales, Scotland and Northern Ireland

Commentary

1280 An outbreak of an infectious animal disease may pose threats over a wide geographical area and the effectiveness of the response must not be inhibited by purely administrative boundaries. BSE proved to be a UK-wide problem and the lessons to be learned are those which relate to such a problem.

1281 It will usually be desirable where there is a problem common to the UK threatening animal health, or both animal and human health, that a common solution should be found, that the same legislative measures should be introduced at the same time and that enforcement standards should be similar.

1282 When BSE emerged, the Territories were, in general, content to follow the lead of MAFF and DH. Under devolution a similar attitude cannot be relied upon. SEAC's advice was the critical element in the formulation of policy, but SEAC reported only to MAFF and DH. We do not consider that this was the most satisfactory arrangement then and it certainly would not be satisfactory today. Moreover, information and expertise existed in the Territories that might usefully have informed UK policy-making. It is important that advice and information should be shared by all those who are responsible for animal and human health in the United Kingdom.

Lessons

- Arrangements need to be in place which will facilitate a synchronised approach throughout the United Kingdom to common problems of animal health, or animal and human health.
- Advisory committees set up to advise on problems of animal health, or animal and human health, which are common throughout the United Kingdom should report to the appropriate Departments both in England and in the Territories.
- So far as animal diseases, particularly those which may involve risk to human health, are concerned, a clear understanding should exist as to:
    - i. the identification of those areas where a uniform and synchronised policy and/or implementation is required and who is to take the lead;

ii. the sharing of resources and information;

iii. a structure for consultation and joint decision-making that minimises unnecessary delay.

Lessons from the emergence of vCJD

Commentary

1283 The transmission of BSE to humans was considered most unlikely, but it has happened. The normal incubation period is not yet known, though if that of kuru is any guide, it is likely to be long. It is too early to estimate the number of people who are at present incubating the disease.

Lessons

- Although likelihood of a risk to human life may appear remote, where there is uncertainty all reasonably practicable precautions should be taken.

- Precautionary measures should be strictly enforced even if the risk that they address appears to be remote.

- All pathways by which vCJD may be transmitted between humans must be identified and all reasonably practicable measures taken to block them.

- The needs of victims of vCJD and their families have special features. Consideration should be given to how best the health and welfare services can meet them. Patients for whom a care plan has been carefully arranged have received better management than those for whom this is lacking.

Lessons from the handling of non-food routes of transmission to humans

Commentary

1284 The widespread use of bovine material for a whole range of food and non-food purposes created a large number of potential pathways of infection of BSE to man. The same is true of any potentially zoonotic disease. Handling of the risks to humans calls for the identification of all such pathways, availability of appropriate powers to address the risks and clear allocation of responsibility for doing so.

Lessons

- A comprehensive review to identify all the potential pathways of infection to humans, including those from waste disposal, for a potentially zoonotic disease should be undertaken as a basis for taking steps to prevent transmission. This review should involve all relevant Departments and draw on outside expertise as necessary.

- An overall handling plan with consistent objectives and a timetable should be drawn up and lead responsibility for dealing with each pathway clearly allocated.

- The legislation applicable to different types of product may provide differing and sometimes inconsistent powers for dealing with similar risks or raw materials. Consideration should be given to the need for a power to cut off supply of a widely used but potentially toxic raw material at source.

- Occupational health risks should be considered in relation to each of those pathways and advice or warnings be promptly provided.

Lessons from the approach to BSE and medicines

Commentary

1285 A potential zoonosis with a long incubation period throws up particular problems for the systems that exist to ensure the safety of human and veterinary medicinal products. While Medicines Act licensing decisions need to be insulated from undue pressures, they also need to be taken on a fully accountable basis.

Lessons

- Reliance on reported adverse reactions will not result in the timely identification of problems arising from a disease with a long incubation period. A database of concerns other than those resulting from adverse reactions should be considered.

- The licensing authorities, their advisory committees and others involved in the medicines licensing system each have information and expertise in relation to potential zoonoses that will be of use to the other. Effective action in respect of such diseases depends on this being shared. MAFF, DH and the Medicines Commission should consider what improvements might be needed to existing collaborative arrangements.

- It is not always clear in practice where responsibility rests as between Ministers, officials and advisory committees for advising, determining policy and taking key decisions on medicines. This should be clarified, so as to ensure that important policy decisions are taken by, or approved by, Ministers, whether those decisions are to take action or to take no action.

- The extent of the requirements of confidentiality in relation to the licensing of medicines should be reviewed.

- Medicines Advisory Committees should make clear what is a scientific assessment and what is a value judgement, so that value judgements are not treated as expert assessments of risk.

- Ring-fencing of medicines decisions to insulate them from outside pressures can reduce accountability. There should be properly reasoned and recorded decision-taking, and the criteria being applied should be made openly available.

- Thought should be given to ways of ensuring that those licensing animal-derived medicinal products are properly informed about the sources and collection of materials.

Lesson from the approach to BSE and cosmetics

Commentary

1286 Addressing the possible risks posed by BSE in relation to cosmetics was impeded by lack of knowledge about the cosmetic products available, their composition and uses.

Lesson

- DTI should review the need to maintain data on products which offer a potential pathway of infection.

Lesson from the approach to BSE and occupational risk

Commentary

1287 Delays in drafting and issuing guidance in respect of occupational risks posed by BSE were inordinate.

Lesson

- The Health and Safety Executive (HSE) should consider means of ensuring that the issue of guidance in respect of risks impacting on different occupations is carried out in a manner which is coordinated and expeditious.

Lesson in relation to pollution and waste control

Commentary

1288 The pathways by which the BSE agent might come into contact with humans and animals as a consequence of the disposal of waste did not receive adequate consideration prior to March 1996.

Lesson

- The disposal of waste from any processing of material that may contain the BSE agent should be reviewed to ensure that it does not involve risk of infection of humans or animals.

Lessons in relation to research

Commentary

1289 We have noted the very large number of research projects that were undertaken in response to BSE. We have also drawn attention to a number of areas where, with hindsight, we can see that research could profitably have been started earlier or pursued with more vigour. Had an improved structure for research coordination been in place, many of these deficiencies might have been avoided.

Lessons

- Where a problem in animal and human health arises that leads to demands for research of the scale and diversity required by BSE, it is desirable that Government Departments and Agencies coordinate their efforts.

- Coordination of the research effort is desirable in order to achieve:

    - identification of gaps in research;
    - determination of research priorities;
    - identification of the best sources of expert assistance;
    - a well-constructed plan for funding from the outset;
    - competition for research projects;
    - peer review of projects; and
    - efficient arrangements for provision of clinical material to researchers.

- The progress of research and the implications of any new developments must be kept under continuous and open review.

- Our conclusion that BSE was probably present in the cattle herd in the 1970s may have implications for past and current assessments of risk which have assumed that the earliest date of infection was around 1980. This illustrates the importance of setting out assumptions and keeping them under review.

- What is now known about the relative sensitivity of mouse bioassay compared with calf bioassay may have implications for the conclusions drawn from mouse bioassays. These need to be reconsidered systematically.

The use of scientific advisory committees

Commentary

1290 Volume 4 of our Report deals in detail with the assistance provided by the Southwood Working Party and Volume 11 with the assistance provided by the

Tyrrell Committee and SEAC. The Government relied heavily on the advice of SEAC during most of the period with which we are concerned, and in Volume 11 we discuss, with commentary, the lessons to be learned from the use of this Committee. We shall not repeat that commentary here, but briefly itemise the lessons which apply to such committees.

Lessons

Setting up the committee

- The areas of advice that are required from the advisory committee should be identified as precisely as possible before the committee is set up.

- The terms of reference should specify with as much precision as possible the role of the committee.

- The composition of the committee should include experts in the areas of the advice that is likely to be required.

- Those invited to join a committee should be given a realistic estimate of the commitment required.

- A lay member can play a valuable role on an expert committee.[108]

- Government should seek advice from the professional or other body best qualified to advise on suitable candidates for membership.

- Potential conflicts of interest should not preclude selection of those members otherwise best qualified, but conflicts of interest should be declared and registered.

- Where any item of business involves an apparent conflict of interest on the part of a member, that should be declared.

- Where the workload of a committee is considerable, it is reasonable that members who are not public servants should be remunerated.

- It will often be desirable to draw the secretariat from the commissioning Department(s) in order to provide a two-way channel of communication.

- In such cases, as in all cases, the secretariat must be careful to respect the independence of the committee.

The role of the committee in relation to policy

- Where a policy decision is urgent, consideration should be given to whether delaying the decision pending advice from an advisory committee is the best course.

- Consideration should be given at the outset to the manner in which the committee will contribute to deciding policy.

- Government should recognise that if a committee is asked to advise which policy option to adopt, there may be little alternative but to follow the advice given.

---

[108] See the section below on 'Dealing with uncertainty and the communication of risk'

- Where the policy decision involves the balancing of considerations which fall outside the expertise of the committee, it will normally not be appropriate to ask the committee to advise which policy option to adopt.

- It may be appropriate to ask the committee to set out a range of policy options, together with the implications of each.

- Where advice is sought on the implications of policy options, this may best be achieved by dialogue between government and the committee.

- Where advice is required only on those ingredients of a policy decision which fall within the particular expertise of the committee, questions should be formulated with precision to achieve that result.

- Where a Department has concerns about the practical implications of advice that a committee may give, these should be placed openly before the committee.

- Where a committee is asked to advise on risk management, it will normally be helpful for the committee to follow a formal structure based on recognised principles of risk assessment.

The form of the advice

- Advice should normally be given in writing.

- Advice should be in terms that can be understood by a layperson.

- Advice should clearly state the reasons for conclusions.

- Assumptions underlying advice should be made clear.

- Advice should identify the nature and extent of any areas of uncertainty.

- Where appropriate, the advice should set out the different policy options and the implications of each.

Communication of the advice

- The advice of the committee, together with any papers necessary for the full understanding of that advice, should be circulated to all within government with responsibility for policy decisions in respect of which the advice is relevant.

- The advice of the committee should normally be made public by the committee.

- The proceedings of the committee should be as open as is compatible with the requirements of confidentiality.

Review of the advice

- Departments should retain 'in house' sufficient expertise to ensure that the advice of advisory committees, and the reasoning behind it, can be understood and evaluated.

- Advice given by a committee should be reviewed by those to whom it is given to ensure that the reasons for the advice are understood and appear sound.

- Where the reasoning of the advice of a committee is unclear, clarification should be obtained from the committee.

Dealing with uncertainty and the communication of risk

Commentary

1291 Some argue that it is not the task of government to protect the public against risk in circumstances where the individual can accept or avoid the risk by making his or her own informed choice. Where the hazard is transparent and one that the individual can readily avoid, this argument has force. Most people believe, however, that government has an important role to play in reducing the extent to which the consumer is exposed to hazard. They believe, for instance, that the Government should do all that is reasonably practicable to see that the food that they eat and the medicines that they take are reasonably safe.

1292 The Government adopted this approach in seeking to protect the public from the possibility that BSE might pose a hazard to human health. We have already considered the extent to which the way that it set about achieving that objective was an adequate response to the emergence of BSE. At this point we are concerned with the lessons to be learned from one aspect of the response that proved particularly unsatisfactory – communication of risk to the public. Although we have made a number of individual criticisms in respect of risk communication, the lessons to be learned are based on hindsight and relate to the overall approach of reassurance that was adopted. We do not consider that individuals should be criticised for following that approach.

1293 The problem is not an easy one. The public are anxious to understand the basis upon which the Government's decisions on risk management are taken. The Government does not set out to achieve zero risk, but to reduce risk to a level which should be acceptable to the reasonable consumer. The individual consumer wishes to be satisfied that the Government has drawn the line in the right place. How can the Government best satisfy the public that this aim has been achieved? We discussed this question with a number of witnesses.

1294 Throughout the BSE story, the approach to communication of risk was shaped by a consuming fear of provoking an irrational public scare. This applied not merely to the Government, but to advisory committees, to those responsible for the safety of medicines, to Chief Medical Officers and to the Meat and Livestock Commission. All witnesses agreed that information should not be withheld from the public, but some spoke of the need to control the manner of its release. Mr Meldrum spoke of the desirability of releasing information 'in an orderly fashion' – of ensuring that the whole package of information was put together, taking care in the process not to 'rock the boat'.

1295 Mr Brian Dickinson, who was a member of MAFF's Food Safety Group, put the matter in this way:

> Given the strength of public debate on the matter at the time one was aware of slightly leaning into the wind. You could not just stand upright and give a totally impartial, objective view of what was the situation. There was a stronger danger of being misinterpreted one way rather than the other, and we tended to make more reassuring sounding statements than might ideally have been said.

1296 We felt that this was an accurate description of the general approach to risk communication. We have seen that it provoked increasing scepticism and, on 20 March 1996, the reaction that the Government had been deceiving the public.

1297 In discussing this topic with us, Sir Robert May, Chief Scientific Adviser, expressed the following view:

> You can see the temptation on occasion to wish to hold the facts close so that you can have internal discussion and the formation of a consensus so that a simple message can be taken out into the market place. My view is strongly that that temptation must be resisted, and that the full messy process whereby scientific understanding is arrived at with all its problems has to be spilled out into the open.

1298 This view received strong support from representatives of the consumer organisations. They emphasised the need for open scientific debate. Ms Sheila McKechnie, the Director of the Consumers' Association, emphasised the need to develop a culture of trust. She commented that:

> There is nothing more nanny-ish than withholding information from people on the ground that they may react irrationally to that information.

1299 She made the point that organisations build up credibility by openness. She expressed the hope that the Food Standards Agency would achieve this.

1300 Everyone agreed that the Government had a problem with credibility. A number of Government Ministers told us that they had lost credibility with the public, so that it was necessary to get independent experts to lend credibility to public pronouncements about risk. Mrs Bottomley spoke of the need for the public to receive information free of 'political overtones'. She told us that she did all that she could to promote the Chief Medical Officer as an independent expert who could be trusted by the nation.

1301 Our experience over this lengthy Inquiry has led us to the firm conclusion that a policy of openness is the correct approach. When responding to public or media demand for advice, the Government must resist the temptation of attempting to appear to have all the answers in a situation of uncertainty. We believe that food scares and vaccine scares thrive on a belief that the Government is withholding information. If doubts are openly expressed and publicly explored, the public are capable of responding rationally and are more likely to accept reassurance and advice if and when it comes. We note, by way of example, that SEAC and MAFF have made public the fact that an investigation is being carried out into the question of whether BSE has passed into sheep. We do not understand that this has led to a boycott of lamb.

Lessons

- To establish credibility it is necessary to generate trust.
- Trust can only be generated by openness.
- Openness requires recognition of uncertainty, where it exists.
- The importance of precautionary measures should not be played down on the grounds that the risk is unproved.
- The public should be trusted to respond rationally to openness.
- Scientific investigation of risk should be open and transparent.
- The advice and the reasoning of advisory committees should be made public.
- The trust that the public has in Chief Medical Officers is precious and should not be put at risk.
- Any advice given by a CMO or advisory committee should be, and be seen to be, objective and independent of government.
- The role, if any, of the Chief Veterinary Officer in making public statements in relation to risk to human health from a zoonosis or potential zoonosis should be clarified.
- The activities of the Meat and Livestock Commission (MLC) in the period up to 20 March 1996 do not appear to have represented all its statutory objectives. The MLC has submitted to us proposals in relation to its future role. We recommend that these receive consideration in the light of our Report.

The legislative framework

Commentary

1302 The Government's response to BSE adopted different approaches to dealing with the risk that the BSE agent in cattle incubating the disease or showing signs of it might be transmitted to other animals or to humans.

- Cattle showing clinical signs were compulsorily slaughtered and destroyed.
- The incorporation of high-risk tissues from apparently healthy cattle in human food was forbidden.
- The incorporation of ruminant protein in feed for ruminant animals was banned.
- The incorporation of high-risk tissues from apparently healthy cattle in animal feed was banned.

- The disposal of high-risk tissues was regulated so that, in effect, they could only be disposed of as waste.

- The use of bovine products or by-products of UK origin in the manufacture of medicinal products was phased out in compliance with guidelines.

- Recovery of mechanically recovered meat (MRM) from the spinal column of cattle was forbidden.

The problem

1303 The statutory powers relied on in adopting these measures were enacted in order to deal with known hazards. However, while it was established that BSE was a major disease threat to cattle, it was for several years unknown whether it was a hazard to human beings and other animals, and if so, how great a risk it posed. The generally held belief of the Government's scientific and veterinary advisers was that BSE probably did not pose a risk to human beings, pigs or poultry. Moreover, even the risk to cattle was not fully established; it was unknown whether BSE could infect cattle other than by some form of ingestion. Thus an unusual feature of the BSE story was that the Government imposed Regulations to address risks that scientists believed probably did not exist, or at least could not confirm as probably existing.

1304 The Government had to take action on BSE in the face of two other significant uncertainties. First, in the absence of a diagnostic test for BSE in live animals, it was impossible to know which animals might be incubating the disease. It could be statistically demonstrated, in the case of any individual animal at the time of slaughter, that that animal was very unlikely to be incubating BSE. Second, it was probable that not all parts of an infected animal might carry infectivity sufficient to transmit the disease to other animals of its own or other species.

1305 The evidence disclosed a number of occasions on which lawyers in MAFF's Legal Department expressed concern as to whether precautionary measures which were being proposed fell within the powers conferred by the legislation under which they were to be introduced. We consider it desirable that legislation should clearly empower Ministers to take precautionary measures in a situation where the existence of a hazard is uncertain. We believe that there are areas where this may not be the case. We have not attempted a detailed analysis of the law in these areas, for this is not part of our task. We draw attention to them so that they may receive further consideration.

Power to order the slaughter of animals

1306 Section 32(1) of the Animal Health Act gives the Minister power, if he thinks fit, to order the slaughter of 'any animal which is affected or suspected of being affected with any disease to which this section applies, or has been exposed to the infection of any such disease'.

1307 Mr MacGregor used this power when introducing the slaughter and compensation scheme in August 1988. The primary reason why he did so was in order to address what was considered to be the remote possibility that BSE was transmissible to humans.

1308 MAFF lawyers expressed doubts as to whether s.32(1) could be used in these circumstances. We do not know whether these doubts were resolved or, if they were, on what basis. We consider that there was certainly scope for doubt as to the extent of the Minister's powers under s.32(1), having particular regard to the fact that:

  i. scientists considered it unlikely that BSE was transmissible to humans; and

  ii. BSE had not at that time been designated a zoonosis under S.29 of the Act.

1309 Consideration was given to a policy of slaughtering animals in the same herd as a BSE victim, or slaughtering the offspring of BSE victims, because of the possibility that BSE might be vertically or horizontally transmissible. Again we think that there would have been some doubt as to the power of the Minister to introduce such a policy under s.32(1) of the Act, having regard to the uncertainty as to the manner in which BSE might be transmitted.

1310 An animal which was not showing clinical signs of BSE would not, ordinarily, be said to be 'affected with the disease'. Furthermore, even if the word 'affected' in section 32(1) included pre-clinical infection, it would be difficult to say of any such animal that it was 'suspected of being affected with BSE', since statistically this would be highly improbable in the case of any individual animal. Nor is it clear that an animal could properly be described as 'exposed to infection' in circumstances where it was uncertain whether transmission of infection was possible.

Power to order the destruction of parts of an animal

1311 Section 1 of the Animal Health Act 1981 gives Ministers power to make 'such orders as they think fit . . . for the purpose of in any manner preventing the spread of disease', and section 8 gives them power to make 'such orders as they think fit' for prohibiting and regulating the removal of 'carcasses, fodder, litter, dung and other things'. Section 35(1) of the Act also gives Ministers power to order the seizure, and impose requirements for the destruction, burial, disposal or treatment, of 'anything, whether animate or inanimate, by or by means of which it appears to them that any disease to which this subsection applies might be carried or transmitted'.

1312 The powers under section 35(1) were used in 1991 to give MAFF the power to seize, destroy and dispose of the carcasses of animals suspected of having died from BSE. The powers under sections 1 and 8 were used to protect human health by ordering the destruction of milk from cows affected by BSE, after BSE had been designated a zoonosis.

1313 These sections of the Animal Health Act are in very wide terms. The question arises of whether they could have been used to order the destruction of SBO as a precautionary measure to safeguard human health, whether through foodstuffs or any other consumer product. We consider that had such a course been adopted, a challenge might have been anticipated on the grounds that:

i. it was statistically highly unlikely that any individual animal was incubating the disease; and

ii. scientists believed it unlikely that tissues from an animal incubating the disease posed any risk to humans.

1314 We do not suggest that such a challenge would necessarily have succeeded.

Power to ban the use of material for specified purposes

1315 Apart from the slaughter and compensation policy, which related only to cattle diagnosed as showing clinical signs of BSE, and the power to seize and destroy carcasses of animals suspected of having died of BSE, the Government did not order the compulsory destruction and removal from circulation of any animals, parts of animals or material derived from or connected with animals which might have been incubating or exposed to BSE. Instead, the Government adopted the alternative approach of banning the use of potentially infective material for particular purposes. Thus the ruminant feed ban prohibited the use of ruminant protein in feed for ruminants; the human and animal SBO bans prohibited the use of particular bovine tissues in food for human and animal consumption, and subsequently prohibited the movement of MBM derived from SBO material to any unlicensed destination; and MRM derived from bovine vertebral columns was banned from use in human food. The question arises whether Ministers had adequate powers to adopt the approach of banning suspect material for particular purposes in the face of the uncertainties about BSE which we have outlined above.

1316 Under a range of different statutes, Ministers had power to take action to block potential routes of transmission of animal diseases by imposing requirements as to the manufacture, sale or supply of products which might incorporate animal material. Thus:

- the Animal Health Act, the Food Act 1984 and its successor, the Food Safety Act 1990, gave the relevant Ministers power in certain circumstances to ban animals and animal tissues from incorporation in food for animal and human consumption;

- the Consumer Protection Act 1987 gave the Secretary of State for Trade and Industry power to make provisions for the purpose of securing that goods were safe, and for the purpose of securing that goods which were unsafe were not made available to persons generally;

- the Environmental Protection Act 1990 gave power to regulate the release of harmful substances into the environment; and

- the Medicines Act 1968 gave the licensing authorities power to impose requirements as to methods of manufacture or as to product ingredients as a condition of granting product licences for human and veterinary medicines.

1317 In the case of the powers granted by each of these statutes, questions were liable to arise as to whether they empowered action on a precautionary basis in circumstances where the existence of risk was not merely uncertain, but considered very unlikely. Thus, when it was proposed to introduce the human SBO ban in June 1989, MAFF lawyers advised the administrators that 'given that it is not possible to prove that the offal to be banned is in fact "unfit" for human consumption, it will be necessary to be able to justify the reasonableness of provisions made as to use'.[109] They recognised the possibility of a challenge to a ban introduced under the Food Act in order to protect humans from a risk which was far from established, and in fact considered to be remote.

Legislative constraints in relation to medicines

1318 The Food Act 1984 and the Food Safety Act 1990 contained powers to prohibit the sale or use of any specified substance or any substance of a specified class in or as food intended for sale for human consumption. As MAFF lawyers pointed out when the provisions of the human SBO ban were being considered, this power did not enable prohibition of the use of these substances in or in the production of medicines. The legislative scheme for regulating the safety of medicines was very different.

1319 In granting or renewing any product licence under the licensing regime established by the Medicines Act, the licensing authorities could have made it a condition that material from BSE-affected cattle, and SBO from any cattle, should not be used in the manufacture of a product. However, it does not appear that the licensing authorities could have made this a general requirement to cover all human and veterinary medicinal products. They could only have acted on a case-by-case basis by including such a requirement in every licence for a product which might include such material, as and when an application was made for the grant or renewal of a product licence.

1320 As for existing licences, the statutory power to suspend, revoke or vary a licence was subject to a requirement that the licence holder should be given notice of the intention to revoke or vary the licence and afforded an opportunity to appear before the relevant section 4 committee or to make representations in writing as to the proposed revocation or variation. While the licensing authority had power to suspend an individual licence with immediate effect in the interests of safety, such suspension could not exceed a period of three months pending consideration as to whether the licence should be varied or revoked, and the licence holder was entitled to appear before the relevant committee and to make representations on the matter.[110]

1321 If, in response to BSE, the licensing authority had wished to use its statutory powers to ensure that UK bovine material was not used in the manufacture of medicinal products, it seems that it would have had to revoke or vary every relevant product licence (possibly after a suspension of up to three months), and in doing so it would have had to give each current licence holder or applicant the opportunity to appear before the relevant section 4 committee to argue against the proposed revocation or variation. This would have been an administrative nightmare. In these circumstances it is not surprising that the decision was taken to issue guidelines rather than attempt to use formal statutory powers.

1322 We consider that it might be of value if licensing authorities had a statutory power under the Medicines Act to impose a general prohibition on the use of substances which are considered to be unsafe in the manufacture of any human and

[109] YB89/6.12/3.1
[110] Section 29 and schedule 2, paras 1–14

veterinary medicines. We appreciate, however, that this suggestion may not be compatible with a regulatory regime which is now governed by European law.

Legislative constraints in relation to cosmetics

1323 Cosmetics is another area where the regulatory regime is governed by European law. The Cosmetic Products (Safety) Regulations 1989 give effect to the 1976 EC Cosmetics Directive. Little scope is left for independent regulatory action by the UK Government, and effecting changes to European Regulations can be a lengthy business.

1324 In these circumstances we were told that in practice the regulation of the cosmetics industry operated on an informal and voluntary basis, under which guidance was given to and implemented by the industry. This was the course adopted in relation to BSE. It does not seem to us that this regulatory regime caters satisfactorily for a situation such as the emergence of BSE.

General constraints of European law

1325 When a manufacturer of MRM sought judicial review of the Specified Bovine Offal (Amendment) Order 1995, one of the arguments put forward was that once definitive measures for a relevant outbreak of disease had been adopted by the European Commission at the EU level, individual Member States were no longer entitled to adopt unilateral measures to deal with the risks posed by the disease.

1326 The High Court granted leave to seek a judicial review of the Order, thereby indicating that it considered the matter to be at least reasonably arguable. However, the judicial review was abandoned after 20 March 1996, and so this argument was not tested at the time. It may well remain open to those who object to actions taken by the Government to deal with zoonoses generally, and BSE in particular, where those actions go beyond EU measures taken under Directives 89/662/EEC and 90/425/EEC.

1327 If the argument is correct, the consequences are worrying. First of all, it calls into question the lawfulness of the Specified Bovine Offal (Amendment) Order 1995. The Government decided to act speedily to ban the use of bovine vertebral columns in the manufacture of MRM on the advice of SEAC. We believe that such action was clearly desirable in the interests of human health. However, this important measure could have been open to challenge under European law, at least until it was adopted by the Commission in July 1997 by Decision 97/534.

1328 The argument also has implications for the future handling of other zoonoses or potential zoonoses. It suggests that in matters governed by the Directives we have cited, the Government may not be able to take unilateral action in the event of a reassessment of the risks associated with a particular disease outbreak.

1329 We understand that a similar point is currently before the European Court of Justice. We expect that this issue will be reviewed by MAFF when the decision of the Court is known. If, in the light of that decision, there remains any danger that measures for the protection of human or animal health may be readily susceptible to challenge, consideration will need to be given to steps to minimise this danger.

Lessons

- Where an animal disease is identified, which could be transmitted to animals or humans via a range of possible routes, powers under UK and European law which enable Ministers to order the slaughter of animals, and the destruction of animal tissues or anything which might carry infection, should not be restricted merely because it cannot be established as a reasonable probability, as opposed to a mere possibility:

    i. that the disease is transmissible; or

    ii. that a particular animal may be infected by the disease in question; or

    iii. that particular organs or tissues in an animal may carry infection.

- Similarly, any powers under UK and European law which enable Ministers to adopt an alternative approach of banning the use of any substances for particular purposes in order to protect human or animal health should not be restricted merely because one or more of the matters referred to above cannot be established as a reasonable probability, as opposed to a mere possibility.

- Current medicines and consumer protection legislation should be reviewed with a view to giving the Government power to act swiftly and comprehensively to ban the use of any substances or processes which might pose a risk to human or animal health.

- The Government should review and clarify its powers under European law to introduce emergency measures for the protection of public and animal health in relation to outbreaks of disease where measures have previously been taken by the European Commission.

The experience of vCJD victims and their families

Commentary

1330 Members of the families of 15 young victims of vCJD came to tell us of what they had experienced. Many more provided us with statements. The description of the clinical treatment of the disease that has been set out in Volume 8 does not fully bring home the horror of what in each case was a harrowing personal tragedy. It is particularly hideous to see young people struck down by a destructive neurological disease of the kind that more usually strikes those who have enjoyed something close to a full life-span.

1331 The start of the nightmare is an inexplicable change of personality. A happy, outgoing and confident young person develops mood swings, depression and lapses of short-term memory. Worried parents or relatives consult their GP, who can find no clinical signs and prescribes an anti-depressant. As the symptoms worsen a referral to a psychiatrist follows.

1332 The psychiatrist finds no sign of organic disease and treats the patient for psychiatric illness, sometimes as an inpatient in a psychiatric ward, where both the environment and the treatment are inappropriate. No improvement follows. For the victim and the relatives this is a time of acute anxiety, but worse is to follow. Neurological symptoms supervene: pins and needles and pains in the limbs, unsteadiness of gait, failures of muscle coordination. A referral is made to a neurologist. A neurological condition is diagnosed – the nature of it may not be. There are other conditions that have similar signs and symptoms to those of vCJD.

1333 Different tests are carried out, some invasive and unpleasant. Sometimes vCJD is suspected, sometimes it is not. The symptoms worsen: speech difficulties, impairment of intellect, involuntary movements, incontinence, progressive immobility until the victim is bedridden. It becomes plain that there will be no recovery.

1334 Some families want to care for their loved one at home until the end comes. Others seek a suitable hospital or hospice. In either case their anxiety is that the patient's final days should be spent in a caring, secure and comfortable environment.

1335 The victims of vCJD and their families have special needs. Degenerative neurological diseases of the young are rare. The structure of the health service makes no special provision for them. Hospital facilities for the elderly who are terminally ill are seldom the place for young people. Hospices that care for those whose days can be numbered may be reluctant to accept patients for whom it is impossible to predict when the end will come.

1336 The evidence that we received showed widely varying standards of management and care of victims of vCJD and of support for their families.

Lessons

1337 What is needed includes:

- as speedy as possible a diagnosis of vCJD;
- informed and sympathetic advice to relatives about the future course of the disease and the needs of the patient;
- speedy assistance for those who wish to care for the victim at home. Needs often include aids for the care of the disabled, modification to the home, financial assistance and respite care;
- a coordinated care package which addresses the needs of the victims and their families; and, if requested;
- a suitable institutional environment for a young person, incapacitated and terminally ill.

1338 It should occasion neither surprise nor individual criticism that these needs were frequently not met in the early days of the disease. We are now able to look back with hindsight. The lesson is clear: the needs of vCJD victims call for a different approach by the health service and the social services departments of local authorities.

Annex 1: Procedures adopted by the BSE Inquiry

1339 In this annex we describe how we sought to achieve our aims of being thorough, open and fair.

Thoroughness and openness

1340 At our preliminary hearing in January 1998 we asked anyone who thought they had relevant evidence to contact the Secretary to the Inquiry.

1341 To assist us in understanding the evidence we would be hearing, we pursued a course of education in order to acquire the necessary background knowledge. We attended a series of lectures on topics including microbiology, epidemiology and toxicology. We also went on a series of visits which we describe below. Government Departments set up 'Liaison Units' to assist us. The first of many tasks these Units undertook was to assemble a set of initial background documents which we published as our Initial Background Documents (IBD) series of bundles.

1342 With the assistance of the Liaison Units, the Inquiry Secretariat identified civil servants who appeared likely to have had an involvement with BSE and variant CJD. These civil servants were then divided into two groups. Witnesses identified as probably having only a peripheral involvement in matters of interest to the Inquiry were initially asked to provide general information about the posts they held and the nature of the dealings they had with BSE or vCJD between 1985 and 20 March 1996. Civil servants identified as probably playing a more central role were asked to provide a thorough statement of the part they played, their responsibilities as they understood them at the time, the information they received, the actions and decisions they took and the reasons for them.

1343 A consultation document on our procedures was circulated in January 1998. This explained that we would be seeking evidence from scientists, those who could give evidence of fact relating to the period prior to the outbreak of BSE (including evidence as to the manufacture of cattle feed and the rendering processes involved), administrators, families of victims of vCJD, the farming industry and other commercial interests, consumer representatives, former Ministers and others. We invited people to suggest names of witnesses for the Inquiry. As the Inquiry proceeded, we requested many individuals to provide supplemental statements, clarifying evidence or addressing further issues. In total, we have published over 1,000 witness statements from over 630 different individuals.

1344 Many of those who played a more central role in events were invited to participate in oral hearings. We heard oral evidence on 138 days. Each hearing was in public and we tried to make the atmosphere at these hearings as informal as possible. We permitted a live radio broadcast of our proceedings and television cameras were permitted when witnesses were not giving evidence.

1345 The witness statements provided by those scheduled to give oral evidence were published prior to the relevant hearing. We invited comments from relevant individuals on the content of these statements, and where appropriate these were raised with the witness at the oral hearing.

1346 We took full advantage of information technology to make transcripts of these hearings available over the Internet, usually within a few hours of the witness giving evidence. We also provided free access to all witness statements, timetables, and background information on our website. This website was extremely popular. Over 160,000 witness statements and almost 86,000 transcripts were accessed from our website, which received over 1.5 million page requests. In April 1998 the Inquiry was awarded a Freedom of Information Award by the Campaign for Freedom of Information for its innovative use of the Internet. Modern technology was used in other ways – during one hearing we discussed epidemiological evidence via a video link with scientists in Canada and New Zealand.

1347 A less glamorous, but essential, part of the process of the Inquiry was the analysis of documentation. Members of the Secretariat went in teams to Government Departments to conduct a trawl of their files. Most information came from the Ministry of Agriculture, Fisheries and Food and the Department of Health. Other Departments which supplied information included the Health and Agriculture Departments in Wales, Scotland and Northern Ireland. These teams examined about 3,000 files, and identified approximately 75,000 pages of documents as being of interest to the Inquiry. Documents were also supplied to the Inquiry by companies, trade associations, scientists, and other individuals. Analysis of the documents we received, and requests for further material on points arising from them, was a continuing process.

1348 Our Inquiry was unusual in beginning oral hearings before completion of the task of finding and collating relevant documentary evidence. In the early stages we were necessarily reliant on witnesses to point us to relevant material. As our documentary trawl proceeded we were able to check whether relevant avenues of investigation had been sufficiently covered.

1349 Throughout the Inquiry, we sought to make available to the public the contemporaneous documents we considered relevant to our work. A reference room containing a full set of all materials was available for use by the press and public. In addition to all published witness statements and transcripts of oral evidence, these included:

- a mass of shorter documents (such as letters and minutes) arranged in chronological order (the Year Book, or YB, series). This series grew considerably during the Inquiry and ended up with nearly 16,000 separate documents;
- bulky materials, such as book chapters and reports (the Materials, or M, series);
- articles from scientific journals, telling much of the scientific story (the Journal series); and
- the selection of 'initial background documents' provided by the Liaison Units referred to above (the IBD series).

1350 In addition to learning about scientific topics, we went on several visits. These included tours of an abattoir, a rendering plant, a Veterinary Investigation Centre, and two farms in Wales. We went to Weybridge to visit the Central Veterinary Laboratory and to Edinburgh to visit the CJD Surveillance Unit and the Institute for Animal Health's Neuropathogenesis Unit. We also visited a livestock market in Northern Ireland to see the cattle-tracking system in operation and were shown the Animal Health Computerised Traceability System at the headquarters of the Department of Agriculture for Northern Ireland.

1351 Early in the Inquiry, we issued a number of working documents, including a glossary, a dramatis personae and a time-line setting out some of the main events in chronological order. More ambitiously, in December 1998 the Inquiry began to publish draft factual accounts (DFAs) of aspects of the history of BSE and vCJD. The DFAs were placed on our website and sent to witnesses. They were intended to help us clarify the overall picture and to enable all those who were concerned or interested to draw attention to any errors or significant omissions in the drafts. The DFAs were not definitive. We recognised prior to their publication that they could contain errors or omissions. We stressed that DFAs should be treated as no more than working documents, intended to set out relevant evidence in a neutral manner.

1352 Following the publication of the first tranche of DFAs, some witnesses raised concerns with us. They were very concerned that substantial amendment was required and that the original drafts were in places inaccurate or misleading. After considering what they said, we produced revised versions of many of the DFAs, taking account of the comments and additions which witnesses had, as we expected, proposed. The revised versions (RFAs) produced with the help of witnesses and others were considerably improved and this assisted us greatly in establishing the course of events. Further DFAs were published as the Inquiry proceeded. Updates to both the DFAs and RFAs were produced in some cases to deal with comments and to draw attention to further relevant evidence.

1353 We believe that the DFAs, RFAs and updates assisted many of those who were taking an interest in the Inquiry's work. We could not produce DFAs for all aspects of the story, but where they were produced, they collated a mass of relevant information in a way which enabled it to be digested and reviewed. They also enabled witnesses to refresh their memory of events and identify evidence upon which they wished to comment.

1354 In June 1998 we published a document setting out our understanding of government structures for scientific research. This was followed in 1999 by discussion papers inviting comments on issues relating to the role of the advisory committees, particularly SEAC, and on epidemiology. When the hearing of oral evidence drew to a close, we issued a more general invitation to supply any further comments anyone wished to make.

1355 In all we received over 11,700 letters, e-mails and faxes in relation to our work during the course of the Inquiry.

Fairness

1356 We gave an indication of the procedures we proposed to adopt at our Preliminary Hearing in January 1998. We considered it important to receive comments on our proposed procedures and therefore set these out in more detail in a consultation document issued by the Inquiry Secretariat at the end of that month. After taking account of comments on the consultation document, we issued a statement of our intended procedures.

1357 Further statements on our procedures in relation to later aspects of our work were issued during the course of the Inquiry. We did not regard our Statements on Procedures documents as an inflexible account of our procedures. We were prepared to, and did, vary our procedures in the light of representations and changing circumstances. The Statements were intended merely as a helpful guide to those participating in and following our work. Anyone wishing to learn more about the detail of the procedures we adopted may wish to refer to those Statements.[111]

1358 As we had proposed in our original consultation document, we adopted a two-phase approach to our work. The first phase, 'Phase 1', was confined to fact-finding. In 'Phase 2', we moved on to examine questions which required clarification, issues on which there were conflicts of evidence, and potential criticisms which might be made of individuals. This description seems to have given rise to a misunderstanding: some thought that there would be no, or no substantial, further fact-finding in Phase 2. In a revised Statement on Procedures for Phase 2, we made it clear that during Phase 2 we would continue to seek further evidence of the facts as we thought appropriate having regard to our Terms of Reference.

1359 As with Phase 1, our procedures for Phase 2 were the subject of a consultation process. Our consultation document explained that the Secretariat would write to individuals identifying potential criticisms. (Letters of this kind were recommended by the Royal Commission on Tribunals of Inquiry chaired by Lord Justice Salmon in 1966,[112] and are known as 'Salmon letters'.) All who received such letters would be asked to respond in writing and would be entitled (if they wished) to answer any remaining concerns at an oral hearing.

1360 A number of concerns were expressed by witnesses. In particular, concerns were expressed in relation to the confidentiality of the potential criticisms we wished to explore with witnesses and of the response of those witnesses. We concluded that we could not guarantee to keep potential criticisms or the replies confidential. We stated in the relevant Statement of Procedures document that neither our letter notifying our concerns nor the response from the witness would be treated as documents over which the individual concerned had a right of confidentiality. Material from either document could be disclosed where such disclosure was considered necessary for the fair and proper conduct of the Inquiry.

1361 Those facing potential criticism are naturally concerned to be aware of any information which might be in conflict with the potential criticism. We used DFAs, RFAs and updates to ensure that witnesses were kept informed of relevant evidence. Our Secretariat undertook to consider whether there was any evidence of this kind

---

[111] Inquiry Announcements bundle 2, tabs 1, 10, 15 and 23 (IA2 tabs 1, 10, 15 and 23)
[112] Cmnd 3121

which had not been referred to in a DFA (or comments on a DFA) sent to an individual facing potential criticism, and to inform that person of any such evidence. We added in our Statement of Procedures for Phase 2 that if material were supplied to the Inquiry in confidence, and the confidentiality were maintained, we would pay no regard to anything in that material supporting a potential criticism. If confidential information could reasonably enable an individual to contradict an issue arising out of a potential criticism, we would discuss with that individual what procedures should be adopted to deal with the material.

1362 In order to ensure that all relevant information was in the public domain, we requested that responses to potential criticisms be accompanied by a statement for the purpose of publication, which set out all factual matters on which the recipient of a letter of potential criticism wished to rely in addition to any evidence already provided in material published by the Inquiry. Not all those involved followed this course. The Inquiry Secretariat had to devote substantial resources to going through responses, identifying new evidence of fact and putting forward a proposed statement for publication. On occasion, to ensure that new evidence of fact was put in the public domain, it was necessary for the Inquiry to publish a 'statement of information provided by a witness' in the absence of approval from the relevant witness.

1363 At first, we had envisaged a 'final stage' of our Inquiry when those participating in the Inquiry would be given a relatively short time in which to make written submissions on relevant aspects. As Phase 2 progressed, we thought it would be more useful, once the main evidence relevant to a particular area was complete, to write to those facing potential criticism identifying anything which no longer needed to be pursued, and suggesting a time within which additional comments on extant potential criticisms should be supplied.

1364 We also concluded in November 1999 that the time had come to reduce the burden on Inquiry resources and change our procedures. It seemed to us that new factual evidence in Additional Comments would not necessarily require a new statement for publication. Our Statement on Procedures for Additional Comments said that we did not propose to publish any Additional Comments we received. We recognised that it was possible that such comments could contain fresh evidence on matters of fact tending to contradict an extant potential criticism, and proposed that in such circumstances we would make arrangements to ensure that anyone notified of the potential criticism in question was informed. This appeared uncontroversial, but when Additional Comments were submitted, there were some who took issue with this. In contrast to the stance adopted at the time of receipt of Salmon letters, a number of those facing potential criticisms said that they wanted their Additional Comments to be published. We considered, in each case, whether we should depart from the procedures we had envisaged for Additional Comments, but concluded that we should not.

Annex 2: Individual criticisms

1365 We have given anxious consideration to whether individuals should be criticised in relation to their response to BSE and vCJD. It is a necessary part of our Terms of Reference – but it is not the most important. We would put the lessons to be learnt from BSE at the forefront. Nevertheless, we recognise that the identification of individual criticisms is an important part of our remit, and we have therefore set out this information in this annex. We draw attention to the fact that the areas where we have criticised individuals are relatively few. We have listed the individual criticisms below so that their nature and limitations can be clearly seen. Cross-references are given to locations in the Report where precise details will be found, along with information needed to set the matter in context.

1366 The Report comments on the response of Government Departments and others, and identifies inadequacies. The mere fact that a response on a particular issue was inadequate, or that some part of the response was regrettable or unfortunate, does not mean that individuals are criticised. Only on those occasions when we consider that somebody should have acted differently, in the light of knowledge at the time, have we criticised that individual. In this volume we point out that these criticisms must be set in context. At this point we would invite the reader to turn to paragraphs 1245–59 in Chapter 13, for what is said there is highly relevant to the remainder of this annex. If those criticised were misguided, they were nonetheless acting in accordance with what they conceived to be the proper performance of their duties. The overall value of the contributions that they have made should not be lost from view. Those who were most active in addressing the challenges of BSE are those who are most likely to have made mistakes. It is in that context that the Report makes the following criticisms:

The early years

- Dr Watson should have sought the assistance of the NPU from the outset (Volume 1, paragraph 175; Volume 3, paragraphs 2.137–2.148).

- Dr Watson and Dr Williams should have urged the merits of publication of information about BSE, and Mr Rees should have permitted it (Volume 1, paragraphs 176–178; Volume 3, paragraphs 2.137–2.194).

- Mr Rees should have permitted publication of a proposed article which compared BSE with scrapie (Volume 1, paragraph 179; Volume 3, paragraphs 2.137–2.194).

- Mr Meldrum should have ensured that proper consideration was given to the impact of cross-contamination on the ruminant feed ban (Volume 1, paragraph 214; Volume 3, paragraphs 4.116–4.157).

- Dr Watson, Mr Rees and Mr Cruickshank should have sought to involve the Department of Health in consideration of the risk to human health from BSE prior to March 1988 (Volume 1, paragraph 234; Volume 3, paragraphs 5.114–5.159).

The Southwood Working Party

- The Working Party should have made it plain that the section of their report dealing with epidemiology had been provided by Mr Wilesmith and was based on data which they had not been able to review (Volume 1, paragraph 260; Volume 4, paragraph 10.28).

- The Working Party should have made it clear that, in describing the risk as remote, they were intending to indicate that steps should be taken to reduce the risk so that it was as low as reasonably practicable (Volume 1, paraagraph 272; Volume 4, paragraphs 10.35 and 10.36).

- The Working Party should have pointed out the possible risk to the human food chain from cattle incubating BSE, and pointed out the need to consider identifying such steps as were reasonably practicable to prevent potentially infective tissue being eaten by humans generally, not just babies (Volume 1, paragraphs 273 and 275; Volume 4, paragraphs 10.53–10.82).

- The Working Party should not have allowed their Report to give the reader a false impression of their assessment of the risk relating to medicinal products and occupational exposure (Volume 1, paragraphs 278–9; Volume 4, paragraphs 10.83–10.109).

Protection of animal health, 1989–96

- In May 1990 Mr Gummer was informed of a cat that had come down with FSE, and understood from Mr Meldrum that there was no likely connection between this cat and BSE. Mr Meldrum should not have given Mr Gummer that impression (Volume 1, paragraphs 363 and 650; Volume 6, paragraphs 4.687–4.702).

- While we do not say that Mr Meldrum and Mr Lowson should have identified all the answers to the considerable problems posed by the ban on SBO in animal feed, they should at least have identified that serious problems existed (Volume 1, paragraphs 415–16; Volume 5, paragraphs 4.789–4.853).

Protection of human health, 1989–96

- Sir Donald Acheson and Mr Clarke should have ensured that the Department of Health reviewed the *Southwood Report*, and in particular considered the question why, if offal was not safe for babies, it was nevertheless safe for adults (Volume 1, paragraphs 542 and 550; Volume 6, paragraphs 3.63–3.134).

- Mrs Attridge should have pursued the question 'Why should we take action on baby food and not on hamburgers?'; Mr Cruickshank should have taken steps to find out why the *Southwood Report* drew a distinction between babies and others and between clinical and subclinical animals; and

FINDINGS AND CONCLUSIONS

Mr Meldrum should have pursued these questions (Volume 1, paragraph 552; Volume 6, paragraphs 3.102–3.116).

- Mr Andrews should have raised with Mr MacGregor the need to have an answer to the question why action should be taken on baby food and not other food, and Mr MacGregor himself should have seen that the question was pursued (Volume 1, paragraph 553; Volume 6, paragraphs 3.63–3.124).

- Mr MacGregor is commended for introducing the SBO ban, but he should not have agreed to a presentation of that ban which played down its importance as a protection for human health (Volume 1, paragraph 569; Volume 6, Chapter 3, paragraphs 3.358–3.320).

- Mr Colin Maclean was responsible for inaccurate statements to the public in material prepared on behalf of the MLC in 1990. These statements, which exaggerated the safety of beef and suggested that precautions that had been put in place were unnecessary, were capable of misleading and Mr Maclean should have been more careful (Volume 1, paragraphs 645 and 654; Volume 6, Chapter 4, paragraphs 4.729–4.743).

- Sir Donald Acheson should have appreciated that his public statement about the cat was likely to give false reassurance about the possibility that BSE might be transmissible to humans; the possibility of BSE having being transmitted to a cat was cause for concern and needed to be investigated by scientists (Volume 1, paragraph 660; Volume 6, Chapter 4, paragraphs 4.710–4.724).

- Dr Metters told colleagues they should avoid the implication that 'somehow the disease poses a risk to human health'; he should not have adopted this approach (Volume 1, paragraph 672; Volume 6, Chapter 4, paragraphs 4.725–4.728).

- Sir Kenneth Calman should not have made statements in 1993 and 1995 without ensuring that they fairly reflected his appraisal of the risk posed by BSE (Volume 1, paragraphs 721–4 and 770; Volume 6, paragraphs 5.337–5.349 and 6.341–6.351).

- Dr Kendell should not have made a public statement in 1995 which did not make it plain that the safety of eating beef was dependent on strict compliance with the precautionary measures introduced by the Government (Volume 1, paragraph 773; Volume 9, paragraphs 11.40–11.53).

- Mr Colin Maclean, as Director-General of the MLC, was responsible for the vigorous advertising campaign that the MLC ran in 1995. In the course of that campaign there were occasions when hyperbole displaced accuracy. Mr Maclean should not have allowed this (Volume 1, paragraph 781; Volume 6, paragraphs 6.370 and 6.354–6.377).

- Mr Colin Maclean sent Dr Kimberlin a list of model answers which the MLC would have liked SEAC to give to questions which Mr Hogg had posed to the Committee. Dr Kimberlin was both a consultant to the MLC and a member of SEAC. Mr Maclean should not have asked Dr Kimberlin to provide this assistance; Dr Kimberlin should have told the members of SEAC of the request that the MLC had made (Volume 1, paragraphs 784–788; Volume 6, paragraphs 7.5–7.52).

- Dr Wight sent minutes to Sir Kenneth Calman of SEAC's meetings on 5 January and 1 February 1996 which were inadequate in certain respects. Her January minute should have communicated the concerns expressed at the SEAC meeting by Dr Will. Her minute of the February meeting should have communicated the concerns expressed by Professor Pattison and Professor Collinge (Volume 1, paragraphs 798–800 and 805; Volume 6, paragraphs 7.100–7.107 and 7.160–7.164).

- Mr Eddy circulated a minute about the SEAC meeting on 1 February to Mr Hogg, Mrs Browning, Mr Packer, Mr Carden and Mr Meldrum. He should have included a clear warning of the concerns that had been expressed about the young cases of CJD and the possibility that they might prove to be linked to BSE (Volume 1, paragraph 804; Volume 6, paragraphs 7.139–7.159).

- Despite the shortcomings in Mr Eddy's minute, on reading that minute Mr Hogg and Mrs Browning should have sought to discuss its implications with Mr Packer, Mr Carden and Mr Meldrum. Similarly, on reading that minute, those officials, after discussion among themselves, ought to have raised its implications with Mrs Browning and Mr Hogg. Each of these five individuals should have considered the action that might be required should the scientists advise that BSE had probably been transmitted to humans, and they should have recognised the need for MAFF and DH to address the implications in conjunction, for example by seeking the views of Sir Kenneth Calman and by discussion between Mr Hogg and Mr Dorrell (Volume 1, paragraph 837; Volume 6, paragraphs 7.390–7.482).

- When Sir Kenneth Calman and Dr Metters received Dr Wight's minute about SEAC's meeting of 1 February 1996, albeit that it was couched in sedative terms, they should have initiated discussions with MAFF officials to discuss the implications of the new evidence and Sir Kenneth should have alerted Mr Dorrell (Volume 1, paragraph 842; Volume 6, paragraphs 7.390–7.482).

- Mr M B Baker and, to a lesser degree, Mr Jacobs should have taken steps to avoid the delay that occurred during parts of 1991 and 1992, in circulating advice to schools about dissecting bovine eyeballs (Volume 1, paragraph 1045; Volume 6, paragraphs 9.141–9.151).

Medicines and cosmetics

- Dr Gerald Jones was responsible for deciding the priority to be accorded to BSE in relation to other work within Medicines Division and setting in hand appropriate action. He should have asked for the paper for the Biologicals Sub-Committee (BSC) to be prepared for the September rather than the November meeting (Volume 1, paragraphs 890–1; Volume 7, paragraphs 4.127–4.141).

- Dr Pickles and Mr Lowson should have alerted DTI in 1989 to the need to consider cosmetics products in relation to BSE (Volume 1, paragraphs 1006–8; Volume 7, paragraphs 8.147–8.159).

Potential pathways of infection

- There was a need for an overview of the uses of bovine tissues. Mr Lowson should have ensured that this matter was promptly and properly addressed (Volume 1, paragraph 1078; Volume 7, paragraphs 9.124–9.173).

Glossary

For fuller explanations of these terms, and of others elsewhere in the Report, see the main Glossary in vol. 16: *Reference Material* (or via an electronic link in the website or CD-ROM versions).

ACDP	Advisory Committee on Dangerous Pathogens
ADAS	Agricultural Development and Advisory Service
ACDPWG	ACDP Working Group
ACVO	Assistant Chief Veterinary Officer
AFRC	Agricultural and Food Research Council
ALARP	As Low As Reasonably Practicable
AMI	Authorised Meat Inspector
Ante-mortem	Before slaughter
BABs	Cattle that are verified as having been Born After the ruminant feed Ban, and which are confirmed to be suffering from BSE
Biologicals	Medicinal and other products made from biological materials
BSC	Biologicals Sub-Committee
BSE	Bovine Spongiform Encephalopathy
BSEWG	BSE Working Group
BVA	British Veterinary Association
CAP	Common Agricultural Policy (of the European Union)
CDSM	Committee on Dental and Surgical Materials
CJD	Creutzfeldt-Jakob Disease
CJDSU	CJD Surveillance Unit, Edinburgh
CMO	Chief Medical Officer
CNS	Central nervous system
CRM	Committee on Review of Medicines
CSM	Committee on Safety of Medicines
CTC	Clinical Trial Certificate
CTPA	Cosmetic, Toiletry and Perfumery Association
CVL	Central Veterinary Laboratory, Weybridge
CVO	Chief Veterinary Officer
CWD	Chronic Wasting Disease (in mule deer and elk)
DANI	Department of Agriculture for Northern Ireland
DES	Department of Education and Science (later the Department for Education and Employment)
DFA	Draft Factual Account

DH	Department of Health (until 1988 part of the Department of Health and Social Security, DHSS)
DoE	Department of the Environment (now the Department of the Environment, Transport and the Regions)
DTI	Department of Trade and Industry
Dura mater	The outermost and strongest of the three membranes which envelop the brain and spinal cord
EC	European Community (*see* EU)
EHO	Environmental Health Officer
ELISA	Enzyme Linked ImmunoSorbent Assay
EU	European Union. When the EU came into existence on 1 November 1993 as a result of the Maastricht Treaty, it incorporated but did not replace the EC. In this Report the term EU is more generally used for consistency's sake (even if sometimes chronologically incorrect), except where specific reference is made to the functions conferred by the European Community Treaty or to its legal effect.
Fell	An animal's hide or skin with its hair
FFI	Fatal Familial Insomnia
FSE	Feline Spongiform Encephalopathy
GSS	Gerstmann-Sträussler-Scheinker syndrome, also known as Gerstmann-Sträussler Syndrome
HMI	Her Majesty's Inspector (of Schools)
HSE	Health and Safety Executive
Index case	A first case in a specified group
JCVI	Joint Committee on Vaccination and Immunisation
LA	Local authority
Lamming Committee	Expert Group on Animal Feedingstuffs
LRS	Lymphoreticular system
LVI	Licensed (or Local) Veterinary Inspector
MAFF	Ministry of Agriculture, Fisheries and Food
MAIL	Medicines Act Information Leaflet
MBM	Meat and bone meal
MD	Medicines Division (DH)
MCA	Medicines Control Agency, which became a separate body from DH in 1989 and an Executive Agency in 1991
MH1	A Meat Hygiene Inspection form
MHS	National Meat Hygiene Service
MLC	Meat and Livestock Commission
MP	Member of Parliament
MRC	Medical Research Council
MRM	Mechanically recovered meat

MSSR	The Meat (Sterilisation and Staining) Regulations 1982
NFU	National Farmers' Union
NHS	National Health Service
NI	Northern Ireland
NIBSC	National Institute for Biological Standards and Control
NPU	Neuropathogenesis Unit, Edinburgh
OIE	Office International des Epizooties
OPs	Organophosphates or organophosphorus insecticides
OVS	Official Veterinary Surgeon
PD	Procurement Directorate (for the NHS)
PHLS	Public Health Laboratory Service
Q&A	Question and Answer
RFA	Revised Factual Account
SBO	Specified Bovine Offal (brain, spinal cord, spleen, thymus, tonsils and intestines)
SBO bans	The 'human SBO ban' banned the sale or use of SBO in food for human consumption. The 'animal SBO ban' banned its use in animal feed and the feeding of SBO to animals
SEAC	Spongiform Encephalopathy Advisory Committee
SEAR	Safety, Efficacy and Adverse Reactions (a subcommittee of the CSM)
SEs	Spongiform Encephalopathies
SVO	Senior Veterinary Office
SVS	State Veterinary Service
Territorial Departments	A collective term used by officials in Wales, Scotland and Northern Ireland, as well as in Whitehall, to the Government Departments in the Welsh Office, the Scottish Office and Northern Ireland (hence also 'Territories')
Titre	A measure of concentration of a substance
TME	Transmissible Mink Encephalopathy
Tolworth (Surrey)	Part of MAFF: location of the Animal Health Group and of the headquarters of the SVS
TSE	Transmissible Spongiform Encephalopathy
UK	United Kingdom (of Great Britain and Northern Ireland – Great Britain comprises England, Scotland and Wales)
UKASTA	UK Agricultural Supply Trade Association
UKRA	UK Renderers' Association
vCJD	Variant Creutzfeldt-Jakob Disease
VFS	Veterinary Field Service
VI	Veterinary Investigation
VIC	Veterinary Investigation Centre of the VI Service

VIO	Veterinary Investigation Officer
VI Service	Veterinary Investigation Service
VMD	Veterinary Medicines Directorate
VO	Veterinary Officer
VPC	Veterinary Products Committee
Zoonosis	Animal disease that can be transmitted to humans

Who's who

For fuller descriptions of many of the following people, see the main Who's who in vol. 16: *Reference Material* (or via an electronic link in the website or CD-ROM versions).

Sir Donald Acheson	Chief Medical Officer, 1983–91
Dr Paul Adams	DH Senior and then Principal Medical Officer, Medicines Division (later MCA)
Professor Jeffrey Almond	Member of SEAC from December 1995
Mr (later Sir) Derek Andrews	MAFF Permanent Secretary, 1987–93
Mr Mike Ansfield	MAFF senior scientific officer
Professor Sir James Armour	Chair, Veterinary Products Committee
Professor Sir William Asscher	Chair, Committee on Safety of Medicines, 1987–92
Mrs Elizabeth Attridge	MAFF Grade 3 responsible for the Animal Health and Veterinary Group, 1989–91 (and previously for the Emergencies, Food Quality and Pest Controls Group)
Sir John Badenoch	Chair, Joint Committee on Vaccination and Immunisation
Dr Harry Baker	MRC Clinical Research Centre
Mr Keith Baker	MAFF ACVO responsible for Red Meat Hygiene sections, 1988–96
Mr M B Baker	DES Head of Schools Branch 3
Professor R M Barlow	Royal Veterinary College; member of SEAC, 1990–96
Dr A J Beale	Wellcome
Professor Sir Colin Berry	Chair, Committee on Dental and Surgical Materials, 1982–92; member, BSEWG, 1989–92
Professor Peter Biggs	Director, Institute for Animal Health, 1986–88
Mrs Virginia Bottomley MP	DH Minister of State (Commons), 1989–92, and Secretary of State, 1992–95
Professor John Bourne	Director, Institute for Animal Health, from 1988; member of the Tyrrell Committee
Mr Raymond Bradley	Head of the Pathology Department, CVL, 1983–95, and the CVL's BSE research coordinator, 1987–95
Mrs Jane Brown	MAFF Head of Meat Hygiene Division, 1990–96
Mrs Angela Browning MP	MAFF Parliamentary Secretary (Commons), 1994–97

Dr (later Sir) Kenneth Calman	Chief Medical Officer, 1991–98
Mr Richard Carden	MAFF Deputy Secretary, Head of Food Safety Directorate, from 1994
Mr Peter Carrigan	Director, Specialpack Ltd
Mr Richard Cawthorne	MAFF Head of Veterinary Investigation Section, SVS, 1987–91; Head of Animal Health (Zoonoses) Division, 1991–95; Veterinary Head of Notifiable Diseases Section from 1995
Mr Christopher Clarke	Authorised Meat Inspector
Mr Kenneth Clarke MP	DH Secretary of State, 1988–90
Professor Gerald Collee	Chair, BSE Working Group of the CSM; and Chair, the Biologicals Sub-Committee of the CSM
Professor John Collinge	Professor of Molecular Neurogenetics, Imperial College School of Medicine at St Mary's; member of SEAC since December 1995
Dr E M (Mary) Cooke	Deputy Director, Public Health Laboratory Service; member of the Lamming Committee
Mr R Cooper	Director, Sainsburys
Mr Mike Corbally	Institute of Environmental Health Officers
Mr Philip Corrigan	Meat Hygiene Service, Head of Operations, January to August 1995
Mr M Cranwell	Starcross Veterinary Investigation Centre, Exeter
Mr Iain Crawford	MAFF Director of the Veterinary Field Service, 1988–98
Mr J Creedy	Her Majesty's Inspector (of Schools)
Mr Alistair Cruickshank	MAFF Grade 3 responsible for the Animal Health Group, December 1986 to December 1989
Mr P W Cunliffe	Joint author of the Evans/Cunliffe management report on control of medicines
Mrs Edwina Currie MP	DH Parliamentary Under-Secretary of State, 1986–88
Mr David Curry MP	MAFF Parliamentary Secretary (Commons), 1989–92, and Minister of State, 1992–93
Mr Don Curry	Chair, Meat and Livestock Commission, since 1993
Mr Ron Davies MP	Opposition (Labour) spokesman on Agriculture
Dr Stephen Dealler	Consultant Microbiologist, Burnley General Hospital
Dr Alan Dickinson	Spongiform encephalopathy researcher; Director of the Neuropathogenesis Unit, 1981–87

Mr Brian Dickinson	MAFF Head of Food Safety Group (later Food Safety and Science Group), 1989–96
Mr Jonathan Dimbleby	Television interviewer
Mr Stephen Dorrell MP	DH Parliamentary Under-Secretary (Commons), 1990–92, and Secretary of State, 1995–97
Mr Thomas Eddy	MAFF Head of Animal Health (Disease Control) Division, and SEAC Secretariat, from June 1993
Professor Anthony Epstein	Professor of Pathology; member of the Southwood Working Party
Dr Diana Ernaelsteen	DH Senior Medical Officer and Senior Medical Adviser to DES, 1984–95
Mr John Evans MP	Opposition (Labour) Member of Parliament
Dr N J B Evans	Joint author of the Evans/Cunliffe management report on control of medicines
Dr R J Fielder	DH Senior Principal Scientific Officer (chemical toxicology, including cosmetics), 1988–93
Mr Andrew Fleetwood	MAFF Veterinary Investigation Officer, SVS, 1987–91; Senior Veterinary Officer, Animal Health (Zoonoses) Division, 1991–95
Dr Gerald Forbes	Senior Medical Officer, Scottish Home and Health Department, 1985–89; Director of the Environmental Health (Scotland) Unit, 1989–93
Sir Christopher France	DH Permanent Secretary, 1988–92
Sir Michael Franklin	MAFF Permanent Secretary, 1983–87
Mr Duncan Fry	MAFF Meat Hygiene Division
Dr A J M Garland	Wellcome
Dr John Godfrey	Member of the MAFF Consumer Panel
Sir Simon Gourlay	President of the National Farmers' Union
Dr Helen Grant	Consultant Neuropathologist
Mr Gordon Gresty	North Yorkshire County Council
Mr John Gummer MP	MAFF Minister of State (Commons), 1985–88, and Minister of Agriculture, Fisheries and Food, 1989–93
Mr David Hagger	DH Head of Medicines Branch 1, 1984–90
Dr E L Harris	DH Deputy Chief Medical Officer, 1977–89
Mr Graham Hart	Permanent Secretary, Scottish Home and Health Department, 1990–92; DH Permanent Secretary, 1992–97
Mr Michael Heseltine MP	President of the Board Trade (ie, Secretary of State), DTI, 1992–95; Deputy Prime Minister, 1995–97

Mr Peter Hewson	MAFF Superintending Meat Hygiene Adviser, 1992–95; Meat Hygiene Veterinary Section, 1995–96
Dr Deirdre Hine	Chief Medical Officer, Welsh Office, 1990–97
Mr Douglas Hogg MP	Minister of Agriculture, Fisheries and Food, 1995–97
Mr John Horam MP	DH Parliamentary Under-Secretary (Commons), 1995–97
Mr Stephen Hutchins	MAFF Senior Veterinary Officer, Meat Hygiene Veterinary Section, 1987–91
Dr James Ironside	Senior Lecturer in Pathology; neuropathologist, CJD Surveillance Unit, since 1995
Mr Ron Jacobs	DES Schools Branch 3
Dr David Jefferys	DH Principal Medical Officer, 1986–90; Medicines Control Agency, 1990 onwards
Dr Martin Jeffrey	Veterinary Research Officer, Pathology, CVL, 1985–87; Senior Research Officer, Lasswade, 1987 onwards
Mr A M (Mac) Johnston	Royal Veterinary College
Dr Gerald Jones	DH Medicines Division, 1984–89; Health Aspects of the Environment and Food Division, 1992–95
Dr Keith Jones	Chief Executive, Medicines Control Agency, from 1989
Ms Marion Kelly	Director-General, Cosmetic, Toiletry and Perfumery Association
Dr Robert Kendell	Chief Medical Officer, Scottish Office, 1991–96
Mr Alastair Kidd	MAFF, Secretary of the Veterinary Products Committee, 1985–91; Director of Licensing, Veterinary Medicines Directorate, 1989–91
Dr Richard Kimberlin	TSE research scientist at the NPU, 1981–88; independent TSE consultant since 1988; member of the Tyrrell Committee and SEAC
Professor Richard Lacey	Emeritus Professor of Clinical Microbiology, University of Leeds
Professor Eric Lamming	Professor of Animal Physiology; Chair, Lamming Committee (Expert Group on Animal Feedingstuffs)
Mr Norman Lamont MP	Chief Secretary to the Treasury, 1989–90; Chancellor of the Exchequer, 1990–93
Mr Ian Lang MP	Minister of State, Scottish Office, 1987–90; Secretary of State for Scotland, 1990–95
Mr Alan Lawrence	MAFF Animal Health Division, 1979–95; Joint Secretariat, Southwood Working Party

Professor David Lawson	Chair, Committee on Review of Medicines
Dr A Lee	Veterinary Medicines Directorate
Mr Charles Lister	DH, Health Aspects of Environment and Food Division; SEAC Secretariat, 1993–95
Dr Thomas Little	Deputy Director of the CVL, 1986–90; Director and Chief Executive, CVL (later Veterinary Laboratories Agency), from 1990
Mr Murray Love	Medicines Control Agency
Mr Robert Lowson	MAFF Head of Animal Health Division, 1989–91; Head of Animal Health (Disease Control) Division, 1991–93; SEAC Secretariat until 1993
Mr Peter Luff	Central Veterinary Laboratory
Mr John MacGregor MP	Minister of Agriculture, Fisheries and Food, 1987–89
Ms Sheila McKecnhie	Director, Consumers' Association, since 1995
Mr Colin Maclean	Meat and Livestock Commission, Technical Director, 1988–92, and Director-General from 1992
Mr David Maclean	MAFF Parliamentary Secretary (Commons) (Minister for Food Safety), 1989–92
Mr Johnston McNeill	Chief Executive, Meat Hygiene Service, since 1995
Dr Kenneth MacOwan	MAFF Scientific Liaison Officer responsible for Veterinary Science, Chief Scientist's Group, 1988–95
Mr John Major MP	Prime Minister, 1990–97
Mr Ron Martin	Deputy Chief Veterinary Officer, Northern Ireland, 1985–90, and Chief Veterinary Officer from 1990
Dr William Martin	Director of the Moredun Research Institute, 1971–85; member of the Southwood Working Party
Mr John Maslin	MAFF Animal Health Group; Head, BSE branch, 1990–93
Dr Danny Matthews	MAFF Senior Veterinary Officer, 1988–96
Professor W Bryan Matthews	Professor of Clinical Neurology; specialist in TSEs
Sir Robert May	Chief Scientific Adviser from 1995
Mr Keith Meldrum	Chief Veterinary Officer, 1988–97
Mr David Mellor MP	DH Minister of State (Commons), 1988–89
Dr Jeremy Metters	DH Senior Principal Medical Officer; Deputy Chief Medical Officer, 1989 onwards
Dr Philip Minor	Head of the Division of Virology, National Institute for Biological Standards and Control

FINDINGS AND CONCLUSIONS

Lord Montagu of Beaulieu	Rural landlord
Mr Thomas Murray	DH, Environmental Health and Food Safety Division, 1990–95; SEAC Secretariat, 1990–93
Mr Richard Packer	MAFF Permanent Secretary from 1993
Dr Michael Painter	Member of SEAC from December 1995
Dr Will Patterson	Consultant in Public Health Medicine
Professor Sir John Pattison	Professor of Medical Microbiology; Chair of SEAC from 1995
Dr Hilary Pickles	DH Principal Medical Officer; Joint Secretariat, Southwood Working Party; Joint Secretary, Tyrrell Committee; SEAC observer to 1991; DH lead on BSE, 1988–91
Mr Michael Portillo MP	Chief Secretary to the Treasury, 1992–94
Dr Stanley Prusiner	Professor of Neurology and Biochemistry, University of California School of Medicine; Nobel Prize winner for research into prion proteins
Dr John Purves	DH Senior Principal Pharmaceutical Officer, 1985–90; Unit Manager, Biological Unit, Medicines Control Agency, from 1990
Professor M D Rawlins	Chair, Committee on Safety of Medicines subcommittee on Safety, Efficacy and Adverse Reactions
Mr John Redwood MP	Secretary of State for Wales, 1993–95
Mr James Reed	Director-General, UK Agricultural Supply Trade Association (UKASTA)
Mr William Rees	Chief Veterinary Officer, 1980–88
Miss Gillian Richmond	MAFF Legal Department
Dr Rosalind Ridley	MRC Clinical Research Centre
Mr Chris Rogers	MAFF Meat Trade Adviser
Mr Richard Roscoe	Head of the Chemical Hazards Section of the Consumer Safety Unit (including safety of cosmetics), DTI, 1983–92
Dr Frances Rotblat	DH Senior Medical Officer; Medical Assessor to the Biologicals Sub-Committee of the Committee on Safety of Medicines, until 1990; Principal Assessor to the Safety, Efficacy and Adverse Reactions subcommittee of the CSM, 1990–95
Dr Eileen Rubery	DH Under Secretary (Grade 3), Head of Health Aspects of the Environment and Food Division
Dr James Rutter	Director of Veterinary Medicines and Chief Executive, Veterinary Medicines Directorate
Mr F J H Scollen	MAFF Animal Health Division

Mr James Scudamore	Assistant Chief Veterinary Officer, Scotland, 1990–96
Mrs Gillian Shephard MP	Minister of Agriculture, Fisheries and Food, 1993–94
Mr Alick Simmons	MAFF Meat Hygiene Veterinary Section, 1991–95
Mr Mike Skinner	DH official; SEAC Secretariat from 1995
Dr Roger Skinner	DH Principal Medical Officer from 1989
Mr John Sloggem	DH Pharmaceutical Officer (later with the Medicines Control Agency), 1985–97
Mr Edward (Ted) Smith	MAFF Deputy Secretary until 1989
Professor Peter G Smith	Member of SEAC from January 1996
Mr Nicholas Soames	MAFF Parliamentary Secretary (Commons), 1992–94
Mr Peter Soul	Director of Operations, Meat Hygiene Service, since 1995
Professor Sir Richard Southwood	Professor of Zoology; Chair, Southwood Working Party
Mr John Suich	MAFF Animal Health Division
Dr David Taylor	Principal Research Scientist, Neuropathogenesis Unit
Mr David Taylor	MAFF Veterinary Head of Section, Red Meat Hygiene Section, 1987–97
Mr Kevin Taylor	MAFF Veterinary Head of Notifiable Diseases Section, 1986–91; Assistant Chief Veterinary Officer, 1990–97
Mr Matthew Taylor MP	Liberal Democrat Member of Parliament
Mrs Margaret Thatcher MP	Prime Minister, 1979–90
Professor Philip C Thomas	Scottish Agricultural College; member of the Lamming Committee
Mr Donald Thompson MP	MAFF Parliamentary Secretary (Commons), 1986–89
Sir Bernard Tomlinson	Neuropathologist
Dr David Tyrrell	MRC Common Cold Unit, 1982–90; Chair, Tyrrell Committee; Chair, SEAC, until 1995
Mr William Waldegrave MP	Minister of Agriculture, Fisheries and Food, 1994–95
Mr Peter Walker MP	Secretary of State for Wales, 1987–90
Professor Sir John Walton	Professor of Neurology; member, Southwood Working Party
Dr William Watson	Director of the CVL, 1986–90; member of the Tyrrell Committee and SEAC
Mr Gerald Wells	Head, Neuropathology Section, CVL, since 1985

Mrs Diane Whyte	DH Higher Executive Officer, Environmental Health and Food Safety, 1989–92; Food Hazard Management Unit, 1995–96
Dr Ailsa Wight	DH lead on BSE from 1991; SEAC observer
Mr John Wilesmith	Head of Epidemiology, CVL, since 1986
Dr Robert Will	Director of the CJD Surveillance Unit since 1990; member of the Tyrrell Committee and SEAC
Dr Bernard Williams	Assistant Chief Veterinary Officer, Head of the Veterinary Investigation Service, until 1987
Mr G M Wood	Central Veterinary Laboratory
Dr Mark Woolfe	MAFF Food Science Division II, 1987–95
Mr Ayyildiz Yavash	MAFF Legal Department, 1985–93

Index

A

abattoirs 199
abomasum 115, 230, 247
accidental contamination 220
ACDP 199, 200
ACDP Working Group *see* ACDPWG
ACDPWG 200
Acheson, Sir Donald 13, 46–47,
.................................48, 107–108, 111, 113,
............................129–131, 142, 171, 176,
.........................179, 182, 223, 242, 281–282
Acts of Parliament 29
Adams, Dr Paul 174, 185
Additional Comments 279
Advisory Committee on Dangerous
   Pathogens 240
advisory committees 240–241, 257, 277
Advisory Note, farmers 65, 97
AFF officials 62
AFRC 222–223, 252
Agricultural and Food Research Council *see*
   AFRC
Agricultural Development and Advisory
   Service 227
agricultural industry 98
Agricultural Research Council 226
Agriculture Advisory Panel 102
Agriculture Committee 104–105,
.... 112–113, 123, 124,
....130–131, 134, 241
Agriculture Department 216
air quality 213
ALARP 32, 52–53, 123
ALARP principle 229–230, 232
allergens 185
Almond, Professor Jeffrey 159
Andrews, Sir Derek 13, 44–47, 57,
...............71, 72, 83, 109, 111, 117, 127, 282
animal feed 26–27, 61
animal feed handlers 198
animal feed industry 62
animal feed manufacturers 26
Animal Feedingstuffs Advisory
   Committee 17
animal health 48, 80, 257
Animal Health (Disease Control)
   Division 87
Animal Health (Zoonoses) Division 88
Animal Health Act 1981 ....14, 78, 90,
....92, 101, 267–269
Animal Health and Veterinary
   Group 73, 243
Animal Health Computerised Traceability
   System 277
Animal Health Division 119, 124, 174, 194
Animal Health Group 108

animal protein 23, 94–95
animal protein, recycling 227
animal SBO ban 16–17, 21, 60,
.... 62–64, 67, 80, 83, 85, 88,
.............................90, 94, 127, 133, 136, 164,
.... 212, 229, 236,239, 245, 248, 269, 281
animal SBO ban, breaches 255
animal SBO ban, implementation 18
animal SBO Regulations 87
animal waste 23, 25
Ansfield, Mike 60
aprotonin 28
Arabian oryx 139
Armour, Professor Sir James 168
artificial inseminators 199
As Low As Reasonably Practicable *see* ALARP
Asscher, Professor
   Sir William 173, 176, 179
attack rate experiment 64, 65, 88, 94,
....141, 147, 224, 236
Attridge, Elizabeth 81, 84, 92, 107–109,
.... 111–113, 124, 245, 281
audit 208, 210–211
Australasia 185
Australia 184
Authorised Meat Inspectors 75, 80
autoimmune theory 219, 222

B

BABs 17–18, 21, 39, 61–63,
.... 73–74, 87, 94, 189, 225
BABs, epidemiological investigations 64
baby food 106–109, 111, 281–282
baby food manufacturers 14
baby food recommendations 51, 233
Bacon and Meat Manufacturers'
   Association 110
Badenoch, Sir John 176
Baker, Dr Harry 171
Baker, Keith 73, 81, 83, 85, 123,
....204–205
Baker, Mr M B 283
Barlow, Professor R M 127
BBC Newsnight 129
Beale, Dr A J 171
beef exports 96, 121, 126, 218
beef exports, ban 160
beef market 153, 160
beef, boneless 126
beef, safety 22, 114, 129–133, 142, 145,
.................149–150, 151–153, 234, 239, 282
Belgium 42
benzene 26
Berry, Professor Colin 168
Biggs, Professor Peter 201–202
biological based medicines 171, 173, 175

297

Biologicals Committee 169
Biologicals Sub-Committee *see* BSC
black and white puddings 115
blood 213
bloodmeal 23
bone china 27
bones 117
Bottomley, Mrs Virginia 137, 180, 265
Bourne, Professor John 55
bovine brains 51, 105, 122,
.... 136, 175, 182
bovine eyeballs 235
bovine eyeballs, dissecting 200, 203–205
bovine eyeballs, dissections 240, 242, 283
bovine heart 28
bovine ingredients 182
bovine insulin 171, 186
bovine intestinal mucous 28
bovine intestines 28, 75
bovine lungs 28
bovine material 51, 170, 172, 176,
.... 177, 178, 185, 189, 194,
.... 207, 209–210, 214, 237, 258, 267, 270
bovine material, potentially infective 22
bovine material, wide range of uses 28
bovine materials 184
bovine meat 25, 28
Bovine Offal (Prohibition) (Amendment)
   Regulations 1995 88
Bovine Offal (Prohibition)
   Regulations 1989 .... 75, 79
bovine pancreases 28, 110
bovine prion gene 12
bovine raw materials 110
bovine serous membranes 28
bovine serum 171, 173, 175, 190
bovine serum albumin 28, 191
bovine spinal column 27
Bovine Spongiform Encephalopathy
   (Amendment) Order 1988 101
Bovine Spongiform Encephalopathy
   (Amendment) Order 1996 66
Bovine Spongiform Encephalopathy (No. 2)
   Amendment Order 1990 73
Bovine Spongiform Encephalopathy
   Compensation Order 1988 101
Bovine Spongiform Encephalopathy Order
   1988 39
Bovine Spongiform Encephalopathy Order
   1991 .... 80, 84
bovine vertebral column 150
bovine-based vaccines 107
Bradley, Raymond 34, 40, 63, 87,
.... 94, 114–115, 118, 159,
.... 170, 191, 195, 209, 223, 233
brain 89, 90, 108, 114–117,
.... 172, 177, 189,192, 221
brain and spinal cord 20, 223, 250
brain and spinal cord, removal 16
brain removal 85, 122–123, 237

breeding 96–97
Bristol 34
Bristol University 128
British Diabetic Association 186
British Isles 183
British Medical Journal 47
British Veterinary Association 96
Brown, Jane 135
Browning, Angela 90, 92, 133,
.... 144, 146, 151, 155, 162, 240, 283
BSC 168, 169, 170, 283
BSE 175
BSE agent 20, 36–37, 53, 64,
.... 110, 115, 117, 189, 200,
.... 212, 220–221, 260, 266
BSE agent, inactivation 38, 213
BSE research 14, 49, 54, 56, 121,
.... 222–225, 261
BSE Working Group *see* BSEWG
BSE, ante- and post-mortem tests 224
BSE, compulsory notification 14
BSE, experimental
   transmission 71–72, 121, 233
BSE, identification 227
BSE, infectivity 88, 94–95
BSE, notification compliance 15
BSE, oral transmission 127–128
BSE, sheep 224
BSE, subclinical cases 68
BSE, transmissibility 228–230
BSE, transmission risk 50–55, 157,175,
.... 177, 179–181, 186, 193, 213,
.... 216, 219, 233–235, 238–239,
.... 240, 242, 250, 267, 280, 282–283
BSEWG 168, 176, 181–187, 189, 194
burgers 117
burial 212
butchers 110, 199
BVA 199

C

calf bioassay 261
calf brain 28
calf rations, MBM inclusion rate 41
calf serum 171
Calman, Sir Kenneth 91, 138, 142–143,
.... 145–146, 154, 155, 156–157,
.... 162–163, 216, 282–283
calves, carcasses 116–117
calves, offal 116–117
Campaign for Freedom of Information 276
CAP 7, 23
carcasses 4, 117, 119, 121–122
carcasses, disposal 49, 105, 199, 212–213
carcasses, dressing 25
carcasses, handling 199
Carden, Richard 144, 154, 155,
.... 158, 161–163, 283
Cardiff 216
care plan 258

INDEX

Carrigan, Peter 85
casings 114–116, 217
cats 15, 70–72, 81, 233, 281–282
cattle 121, 212
cattle born after the ban *see* BABs
cattle brains 110
cattle feed 36, 40, 59, 141
cattle feed, accidental contamination 67
cattle feed, contamination 87
cattle feed, cross-contamination 240
cattle glands, handling 198
cattle industry 23
cattle offal, rendering 20
cattle products 211
cattle tracking 277
cattle-tracking, computerised system 96
Cawthorne, Dr Richard 88, 144
CDSM 53, 168, 173, 175, 184, 186
cell cultures 177
central nervous system *see* CNS
Central Veterinary Laboratory *see* CVL
cerebellar dysfunction 100
Charing Cross Hospital, London 110
cheese 115
cheetah 139
Chief Medical Officers *see* CMO
Chief Scientific Adviser *see* SCA
Chief Secretary to the Treasury 103
Chief Veterinary Officer *see* CVO
children 108
civil servants 275
CJD 99–101, 141–142, 147,
.... 147–148, 153–160, 161,
.... 162, 169, 200, 223, 238, 283
CJD surveillance 99–100, 154
CJD Surveillance Unit 243
CJD, monitoring 54
CJD, occupational risk 12
CJDSU 15, 19, 54, 99–101,
.... 141–142, 148, 150, 152,
.... 155, 160, 163, 223, 277
Clarke, Christopher 137
Clarke, Rt Hon
 Kenneth 107–108, 113, 180, 281
Clinical Microbiology 128
Clinical Trial Certificate *see* CTC
CMO 1, 13, 29, 45–46,
.... 180, 216, 234, 244, 252, 264–266
CNS 118, 123–124
Code of Practice for the Handling of SBO at
 Rendering Plants 84
COLIPA 196
collagen, handling 198
collection centres 74, 76, 81, 86,
.... 90, 91, 92
Collee, Professor Gerald 168, 174,
.... 183–184, 186
Collinge, Professor John 140, 148, 154,
.... 155–156, 159, 283
Commission Decision 95/287/EC 65

Committee on Dental and Surgical Materials *see*
 CDSM
Committee on Review of Medicines *see* CRM
Committee on Safety of Medicines *see* CSM
Common Agricultural Policy *see* CAP
Communicable Disease Report
 Review 200
communicable diseases 100
compensation, formulae 101, 105
compliance 145, 149, 282
compliance, voluntary 59
compulsory slaughter and compensation
 scheme 14, 44–47, 101–105,
.... 126, 129, 212, 267
confidentiality 278–279
conflict of interest 262
Conservative Women's Conference 110
Consultative Committee on Research into SEs
 see Tyrrell Committee
Consumer Panel 232
consumer protection 272
Consumer Protection Act 1987 .... 192, 269
Consumers in the European Community
 Group 124
Consumers' Association 124, 265
Consumers' Committee of the Meat and
 Livestock Commission *see* MLC
Cooper, R 103
Corbally, Mike 77, 79, 123
Corrigan, Philip 144
Cosmetic Products (Safety) Regulations
 1989 271
Cosmetic, Toiletry and Perfumery Association
 see CTPA
cosmetics 4, 26–27, 194, 208,
.... 232, 235, 238, 242, 260, 271, 283
Cosmetics Directive 196
cosmetics industry 192
County Councils 59, 78, 80–81, 92, 245
County Trading Standards Officer 102
Cranwell, Mr M 36
Crawford, Iain 83, 86
credibility 265–266
Creedy, J 204
CRM 168, 173, 184
cross-contamination 94
Cruickshank, Alistair 39, 44–45, 109,
.... 118–120, 174, 280–281
CSA 265
CSM 53–54, 167, 170,
.... 172–173, 175, 176–177, 179–182, 234
CTC 170, 171
CTPA 193–196
cull cows 110
Cunliffe, P W 168
Currie, Mrs Edwina 180
Curry, David 103, 112, 121
Curry, Don 145–146

299

CVL 13, 17–18, 40–41, 47,
  .... 88, 114–115, 127, 137,
  .... 141, 169, 170, 171, 191, 195–196,
  .... 209, 223–225, 227, 236, 243, 277
CVL, Animal Health Division 43
CVL, Biological Products and Standards
  Department 167
CVL, Epidemiology Department 13, 36
CVL, Pathology Department 33–34
CVL, veterinary medicines 43
CVO 13–34, 38, 41–42, 244, 252, 266

D

DAFS 216–217
Daily Mail 142–143
Daily Telegraph 149
dairy calves 24
dairy farmers 142, 147–149
dangerous pathogens, handling 200
DANI 118, 277
Davies, Ron 111
Dealler, Dr Stephen 143, 148
death certificates 100
deboning 19, 157, 159–160,
  .... 164, 230, 241
deer 177
dementing illnesses 100
Department of Agriculture and Fisheries for
  Scotland *see* DAFS
Department of Agriculture for Northern Ireland
  see DANI
Department of Education and Science *see* DES
Department of the Environment *see* DoE
Department of Trade and Industry *see* DTI
depression 272
deregulation 30, 245
deregulation initiative 76, 90, 247
DES 29, 203–205, 242
DFAs 277–278
DH 13, 29, 42–43, 45,
  .... 56–57, 99, 100, 107, 108–109,
  .... 112, 113–116, 118, 120, 128,
  .... 131, 132–133, 138, 142, 146,
  .... 152, 153, 155, 156, 169,
  .... 177, 181, 182, 189, 193,
  .... 195, 201, 204, 208–209,
  .... 211, 215, 218, 223, 232, 236,
  .... 246, 252, 254, 256–257, 259,
  .... 276, 280–281, 283
DH officials 235, 240–243
DH, areas of responsibility 242
DH, Medicines Division see DH, MD 167
DH, role 243
DHSS 43
Dickinson, Brian 265
Dickinson, Dr Alan 223
Dimbleby, Jonathan 149
Diseases of Animals (Protein Processing)
  Order 1981 227
distal ileum 137–138, 223

District Councils 75, 76, 78, 80–81,
  .... 92–93, 245
District Councils, enforcement
  responsibilities 25
DoE 29, 242
domestic slaughterhouses 82
Dorrell, Stephen 1, 147, 149, 151,
  .... 158–160, 162–164, 283
dorsal root ganglia 118, 124, 147
Dorset 103
draft factual accounts *see* DFAs
drains 213
DTI 29, 43, 192–197, 208–209,
  .... 232, 242, 260, 283
DTP vaccines 185–186

E

EC 19, 271
EC bone-in beef export ban 19
EC Cosmetics Directive 1976 271
EC Decision 97/534 271
EC Directive 89/662/EEC 271
EC Directive 90/425/EEC 271
Eddy, Thomas 67, 87, 88, 150,
  .... 154, 155–156, 161–162, 283
Edinburgh 34, 170
EEC 23
effluent 212–213
EHOs 122, 134
Eland 139
ELISA test 18, 41, 60–61, 225
Emergencies, Food Quality and Pest Control
  Group 107
enforcement 121, 135, 144, 157,
  .... 192, 217, 255
enforcement of Regulations 120, 133
enforcement, meat hygiene 75–82
enforcement, Regulations 244–245
enforcement, RFB 59–60
England 24, 26, 99, 100
Environment Act 1990 213
Environment Agency 213
Environmental Health (Scotland)
  Unit 124, 216
Environmental Health Departments 76
Environmental Health
  Officers 75, 76, 79, 80, 83, 88, 93
Environmental Protection Act 1990 269
environmental regulation 212–213
epidemiology 33, 56, 99, 100, 141,
  .... 171, 207, 216, 219, 225,
  .... 253, 275–276, 277, 281
Ernaelsteen, Dr Diana 203–205
EU 121, 160
EU cattle export ban 15
EU Cosmetics Directive 192
EU Council of Ministers 16
EU Health Council 196
EU Member States 23–24, 73, 271
EU requirements 213

EU Scientific Veterinary Committee 146
EU Veterinary Inspectors 134
EU Working Party on Cosmetics 196
European Commission 126, 195
European Commission, Scientific Veterinary
    Committee 137
European Court of Justice 271
European guidelines 181, 187
European law 271
European standards, meat hygiene 76
European Union Surveillance 100
Evans Medical 185
Evans, Dr N J B 168
Evans, John 109
Evans/Cunliffe report 172
Eville & Jones 144
expert committees 254
Expert Group on Animal
    Feedingstuffs 16, 60
export-approved slaughterhouses 82
Exports 116–117, 138
exports, SBO 73
eyeballs 90, 283

F

face creams 192
fallen stock 74, 80, 82
families 272–273
farm workers 199
farmers 100, 101, 103, 104,
    .... 199, 226, 251
farmers, Advisory Notes 97, 240
Farmers' Union of Wales 102
farmers' unions 59
farming industry 104, 146
Farming News 46
feather meal 23
feed compounders 116
feed industry 59
feed, accidental contamination 40
feedlines, cleanliness 65
feedmills 27, 160
feedmills, cross-contamination 18, 27,
    .... 40, 41, 60–61, 63, 65, 68,
    .... 141, 225, 236, 239, 253, 280
Feline Spongiform Encephalopathy 60
fell mongers 199
fertiliser 84, 112, 212–213, 229
fertiliser, handling 198
Fertilisers and Feedingstuffs Act 1926 20
fertilisers, unprocessed blood 25
Fielder, Dr R J 192, 193, 195
financial control pressures 30
fishmeal 23
Fleetwood, Andrew 83, 88, 91, 93
foetal calf serum 186, 187, 191
Food Act 1984 .... 24, 75, 82, 83, 113,
    269–270
food chain 149, 158, 217, 250
food industry 114

Food National Interest Group *see* NIG
food safety 114, 121, 146, 164, 237
Food Safety Act 1990 .... 75, 83, 269–270
Food Safety Directorate 121, 154, 232, 243
Food Science Division 107, 124
Food Standards Agency 244, 265
Food Standards Division 124
Forbes, Dr Gerald 124, 216
Form MH6 86
France 187
France, Sir Christopher 243
Franklin, Sir Michael 43, 45, 231
Freedom of Information Award 276
Freeman, Roger 56
fresh meat 27
Fry, Duncan 118
FSE 70–72, 81, 128–129,
    .... 133, 139, 281

G

Garland, Dr A J M 171
gelatine 4, 26, 28, 178, 188,
    .... 196, 211, 238
gemsbok 139
Germany 196
glands 189
Glaxo Wellcome 28
glucagon 28
glue 27
glycerine 26
glycerol beef broth 186
glycosylation 12, 221
goats 177
Godfrey, Dr John 124, 161
Gourlay, Sir Simon 71
Grant, Dr Helen 110, 128
greater kudu 139
greaves 26
green offal 75, 76
Gresty, Gordon 102
growth stimulants 27
Gummer, John 15, 42–43, 56–57,
    .... 60, 61, 62, 69, 70, 70–72,
    .... 79, 81, 103, 112, 114–115,
    .... 120–124, 127–131, 134–136,
    .... 140, 142, 161, 191, 208–209,
    .... 211, 231, 243–244, 248, 281
gut rooms 85

H

Hagger, David 184
haggis 217
hamburgers 109, 143, 281
Harris, Dr E L 107, 108, 169,
    .... 171, 174, 176
Hart, Graham 216
head meat 122–123
head meat, removal 16, 25
head-boning plants 92, 122

heads 74, 90, 91
Health Department 216
health stamp 25, 121–122, 135,
.... 144–145, 245
healthcare professionals, mortuary
    workers 198
heart 108
heart valves 28, 168
Heidelberg 188
heparin 28
Heseltine, Michael 158
Hewson, Peter 89
Hine, Dr Deirdre 216
histopathology 34, 105, 223, 227
HMI 204
Hogg, Douglas 19, 66–67, 90, 91,
.... 96, 133, 144, 144–147,
.... 150–153, 155, 157–160, 162–164, 282
homeopathic medicines 184
Horam, John 156
hormones 166, 220
hospitals 240
House of Commons Agriculture
    Committee 15–16, 96, 217
House of Commons, Agriculture
    Committee 60
HSE 14, 29, 51, 54, 175, 198,
.... 204–205, 208, 242, 260
human cadavers, handling 200
human food chain 42, 51, 76, 110,
.... 116–117, 225, 281
human health 48, 280
human medicines 170, 173
human SBO ban 14, 17–18, 21, 60,
.... 69–70, 76–79, 83, 88, 92,
.... 106, 113–121, 126, 132, 136,
.... 140, 217, 230, 232, 233, 239, 269–270
human SBO ban, ... 245
human SBO Regulations 75, 79, 82, 83,
.... 85, 87, 89, 93, 114, 118–120, 134
humans, BSE transmisisbility 229
humans, infective dose 237
Humberside County Council 128
Humberside Education Authority 15
hunt kennel workers 199
hunt kennels 24, 74, 78, 80–81, 86, 92
Hutchins, Stephen 83, 122, 137,
.... 208, 210–211
Hygiene Advice Teams 144
Hygiene Regulations 25
hygiene standards 134
hygiene standards, slaughterhouses 75, 76

I

IBD 275–276
ileum 110
implants 192
incinerator operatives 199
incinerators 90, 92
incinerators, licensing 212
incubation 148
incubation period 12, 59, 62, 67, 132,
.... 140, 150, 221, 227, 239, 251,
.... 258–259
The Independent 127, 149, 150
inefective tissue, handling 198
infective dose 40, 41, 119, 132, 141
infective dose, minimum 125–126, 224,
.... 236–237, 239
infective tissues 73, 106, 111–112
infectivity 118, 137, 171, 189, 190,
.... 196, 223, 250, 267
Information Note 144
information technology 96
Initial Background Documents *see* IBD
injections 11
inspection, meat 245
Institute for Animal Health 223
Institute for Research on Animal
    Diseases 223
Institution of Environmental Health
    Officers 77, 123
insulin 28, 166
Internet 3, 276
intestines 108, 110, 114–115, 182
intestines, processed 114
intracerebral inoculation 15
Ironside, Dr James 155, 156

J

Jacobs, Ron 204–205, 283
JCVI 176
Jefferys, Dr David 171–172, 174,
.... 184–186
Jeffrey, Dr Martin 34
Johnston, A M 122
Joint Committee on Vaccination and
    Immunisation *see* JCVI
joint guidelines 174, 176
Jones, Dr Gerald 172, 283
Jones, Dr Keith 181
judicial review 120, 271

K

Kelly, Ms Marion 193, 195
Kendell, Dr Robert 150, 216, 282
Kent 34
Kidd, Alistair 184
kidney 108
Kimberlin, Dr Richard 49, 55, 110–111,
.... 115–117, 120, 129, 140,
.... 152–153, 159, 171, 176,
.... 182, 183, 194, 236, 241, 282
knacker's yards 24, 74, 78, 80–82,
.... 86, 87, 92
knackermen 199–200
kuru 258

INDEX

L

lab workers 199
laboratories 240
Lacey, Professor
   Richard 128, 129, 142–143
Lamming Committee 60, 61, 67, 95
Lamming, Professor Eric 16
Lamont, Norman 103–104
Lang, Ian 136
Lawrence, Alan 38–39, 44, 48–49,
   .... 70, 76, 81, 85, 107, 112,
   .... 118, 120, 134, 182, 191,
   .... 195, 208–210
Lawson, Professor David 168
Lee, Dr A 183, 187
Leeds University 128
Liaison Units 275–276
licensed plants 159–160
licensing 178, 186, 259–260
licensing authorities 53, 167, 170, 175,
   .... 184, 259, 269–270
licensing scheme, disposal of SBO
   protein 84
licensing, DH responsibility 232
licensing, medicines 243
Little, Dr Thomas 43, 169, 170, 173
liver 108
local authorities 75, 76, 89, 90, 245
Local Authorities Catering Association 149
Local Authorities, enforcement
   responsibilities 25
local authority inspectors 199
local authority tips 212–213
Love, Murray 184
Lowson, Robert 72, 81–82, 103,
   .... 112, 118–119, 131, 142, 194,
   .... 209, 211, 281, 283–284
LRS 116, 117
Luddington 62–63
Luff, Peter 169
lymph nodes 110, 126, 237
lymphoid tissue 50, 114–115, 127,
   .... 157, 172, 175, 177, 182
lymphoreticular system *see* LRS

M

MacGregor, Rt Hon. John 13–15, 38–39,
   .... 43, 45, 47, 59, 69, 101,
   .... 102, 106–113, 182, 267, 282
Maclean, Colin 127, 129, 151–152, 282
Maclean, Rt. Hon.
   David 15, 72, 112, 114–115,
   .... 121, 122, 128–129, 131
MacOwan, Dr Kenneth 209

MAFF 13, 25, 40–42, 56, 60, 61,
   .... 87, 102, 107, 108–109,
   .... 112–120, 122, 123, 124–126,
   .... 127–129, 131, 138, 146, 152,
   .... 153, 156, 169, 174, 177, 181,
   .... 182, 191–195, 198, 203–205,
   .... 212, 215–216, 218, 222–224,
   .... 246, 252, 254, 256–257, 259,
   .... 266, 268, 271, 276
MAFF Animal Health Division 38, 48,
   .... 54, 208, 211
MAFF Consumer Panel 161
MAFF Meat Hygiene Veterinary
   Section 208
MAFF Meat Trade Adviser 208
MAFF officials 13–14, 29, 39, 43–46,
   .... 48, 57, 59, 65, 72, 80, 91,
   .... 94, 95, 111, 115, 234–235,
   .... 236, 237, 240, 283
MAFF, Animal Health Division 73, 77, 82
MAFF, Animal Medicines Division 167
MAFF, areas of responsibility 242
MAFF, Biologicals Committee 173
MAFF, conflict of interest 231–232
MAFF, Legal
   Department 73, 81, 267, 269–270
MAFF, Meat Hygiene Division 119
MAFF, Meat Hygiene Veterinary
   Section 85
MAFF, structure 243
maggot bait farm workers 199
MAIL 173, 174
Major, John 47, 101, 138, 158
mandatory sampling 63, 66
market handlers 199
marmoset 139
Martin, Dr William 48, 54, 178
Martin, Ron 118
Maslin, John 73, 79, 81, 118, 209
maternal transmission 17, 49, 61–62, 96,
   .... 116, 220, 228
Matthews, Dr Danny 61, 81, 205, 208
Matthews, Professor Bryan 99–100
May, Sir Robert 265
MBM 16, 19–21, 26–27, 33,
   .... 40–41, 54, 59, 60, 61–62, 64,
   .... 66–70, 74, 76, 84, 116,
   .... 117–120, 141, 159, 212,
   .... 217, 220, 228, 236, 245,
   .... 250–251, 253, 269
MBM ban 36, 38, 42
MBM, contamination 13, 87
MC 242, 259
MCA 166, 177, 181, 184,
   .... 188–189, 191, 243
McKechnie, Sheila 265
McNeill, Johnston 91, 136, 144, 145, 151
MD 171–172, 173, 179
MDA 243
measles vaccine 185–186

303

Meat (Sterilisation and Staining) Regulations 1982 .... 74–77, 86
Meat and Livestock Commission *see* MLC
meat hygiene 73
Meat Hygiene Division 123–124, 135, 144
Meat Hygiene Inspection (MH1) form 83
Meat Hygiene Regulations 75, 82
Meat Hygiene Service *see* MHS
Meat Hygiene Veterinary Section 122
meat hygiene, enforcement 76
meat hygiene. 83
meat industry 27
Meat Inspection Regulations 75, 82
meat inspection standards 19
Meat Inspectors 19, 74, 76, 79, 88,
.... 90, 91, 93, 121–122, 135
meat pies 110, 117
meat plants 160
meat products 114, 126, 128
meat products, processed 27
meat trade 199
Medical Devices Agency 181
medical professionals 200
Medical Research Council *see* MRC
medicines 4, 27, 51, 230, 232,
.... 235, 236, 238, 242, 243, 267, 281
Medicines Act 1968 .... 167, 269–270
Medicines Act Information Leaflet *see* MAIL
Medicines Advisory Committees 259
Medicines Commission *see* MC
Medicines Control Agency *see* MCA
Medicines Division, DH 243, 283
medicines licensing 174
Medicines Licensing Authority 51
medicines, injectable 28
medicines, safety 176–177, 182
medicines. 234
Meldrum, Keith 40–41, 44, 60, 61,
.... 61–62, 65, 69, 70, 71, 81,
.... 82, 84, 86–88, 91, 92, 94–96,
.... 102, 105–106, 109–111, 114–116,
.... 120, 123–124, 128–129, 133,
.... 138, 144–146, 154, 155–158,
.... 162, 194, 236, 244, 245, 247–248,
.... 264, 280–283
Mellor, David 47, 135
mesenteric fat 114–115
Metters, Dr Jeremy 111–112, 114–115,
.... 120, 132–133, 145–146, 153,
.... 155, 162–163, 182, 233, 282
MHIs 144
MHS 18–19, 76, 88, 89, 91, 92,
.... 94, 95, 96, 133, 135–136,
.... 144–145, 146, 151, 159, 160, 230
MHS Project Team 134
mice 121
microbiological safety 26
microbiology 100, 275
Middle East 42
milk 268

milk, safety 160
Ministry of Agriculture, Fisheries and Food *see* MAFF
Minor, Dr Philip 169–170, 188
The Mirror 142, 159
MLC 28, 103, 110, 127, 129,
.... 145–146, 151–154, 264, 266, 282
MLC Consumer Committee 124
MMR vaccine 185–186
molecular studies 56
monitor 120
monitoring 99, 119, 121
monitoring, Regulations 245
monitoring, SBO ban 82–86
Montagu of Beaulieu, Lord 44
mood swings 272
Moredun Research Institute, Edinburgh 48
mortuaries 240
moufflon 139
mouse bioassay 224, 261
mouse prion gene 12
movement permits 76–77, 79, 85, 87
MRC 171, 222–223, 252
MRM 16, 27, 53, 122, 123–126,
.... 137, 150, 237, 241, 247–248,
.... 256, 267, 271
MRM, cattle spinal cord ban 19
MRM, contamination 125
MSSR 113, 182
Murray, Thomas 131, 132, 194

N

National Consumer Council 102
National Health Service *see* NHS
National Institute for Biological Standards and Control *see* NIBSC
nerve cells 221
neural tissue 50, 177
neurological symptoms 272
neurologists 99–100
Neuropathogenesis Unit 236
neurophysiologists 199
neurosurgeons 201
neurosurgery 184, 189
New Zealand 184, 186, 276
Newcastle University 149
NFU 71, 102–104
NFU, Milk and Dairy Produce Committee 102
NHS 168, 232
NIBSC 169, 170, 171, 174, 190
NIG 200
NOAH 169
North Africa 42
North Yorkshire County Council 102
Northern Ireland 13, 100, 215, 217–218
Northern Ireland Office 29
notifiable diseases 46, 47, 59, 100, 217
notification 35

INDEX

NPU 34, 40–41, 49, 127–128, 141, 170, 176, 183, 186, 189, 222–224, 252, 277, 280
nyala 139

O

obex section 105
occupational health 49
occupational risk 54, 198, 208, 230, 238, 242, 260, 281
off farm burning 212
offal 25, 112, 114, 149, 281
Office International des Epizooties 42
Official Veterinary Surgeons 75, 79, 88, 89, 90, 93, 245
on-farm mixing 61, 67
oral evidence 277
oral ingestion 15
organophosphate theory 219, 222
Over Thirty Month Scheme 20, 157–160, 162–165
OVSs 134, 144
ox brain 49
ox liver 28

P

Packer, Richard 64, 91, 93, 144, 146, 151, 155, 156–158, 162, 283
parenteral inoculation 50
parenteral products 172–173, 178
Parliamentary Questions 102
Patent Blue V 88, 90
pathology 56, 251
Patterson, Dr Will 148
Pattison, Professor Sir John 147, 150–151, 155–159, 283
PD 168, 190
Pedigree Masterfoods 110, 241
pericardium patches 28, 168
peripheral nervous tissue 118, 247
pet food 26, 28
pet food companies 229
pet food industry 110
pet food manufacturers 16
Pet Food Manufacturers' Association 69
pharmaceutical industry 232
pharmaceuticals 26, 28, 210
PHLS 100–101, 252
photographic chemicals 27
Pickles, Dr Hilary 48–49, 53, 96–97, 107, 111, 113–115, 118, 120, 131, 172, 174–176, 182, 187, 191, 193, 201, 203, 207, 209–211, 243, 283
pig and poultry feed 16, 40, 141, 220, 253
pig eyeballs, dissecting 203–204
pig feed 69, 71, 87, 236, 240
pigs 71–72, 95, 121
pithing rods 24, 200
Pitsham Farm, Sussex 33

pituitary hormone material 169
placenta 229
placental material 192
placental tissue 177
porcine material 187
Portillo, Michael 136
poultry 95
poultry feed 69, 71, 87, 236, 240
prion mutation 220
prion protein 11, 221
prion protein theory 219
Priorities Board for Research and Development in Agriculture and Food 222
Procurement Directorate *see* PD
product licence 167
Product Licences of Right 168–169, 173, 184
production lines, separate 66, 68
production plant, cleansing 84
prosecutions 144, 145
Prosper De Mulder 26, 85
Protein Processing Order 1981 83
proximal colon 110
Prusiner, Professor Stanley 11
puma 139
Purves, Dr John 172, 185, 188

Q

quality assurance schemes 27
questionnaires 100

R

random inspections 105
Rawlins, Professor M D 176
Redwood, John 136
Reed, James 62
Rees, William 13, 34–35, 38–39, 43–44, 170, 280
Regulations 203, 217, 267
Regulations, breaches 101, 145
renderers 74, 76, 80–85, 86, 88, 91, 95, 199–200
renderers, dedicated SBO facilities 91
renderers, licensing 87
renderers, SBO processing lines 92
rendering 25–26, 38, 54, 117, 212, 220, 226, 251, 277
rendering industry, Regulations 1981 .... 26, 37
rendering industry, Regulations 1989 37
rendering plant waste 208, 211, 213
rendering plants 236
rendering, effluent 238
rennet 114–115, 247
Research Councils 29, 222–224
revised factual accounts *see* RFAs
RFAs 277–278
RFB 94, 95, 116, 159, 164, 228, 238, 239

RFB, breaches	67
Richmond, Gillian	81–82
Ridley, Dr Rosalind	171
Rimmer, Vicky	143
risk communication	127, 164, 264–266
risk evaluation	31
risk management	31
risk, perception	232–235
rivers	213
Rogers, Chris	208–209, 211
Roscoe, Richard	192–195, 208
Rotblat, Dr Frances	172, 185
Royal Commission on Environmental Pollution 1979	226
Royal Veterinary College	127
Rubery, Dr Eileen	155, 162
ruminant feed ban	13, 18, 21, 38–42, 47–48, 59, 217–218, 236, 253, 266, 269, 280
ruminant feed ban, breaches	62
ruminant offal	14
ruminant protein	61
ruminant protein, animal feed	63, 64, 225
Russia	129
Rutter, Dr James	181

S

Safety, Efficacy and Adverse Reactions see SEAR	
Sainsburys	103
salmonella	83
sausage casings	247
sausages	110, 114–115, 117, 217
SBO	14, 118
SBO ban	16, 53, 55, 129, 131, 138, 144, 157, 158, 160, 203, 208, 229, 241, 247, 282
SBO Regulations	122, 123, 144, 146, 149, 150, 177, 235, 237
SBO, controls	88
SBO, dedicated facilities	92, 95
SBO, disposal	86, 87, 137, 144, 212, 213
SBO, handling	78–79, 80, 82, 87, 90–91, 92–93, 144
SBO, MBM contamination	236
SBO, staining	87, 88–89, 90, 92–93, 245
school meals	15, 128, 149
schools	203, 283
scientists	252, 268, 276
scimitar-horned oryx	139
Scollen, F J H	174, 188
Scotland	13, 24, 26, 100, 215–216, 218
Scottish Education Department	203
Scottish Home and Health Department	216
Scottish Office	29, 216–217
scrapie	13, 34, 37, 43, 228, 233, 238, 280
scrapie agent	220–221, 224
scrapie infected feed	36
scrapie research	116–117
scrapie, cattle, transmission	49
Scudamore, James	216
SEAC	15–17, 19, 57, 66, 81, 84, 95, 96–97, 99, 110, 119, 121, 123–126, 129, 131–133, 135, 137, 138–142, 146–148, 150–160, 163–164, 183, 194, 201, 203–204, 208–211, 212–213, 216–217, 224, 230, 232, 233–234, 237–241, 244, 247, 253, 256–257, 262, 266, 271, 277, 282–283
SEAR	173, 176
Secretary of State for Health	1, 167
Secretary of State for Trade and Industry	269
Secretary of State for Wales	102
section 4 committees	243
Select Committee on the Environment	213
self-regulation, feed industry	61
SEs see spongiform encephalopathies	
sewage sludge	213
sewers	213
sheep	177
sheep eyeballs, dissecting	203
sheep tissues	210
Shephard, Gerald	136
Shephard, Gillian	17, 62, 64, 138
short-term memory	272
Siamese cat	15
Simmons, Alick	83–84, 86–87, 137
Single European Market	134
Skinner, Dr Roger	162
Skinner, Mike	156
slaughter and compensation policy	236
slaughter, compulsory	17, 21, 47, 106, 229, 254, 266, 269, 272
slaughterhouse practices	16, 25, 123–125, 135, 146, 149, 216–217
slaughterhouse products	205
slaughterhouse waste	208, 211, 213
slaughterhouses	19, 24, 54, 69, 81, 95–96, 101, 104, 105, 116, 117–118, 120, 121–122, 133, 134–137, 144, 146, 156, 160, 199–200, 208, 212, 235, 245, 255, 277
Slaughterhouses Act 1974	24–25
slaughterhouses, compliance with Regulations	91
slaughterhouses, drainage waste	238
slaughterhouses, monitoring	119
slaughterhouses, practices	241
slaughterhouses, SBO handling	74–80, 82–83, 85–90
slaughtermen	100, 199
Sloggem, John	170
Slovakia	100

small intestine, calves 18
Smith, Edward 44
Smith, Professor Peter G 155, 159
SmithKline Beecham 28
Soames, Nicholas 62, 63
solvent extracted tallow 37
solvent extraction 26
Soul, Peter 144
South West England 220
Southwood Report 14, 17, 53, 54,
.... 58, 69, 106–109, 174, 176–180,
.... 198, 205, 215, 233, 234,
.... 238–239, 248, 254, 256, 281
Southwood Working Party 14–15, 38,
.... 46–55, 60, 68, 69, 95, 99,
.................................101, 106, 109, 111, 118, 140,
................................171–172, 175, 178, 180, 207,
.... 213, 233–234, 237–238, 240, 241, 281
Southwood, Professor Sir
  Richard 14, 47, 48, 54, 70,
  .... 107–109, 111, 116, 173, 237
soyabean meal 23
Specified Bovine Offal (Amendment) Order
  1995 271
Specified Bovine Offal Order
  1995 .... 78, 90
spinal 237
spinal columns 74
spinal cord 89, 91, 108, 110,
.... 114–120, 121–122, 123–126, 136,
.... 137, 144–145, 146, 182, 221,
.... 237, 239, 245, 248, 267
spinal cord, removal 25
spleen 108, 110, 114–115, 192
spongiform encephalopathies
  (SEs) 34, 50, 139
staining 74, 85, 86, 203, 210
staining, SBO 87
standards 144
Starcross VIC, Exeter 36
State Veterinary Investigation Centres *see* VICs
State Veterinary Service 78, 82, 95–96,
.... 231, 243, 246
Statement on Procedures for Additional
  Comments 279
Statements on Procedures 278
statutory animal SBO ban 72–74, 82
statutory enforcement, RFB 63
Statutory Instruments, relating to BSE 246
statutory monitoring 255
statutory powers 267–270, 272
statutory regulation 16
stockmen 199
stocks 176–177, 179–180, 184,
.... 187, 197
subclinical cases 14, 33, 51, 53, 55,
.... 106, 108, 110–113, 148, 164,
.... 178, 225, 250, 281
subordinate legislation 29
Suich, John 43

The Sun 129
supermarkets 69, 160, 164
surgical devices 4
surgical instruments, sterilisation 11
sutures 115, 166, 173, 184, 187, 189
SVO 208
SVS 1, 13, 35, 135, 136, 144–145,
.... 251, 253
Switzerland 42

T

tablets, coatings 27
tallow 4, 26, 28, 61, 74, 76,
.... 115, 211–212, 238
Taylor, David 118
Taylor, Dr David 170, 183, 186
Taylor, Kevin 39–40, 59, 63, 81,
.... 83, 102, 119
Taylor, Matthew 122
Terms of Reference 262, 280
Territorial Departments 112, 118,
.... 215–216, 218, 246, 257
Thatcher, Margaret 70, 103, 109, 113
The Guardian 110
The Times 44, 103, 110, 149, 151
Thompson, Donald 43, 45, 110
Thruxted Mill, Kent 213
thymus 108, 114–115, 138, 144,
.... 177, 182, 192
thymus, calves 18
toiletries 192, 232
Tomlinson, Sir Bernard 149
tonsils 110, 114–115, 144
topical products 178, 186, 192
toxicology 275
trade associations 59
Trading Standards Departments 78, 192
Trading Standards Officers 78
transgenic mice 12
Treasury 101, 103, 223
treatments 166
tripe 114–115, 230, 247
TSE, transmission 11
TSEs 2, 34, 49, 119, 171, 177,
.... 196, 199–202, 207, 219–223,
.... 227, 229, 234, 239, 250–252
TSEs, disease patterns 12
Tuberculin PPD 185–186
Tyrrell Committee 14–15, 49, 55–57,
.... 99, 110, 190, 213, 223–224,
.... 233, 237, 262
Tyrrell Report 56, 99, 191, 193, 207, 210
Tyrrell, Dr David 14–15, 55, 57,
.... 116, 125, 129, 132, 140, 183,
.... 194, 200, 216, 253

U

UK Renderers' Association *see* UKRA

FINDINGS AND CONCLUSIONS

UKASTA 16, 39–40, 61, 69–70,
.... 112, 236, 239
UKASTA, Code of Practice on cross
  contamination 27
UKASTA, Scientific Committee 63
UKRA 26, 69–70, 84
undertakers 198
unfit meat 74, 76, 212
Unitary Authorities 59, 75, 78
Unitary Authorities, enforcement
  responsibilities 25
United Kingdom Agricultural Supply Trade
  Association *see* UKASTA

V

vaccines 28, 166–167, 175, 176, 177,
.... 179, 180, 185, 187, 189, 234, 243
vaccines, safety 111, 158
vCJD 1–4, 11, 19, 21, 66, 100–101,
.... 130, 143, 154, 160, 189, 196,
.... 198, 219, 221–223, 230, 249,
.... 256, 258, 275, 277, 280
vCJD, victims 3, 272–273
veterinary epidemiology 253
Veterinary Field Service *see* VFS
Veterinary Investigation Centres *see* VICs
Veterinary Investigation Service *see* VIS
veterinary medicines 167, 169, 173,
.... 188, 229, 234, 242, 243, 269
Veterinary Medicines Directorate *see* VMD
veterinary medicines producers 169
Veterinary Officers *see* VOs
Veterinary Products Committee *see* VPC
Veterinary Record 35, 169
veterinary vaccines 175
VFS 18, 82, 83, 85, 86, 89, 91,
.... 93–96, 120, 122, 136, 144,
.... 239, 245, 255
VIC, Worcester 60, 62
VICs 33, 252, 277
VIS 34, 105, 243
VMD 167, 181, 183–184, 187
voluntary animal SBO ban 69, 77–79, 82
voluntary sampling 63, 64, 65
VOs 82, 83, 137
VPC 53–54, 168, 169, 175,
.... 176–177, 182, 187, 234

W

Waldegrave, William 19, 64–67, 88, 90,
.... 135–136
Wales 13, 24, 26, 99, 100, 215–216,
.... 218, 277
Walker, Peter 102
Walton, Professor Sir John 48
waste disposal 116, 213, 242, 260
waste tip 198
water authorities 213

Watson, Dr William 34–35, 40, 43–45,
.... 55, 170, 280
website 276
weight checks 91
weight checks, slaughterhouses 85
Wellcome 171, 185
Wells, Gerald 34–35
Welsh Office 29, 203–204, 216, 246
West Country 33
West Germany 186
western blotting 12
Western General Hospital, Edinburgh 15, 99
Weybridge 33, 277
Whitehall 215–216, 218
Whyte, Mrs Diane 195
Wight, Dr Ailsa 142, 153–156, 162,
.... 196–197, 283
Wilesmith, John 13, 36–39, 41, 44,
.... 49–50, 63–64, 87, 118, 169,
.... 171, 225, 236, 252, 281
Will, Dr Robert 15, 55, 99–100, 129,
.... 141–142, 147, 150–151, 154–155,
.... 157, 159, 183, 194, 256, 283
William Forrest & Son (Paisley) 26
Williams, Dr Bernard 35, 280
Woman's Farming Union 110
Wood, G W 169
Woolfe, Dr Mark 107, 109
workers at disposal sites 199
Working Group of Advisory Committee on
  Dangerous Pathogens *see* ACDP
World Health Organisation 181, 196

Y

Yavash, Ayyildiz 81
yellow card system 168

Z

Zeneca 28
zoo workers 199
zoonoses 242, 250, 252, 254,
.... 258–259, 266, 268, 271
zoonoses, handling 199

308

Printed in the United Kingdom for The Stationery Office Limited
on behalf of the Controller of Her Majesty's Stationery Office
552400 10/2000 19585